Fuel Cell Technologies: State and Perspectives

T0191740

NATO Science Series

A Series presenting the results of scientific meetings supported under the NATO Science Programme.

The Series is published by IOS Press, Amsterdam, and Springer (formerly Kluwer Academic Publishers) in conjunction with the NATO Public Diplomacy Division.

Sub-Series

I.   Life and Behavioural Sciences   IOS Press
II. Mathematics, Physics and Chemistry   Springer (formerly Kluwer Academic Publishers)
III. Computer and Systems Science   IOS Press
IV. Earth and Environmental Sciences   Springer (formerly Kluwer Academic Publishers)

The NATO Science Series continues the series of books published formerly as the NATO ASI Series.

The NATO Science Programme offers support for collaboration in civil science between scientists of countries of the Euro-Atlantic Partnership Council. The types of scientific meeting generally supported are "Advanced Study Institutes" and "Advanced Research Workshops", and the NATO Science Series collects together the results of these meetings. The meetings are co-organized by scientists from NATO countries and scientists from NATO's Partner countries — countries of the CIS and Central and Eastern Europe.

Advanced Study Institutes are high-level tutorial courses offering in-depth study of latest advances in a field.
Advanced Research Workshops are expert meetings aimed at critical assessment of a field, and identification of directions for future action.

As a consequence of the restructuring of the NATO Science Programme in 1999, the NATO Science Series was re-organized to the four sub-series noted above. Please consult the following web sites for information on previous volumes published in the Series.

http://www.nato.int/science
http://www.springeronline.com
http://www.iospress.nl

Series II: Mathematics, Physics and Chemistry – Vol. 202

Fuel Cell Technologies: State and Perspectives

edited by

Nigel Sammes
Department of Mechanical Engineering,
University of Connecticut,
Storrs, CT, U.S.A.

Alevtina Smirnova
Department of Materials Science and Engineering,
University of Connecticut,
Storrs, CT, U.S.A.

and

Oleksandr Vasylyev
Frantcevych Institute for Problems of Materials Science,
National Academy of Science of Ukraine,
Kyiv, Ukraine

 Springer

Proceedings of the NATO Advanced Research Workshop on
Fuel Cell Technologies: State and Perspectives
Kyiv, Ukraine
6–10 June 2004

A C.I.P. Catalogue record for this book is available from the Library of Congress.

ISBN-10 1-4020-3497-0 (PB)
ISBN-13 978-1-4020-3497-8 (PB)
ISBN-10 1-4020-3496-2 (HB)
ISBN-13 978-1-4020-3496-1 (HB)
ISBN-10 1-4020-3498-9 (e-book)
ISBN-13 978-1-4020-3498-5 (e-book)

Published by Springer,
P.O. Box 17, 3300 AA Dordrecht, The Netherlands.

www.springeronline.com

Printed on acid-free paper

All Rights Reserved
© 2005 Springer
No part of this work may be reproduced, stored in a retrieval system, or transmitted in
any form or by any means, electronic, mechanical, photocopying, microfilming, recording
or otherwise, without written permission from the Publisher, with the exception of any
material supplied specifically for the purpose of being entered and executed on a
computer system, for exclusive use by the purchaser of the work.

Printed in the Netherlands.

Contents

Contributing Authors

G.YA. Akimov

K. Aldas

O. Attia

J.A. Bahteeva

V.Y. Baklan

V.I. Barbashov

I.A. Bashmakov

A.G. Belous

M. Berkowski

U. Bismayer

G. Blaβ

L. Blum

T.N. Bondarenko

M. Carrasco

A.P. Carvalho

E. Chinarro

A. Cramer

I.A. Danilenko

D.B. Dan'ko

V.I. Dekhteruk

A.K. Demin

M. Demiralp

A.S. Doroshkevich

Y. Du

V.A. Dubok

O.M. Dudnik

D.A. Durilin

V.N. Fateev

F. Ferrari

L.I. Fiyalka

J.R. Frade

T.G. Gal'chenko

O.Z. Galiy

A. Gebert

R. Gemmen

V.A. Glazunova

V.O. Golub

Y.M. Goryachev

S.A. Grigoriev

R.W. Grimes

U. Guth

L.G.J. de Haart

J.T.S. Irvine

V.P. Ivanitskii

V.V. Ivanov

C. Johnson

J.R. Jurado

A.A. Kalinnikov

D. Kang

Y. Kaplan

V.V. Kharton

V.R. Khrustov

M. Knapp

G.Y. Kolbasov

I.P. Kolesnikova

Y.A. Komysa

T.E. Konctantinova

M. Konstantinova

N. Koprinarov

I. Kosacki

I.A. Kossko

A.V. Kovalevsky

V.L. Kozhevnikov

W.M. Kriven

V.N. Krivoruchko

A. Kruth

M. Kuzmenko

V.V. Lashneva

F.D. Lemkey

I.A. Leonidov

A.S. Lipilin

M. Marinov

F.M.B. Marques

V.V. Martynenko

M.D. Mat

W.A. Meulenberg

B. Moreno

B.A. Movchan

H. Nabielek

E.N. Naumovich

A. Nicholls

N.V. Nikolenko

A.V. Nikonov

K.V. Nosov

G.P. Oreghova

N. Orlovskaya

E.V. Pashkova

M.V. Patrakeev

C. Paulmann

G. Pchelarov

J.C. Perez

J. Piron-Abellan

Y. Pivak

D.Y. Podyalovsky

T. Politova	Y.M. Solonin
V.I. Porembsky	R. Steinberger-Wilckens
N. Poryadchenko	S. Tao
V.V. Primachenko	F. Tietz
W.J. Quadakkers	A. Tomasi
O.P. Rachek	A.I. Tovstolytkin
J. Remmel	R. Travis
B.R. Rosczyk	B.A. Troshen'kin
I.A. Rusetskii	V.B. Troshen'kin
N. Sammes	E.V. Tsipis
J. Sankar	M.V. Uminsky
M.P. Savyak	I.V. Uvarova
R. Sayers	V. Vashook
D.I. Saytskii	L. Vasylechko
A. Senyshyn	O.D. Vasylyev
G.E. Shatalova	V.G. Vereschak
A.L. Shaula	T.N. Veziroğlu
V. Shemet	I.C. Vinke
S.N. Shkerin	G.K. Volkova
I.G. Shulik	O.I. V'yunov
M.I. Siman	Z. Xu
I. Singheiser	O.Z. Yanchevskii
S.J. Skinner	S. Yarmolenko
A. Smirnova	N. Zakowsky
I.S. Sokolovska	A.V. Zyrin

Preface

Fuel Cells are electrochemical devices that convert the chemical energy of a fuel directly into electricity, without the necessity of an intermediate. Fuel cells are considered an emerging technology that can deliver clean, quiet, and potentially renewable energy for primary, base-load and back-up power; this technology now has to be commercialised. As such, this ARW provided an interactive forum for recent advances in the development and commercialization of SOFC and PEM fuel cell systems, as well as technology improvements in the materials and systems.

The objectives of the NATO ARW were to:

1. Discuss progress and problems in the development of fuel cell technologies for advanced energy systems.

2. Exchange ideas for the development and design of solid oxide fuel cells (SOFC) and proton exchange membrane (PEM) fuel cells for electricity and heat generation, vehicles, reversible fuel cell electrolyzes, and fuel cell microturbine hybrids, for example.

3. Identify directions for future research in SOFC, PEM and their use in developed and developing countries, and finally their commercialization.

4. Provide a forum for communication and exchange of ideas for scientists from different institutions and research sectors from around the world.

The following topics were presented:

- Historical aspects of fuel cells internationally

- Worldwide update of fuel cell status

- SOFC advances

- PEM advances

- Materials for SOFC and PEM technologies

- The hydrogen infrastructure

- Novel processing techniques

N Sammes

O. Vasylev

Co-Directors

11/1/2004

Acknowledgments

We would like to thank NATO for financial support that made this ARW possible (Award number PST.ARW.980364). We would also like to thank the European Office of Aerospace Research and Development (FA8655-04-1-5086).

We would like to thank the following sponsoring organizations: The Center for Advanced Materials and Smart Structures, North Carolina A and T State University, The Connecticut Global Fuel Cell Center, The University of Connecticut, The Frantcevych Institute for Problems of Materials Science, and Zirconia Ukraine Ltd.

A warm thank you goes to all those who helped make the ARW a great success including V. Burdin, M. Brychevs'ky, M. Golovkova, O. Koval I., M. Lugovy and I. Okum'.

We would also like to thank the international organizing committee including, L. Blum (Germany), A. Demin (Russian Federation), W. Kriven (USA), N. Orlovskaya (USA), S. Skinner (UK) and V. Skorokhod (Ukraine).

We would like to thank B.J. Mclaughin (University of Connecticut) for doing an excellent job of handling the budget, as well as Jakub Pusz for doing an excellent job of collating the papers. I would like to give a warm thank you to Dr Nina Orlovskaya for her enthusiasm for the ARW.

Finally I would like to thank all those that I have forgotten, as well as all the speakers, session chairs, and participants who made the workshop the tremendous success that it was.

N. Sammes, O. Vasylyev

Co-Directors, NATO ARW

"Fuel Cell Technologies: State and Perspectives"

HISTORIC ASPECTS OF FUEL CELL DEVELOPMENT IN UKRAINE

O.D. VASYLYEV

Frantcevych Institute for Problems of Materials Science, National Academy of Science of Ukraine; Zirconia Ukraine Ltd. 3, Krzhyzhanivs'koho Str., Kyiv-142, 03680, Ukraine

Abstract: This paper is an historical essay on development of fuel cell technologies in Ukraine and short description of the National Program on Fuel Cells. It is shown that fuel cell technologies have both many years positive experience in Ukraine owing to Oganes Davtjan's efforts since early 50th of the 20th century, which were interrupted by political events in the 70th, and opportune perspectives based on achievements determining these up-to-date high technologies like nanosized zirconia and other oxide powders, materials for interconnects and microturbines, production of hydrogen, gasification of coal as well as own zircon-sand and scandium ore deposits, well-developed manufacture of gas turbines and another energetic components and equipment.

Key words: Fuel Cells/History Of Fuel Cells/Zirconia/Nanopowders/Fuel Cell Program

Fuel cell is an energetic technology, which is revolutionizing energy consumption. People have to understand that taking hydrocarbon fuels for production of heat or electricity with traditional ways they are able to use roughly only one-third part of energy stored in the fuel, and the rest of it is simply lost. It means that people might be satisfied with only one third of all organic fuels used at present. Fuel cells, especially their solid oxide and molten carbonate types, give the unique possibility to use that lost part of energy in a very flexible form namely electricity. Except significant saving of energy the fuel cells diminish seriously harmful pollutions typical for power stations and transport based on internal-combustion engines.

Here, one Ciceronian example is worthy to be mentioned. When Daimler Chrysler was asked about their vision of future, their answer was as the next:

1

N. Sammes et al. (eds.), Full Cell Technologies: State and Perspectives, 1-10.
© 2005 *Springer. Printed in the Netherlands.*

"Anticipating future demands, DaimlerChrysler is already today operating a large number of fuel-cell-powered vehicles in successful practical endurance tests" that was accompanied by the nice symbolic picture of a humming-bird drinking water from the car exhaust pipe [1].

Figure 1. O.K. Davtjan together with his wife and colleague E.G. Mysyuk.

While not offering a specific date, the former US vice-president Al Gore said that advances in fuel cell technology and research led him to believe the gasoline engine could be replaced (*by fuel cell engine – author*) before 2017. He has anticipated that fuel cell technologies will reduce the US consumption of foreign oil, clean air, and ultimately make roads safer [2].

The developed countries are actively elaborating numerous technologies concerning fuel cells and put already them on commercialization way for manifold applications. And what about Ukraine? Strictly speaking about today situation, the expression "near zero" is very generous estimation. And what about the past? What is history of this topic in Ukraine?

The work on the National Fuel Cell Program initiated by me as a person, who came out from the field of structural zirconia ceramics, has brightly displayed the very dramatic and emotive story that is worthy to be put in the base of cliffhanger novel or movie.

My the best knowledge is saying that the beginning of Ukrainian fuel cells was put by eminent famous and esoteric Armenian and Ukrainian scientist Prof. Oganes Davtjan (1911-1990, Fig. 1) in 1953 at Mechnikov

Odesa State University at Chair of Physical Chemistry. Oganes Davtjan is the Father of Ukrainian Fuel Cells; Ukrainian Fuel Cells were born in Odesa.

Oganes Karapetovych Davtjan was born in Armenia, in Akhurjan village, on April 15, 1911. His education he obtained at Yerevan and Moscow Universities. His PhD degree in chemistry Oganes Davtjan obtained in 1938 at Moscow Oil Institute for his work "Theory on Association of Molecules". His activity was continued in Moscow Energetic Institute where Davtjan obtained his doctor's degree on Technical Science in 1944 for work "Problems of direct transformation of chemical energy of fuel into electricity". In 1947, Davtjan published his book under the same title, which was the first book on fuel cell theory and technology in the World [3].

Since 1945, after a few years of knocking about the former Soviet Union (after Moscow, Davtjan worked in Kazan', Yerevan, and Vilnius where he suffered on political persecutions for quantum chemistry, which was "bourgeois" science like "prostituting girl of imperialism" in Soviet Union in that time), Oganes Davtjan got at least the chair for physical chemistry at Odesa University in 1953, July 3, where he initiated the works on fuel cells ("Davtjan's elements"). His activity was strongly supported by Ukrainian government under Petro Shelest by the special act N67 (January 20, 1962) on foundation of special laboratory for fuel cells. In contrary to the today government the funding was huge; it allowed building of special laboratory, design bureau, workshops etc.

Davtjan has dealt with solid oxide, molten carbonate, and alkali fuel cells. Since that the laboratory under Davtjan has developed as follows:

- Cheap catalysts for hydrogen and oxygen electrodes based on oxides of La, Co, Ni, Ti etc;
- Design and technology of hydrogen and oxygen electrodes;
- Design of fuel cells stock and technology of its manufacture;
- Automatic control of fuel cell stock.

It is worthy to be noted, Davtjan's team has made the 200 W hydrogen-oxygen alkali fuel cell stocks in 1962 (Fig. 2). Temperature of working gases was 170 °C; their pressure was 20 atm. They have developed the fuel cells that could operate at 700 °C with a solid ionic conductor as electrolyte too. Davtjan added monazite sand to a mix of sodium carbonate, tungsten trioxide, and soda glass "in order to increase the conductivity and mechanical strength". Broers and Katelaar tested the Davtjan's electrolyte in 1960, and established that this electrolyte has two phases, a molten phase consisting of carbonates, phosphates, tungstates and silicates, and a solid phase based on rare earth oxides [4].

Since that 1 kW and 5 kW stocks, which did not require noble expensive metals, were developed and made for atomic power stations to dump peak loads. The Davtjan's progeny put them in the basement of hybrid engines for vehicles (fuel cells α inner-combustion engine) in cooperation with Yerevan, Armenia, and Cherkasy, Ukraine.

For his work in the field of fuel cells Oganes Davtjan was nominated by the Czech Academy of Science for the Nobel's Prize. Unfortunately their efforts were not understood and not supported.

The reader may find more details about the activity of Oganes Davtjan in the fuel cell field in [5].

Figure 2. The Davtjan's 200 W hydrogen-oxygen alkali fuel cell stock.

Except of Ukrainian fuel cells Prof. Oganes Davtjan is the founder of quantum chemistry in the FSU and the author of the first Soviet manual on quantum chemistry published by him in 1962. Since 1968, when O. Davtjan left Odesa for Yerevan, he had no more possibility to continue his experimental work in the fuel cell field. Having a very wide spectrum of devotions, Davtjan had concentrated on theoretical physics and, as a result, he has created the universal "The Theory of the Fundamental Field", which is finally concluding the model of Universe and existing the God as the Supermind of the Universe [6].

It is very important that O. Davtjan is considering the fuel cell effect as a base of human intellectual and motional activities, supposing that, e.g., phosphorus and oxygen reacting in the adenosine three-phosphate molecules of body cells and neurons produce electricity and mechanical movement. The metastable adenosine three-phosphate molecules decompose into stable adenosine two-phosphate molecules and phosphate ions. "The efficiency of this transformation is the highest like in fuel cells" [7].

And now briefly about modern Ukrainian fuel cell history: Being based on many years positive experience in manufacture of real nano-sized zirconia powders with different stabilizers like yttria, calcia, scandia etc. and zirconia ceramics [8, 9] the first Ukrainian demonstrating model of zirconia fuel cell was made and exhibited by January 22, 2002 [10- 12]. It has realized 0.85 V and 0.5 V of electro motive forces with propane gas and ethanol respectively at their direct burning.

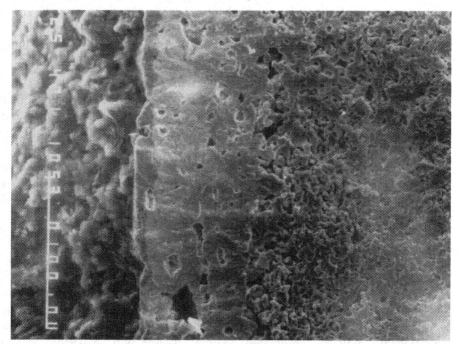

Figure 3. Structure of the first Ukrainian thin film zirconia ceramic fuel cell transformer. Left part of the picture shows the LSM cathode of around 10 μm thickness and its surface followed by the dense solid YSZ electrolyte of 60-70 μm thickness with a few isolated pores; the right part shows the highly porous zirconia–Ni anode. Scanning electron microscopy, Superprobe 733, JEOL.

Now, under auspices of the Ukrainian Parliament the National Program "Fuel cells" for 10 years period is developed [13]. The Program is considering fuel cells based on zirconia as very perspective for Ukraine taking into account its significant natural advantages. They are the most effective and reliable. They do not need in noble metals and may utilize practically all the types of organic fuel. Polymeric proton exchange membrane fuel cells will be developed for a few applications too.

The goal of the "Fuel Cells" Program is consolidation of efforts of Ukrainian society for accelerated development of fuel cell technologies. Its launch is caused by need of significant improvement of efficiency of organic fuel consumption especially natural gas and oil, transition to prevailing use of synthetic gas from coal at production of electric energy; sequestration of pollutions like CO_2, NO_x, SO_x etc. produced with power plants and internal combustion engines (ICE). Fuel cells stimulate a development of related branches of economy like fine technical ceramics, transport etc.

The Program "Fuel Cells" is self-developing. It consists of eight separate interdependent blocks. It is like a scheme of roads-directions, moving along them it is possible to reach the final goal. The first its "building block" is the Scientific & Technological Program "Fuel Cell Technologies", which must determine the next steps of development of the Program as a whole.

Analysis of condition of development of fuel cell technology in the World and Ukraine shows that the difference between Ukraine and other developed countries exists, but this difference is not so significant from the point of view of scientific and technological achievements. The most drastic is the difference in their practical application. But it might be quickly worked out, if scientific and industrial forces would be combined and financially supported in right way. The National Program "Fuel Cells" is directed on removing the gap between different scientific achievements and uniting them in elaboration of fuel cell technologies and their utilization.

The blocks of the National Program "Fuel cells" are as follows (Fig. 4):

Fuel Cell Technologies. This Academic Scientific & Technological Program is the starting block of the National Program, according to which samples of fuel cells, systems for fuel conditioning will be worked out; their test and optimization will be achieved during five years. During this time the other blocks of the National Program will be elaborated and their relationships will be coordinated. Market research will be performed and relationships with other developed countries in the World will be established. Scientific meetings and efforts directed on creation of favorable public opinion will be done routinely.

Fuel Cells Manufacture. The development and manufacture of several types of fuel cells for production of electric and heat energy to meet decentralize and centralize power supply of different power are planned. The special attention is given to zirconia ceramic fuel cells to supply residencies by electricity and heat; to replace internal combustion engines in different transport applications; to work in hybrid pairs with wind and solar generators, and gas turbines; for production of oxygen for medical needs etc.

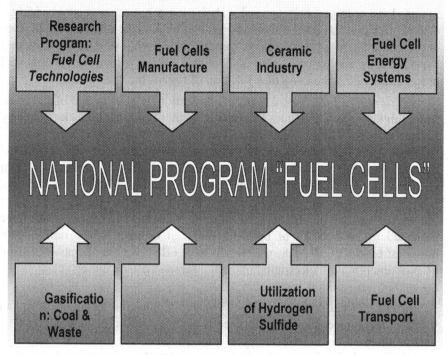

Figure 4. Structure of the Ukrainian National Program "Fuel Cells".

Ceramic Industry. The success in development of zirconia ceramics leads to the development of the whole fuel cell technology, which was taking place during the last 100 years, from time of discovery of "the Nernst's mass", which were zirconia base compounds. The typical strength of Ukrainian "ceramic steel" is around 1 GPa that allows thinning it to micron thickness [8]. Many people are familiar with Ukrainian ceramic scalpels, which not only improves the quality of surgery, but also speeds the healing of wounds. Here Ukraine has the possibilities of supplying the population by surgical tools of new generation, avoiding the stage of production of metal scalpels, which Ukraine never had. Special place in this list of goods takes thermal barrier coatings, without which the improved performance of modern gas turbines, which blades are usually coated by thin layers of protective zirconia ceramic coating, would be impossible.

Fuel Cell Energy Systems for Centralize and Decentralize Power Supply. Several energy systems for residential and centralized power supply of different power must be developed and manufactured. Among them there are systems, which may be used together with solar and wind generators of comparatively small power to avoid the primary weakness of these ecologically pure generators, namely their intermittence that does not allow to produce energy on demand. High temperature reversible SOFC-electrolyzer may be used effectively instead of internal combustion engines and electric accumulators making generators from these renewable energy sources really ecologically pure and reliable. These reversible systems will promote to wider spreading of utilization of renewable energy.

Gasification of Coal and Waste. Gasification of coal is one of the main technologies, which produces the gas for further production of electricity by fuel cells or for other applications. The research should be carried out also for finding the best way to obtain gas from low caloric coal with substantial content of sulfur and also from fuel mixtures, such as biomass or from enriched waste. In Ukraine, scientists of Scientific Center of Coal and Energy Technology and Design Bureau "Pivdenne" have already worked out the technology of gasification of Ukrainian coals. Preliminary tests have indicated the possibility of building fuel cell electric power station with power of blocks of 10 MW, which will have effectiveness of transformation of coal energy into electricity of 65%, compared to existing 34-46% for thermal power stations.

Hybrid of Fuel Cells and Gas Turbines. Today the research efforts in Ukraine are concentrated on fuel cells and gas turbines separately. The combination of these two devices may propose a substantial economic benefit through the increased effectiveness of fuel consumption from 50 to 80% with reducing the emission of nitrogen oxide and CO to 2ppm along with the reduction of the financial investment by 25% over one fuel cell system. The combination of Ukrainian gas turbines manufactures by "Zorya" Plant, Mykolayiv, and coated with Ukrainian Electron-Beam ceramic thermal barrier coating along with zirconia ceramics fuel cells will allow the double efficiency of transformation of fuel energy into electricity and reduce the emission of CO_2 by 30% in comparison with the best gas turbine at simultaneous reduction of cost in production of electric power.

Utilization of hydrogen sulfide. Zirconia ceramic fuel cells are the ideal means to use hydrogen sulfide as a fuel for production of heat, electricity, and value industrial substance – sulfur. Zirconia ceramic fuel cells are reliable in contact with sulfur-containing substances; they may produce electricity in the full analogy with pure hydrogen or organic fuels. Ukrainian scientists from National Polytechnic University (Kyiv) have developed ecologically pure technology of yield of hydrogen sulfide from Black See. This work will promote both production of value energy and substance, and purification of ecologically dangerous Black See.

Fuel Cells for Transport and Transport with Fuel Cell Engines. The wide use of transportation devices with fuel cells would be able to influence positively on reduction of consumption of oil products and minimizing the emissions over dense populated areas. International experience is quite positive here. Fuel cells potentially may substitute the internal combustion engine for all transportation devices, because they have higher effectiveness, much less pollution's and they are able to work on different fuels, such as hydrogen, ethanol, methanol, or natural gas. In Ukraine, as it was mentioned above, scientists have already created the acting samples of electric automobiles, which may be used as a base for testing of fuel cell engines and creation of an industry of transponrt means of new generation.

Finally, the National Program "Fuel Cells" is directed on development in Ukraine the economical and ecologically safe alternative to nuclear power engineering namely based on fuel cells, industry of technical ceramics and transport means including cars of new generation that will allow in general to cut consumption of gas for 30-50% and reduce its price for 20-30% at production of electric energy only, increase significantly consumption of coal to replace natural gas, significantly cut pollution in atmosphere including CO_2 for 50% and create hundreds thousands of new high technological working places.

Advantages from realization of the Program might be as follows:

- Reliable provision of high quality electrical energy;

- Reduction of transportation costs, using electricity and heat for industrial processes and air conditioning at the place of its production;

- Avoiding losses in peak load and voltage;

- Production of electrical energy in areas with very sensitive environment;

- Flexibility in choice of variety of fuel and power;

- Finally, electrical energy will be cheaper;

- Production of doubled amount of electric energy from the same amount of fuel;

- Reducing capital expenses and risks due to flexibility of power and location of station, fast building and starting of module fuel cell system;

- Reducing undesirable expenses at increasing power of the stations;

- Reducing investment in power line and distribution systems;

- Reducing of dependence from gas and oil supplies because of reduction in their use and transition to preferred use of coal, ethanol and biogas;

- Reducing of hazardous emissions, which cause acid rains and warming;

- Increasing of competition of Ukrainian goods in the World market;

- Ukraine could be an exporter of fuel cells to the fast growing markets, which so far have no transmission lines, with requirement of 500 GW of new power production by 2010.

ACKNOWLEDGEMENTS

Author thanks the Ukrainian State Committee for Energy Saving as well as NATO, their Advanced Research Workshop and Science for Peace Programs for their countenance and financial support of efforts directed on Research and Development in fields of solid oxide fuel cells and zirconia ceramics as well as their activity for orientation of Ukrainian economy in direction of Fuel Cell Technologies.

REFERENCES

[1] Schrempp J. DaimlerChrysler: Was the merger a mistake? Business Week, 2003, September 29, p. 44-51.

[2] Alternative Fuel Vehicles. ISATA Magazine, issue 10, June 2000, p.16.

[3] Karamyan G.G. Oganes Davtjan, Lpatu, Yerevan, 2001, 100 p., in Russian.

[4] Molten Carbonate Fuel Cells, http://americanhistory.si.edu/csr/fuel cells/

[5] Baklan V.Yu. This book

[6] Davtjan O.K. Theory of Fundamental Field, Aiyastan, Yerevan, 1995, 313 p., in Russian.

[7] Davtjan O.K. The problem of direct transformation of energy of fuel into electricity. Academy of Science of USSR, Moscow, 1947, in Russian.

[8] Ivaschenko O.V., Vasilev A.D., Peitchev V.G. et al. Fracture mechanisms, and strength of zirconia stabilized partially with yttria. Physical and chemical mechanics of materials, 1992, v.28, N6, p.46-50, in Russian.

[9] Vasylyev O.D., Akymov G.Ya, Koval O.Yu. Zirconia ceramics and their prospects in Ukraine. Refractories and technical ceramics, 2000, N10, p.2-5, in Russian.

[10] Vasylyev O.D., Schokin A.R. Ceramic Fuel Cells: Achievements and Perspectives in Ukraine. Electroinform, 2003, N1, p. 24-27, in Ukrainian.

[11] Vasylyev O.D., Schokin A.R. Ceramic Fuel Cells. In: Energy Saving in Regions. State Committee on Energy Saving, Analytical Issue, Alternative Energetics, Kyiv, 2003, p. 57-61, in Ukrainian.

[12] Vasylyev O.D., Burdin V.V. Fuel Cells. Electroinform, 2004, N1, p. 10-12, in Ukrainian.

[13] Vasylyev O.D. Fuel Cells: The Target Program. Electroinform, 2004, N1, p. 46-47, in Ukrainian.

SOLID OXIDE FUEL CELLS IN RUSSIA

A.K. DEMIN

Institute of High Temperature Electrochemistry, Ural Division of RAS
22, S.Kovalevskoy Str., 620219 Ekaterinburg, Russia, a.demin@ihte.uran.ru

Abstract: The paper gives a review of state-of-the-art and perspectives of Solid Oxide
              Fuel Cells in Russia. R&D in the field of SOFCs is concentrated in institutes of
              Russian Academy of Sciences whereas institutions of Minatom system develop
              SOFC technology. Brief outline of the RAS-Norilsk Nickel program
              "Hydrogen Energy and Fuel Cells" is presented.

Key words: Solid Oxide Fuel Cells

Fundamental research in the field of SOFCs started in Russia in the end of 1950s at the Institute of Electrochemistry in Sverdlovsk (now the Institute of High Temperature Electrochemistry, IHTE, Ekaterinburg). From the very outset, the works included study of solid oxide electrolytes (SOEs), electrode materials and electrode kinetics, other components of SOFC: interconnects, seals, etc.

The investigations of SOEs with oxygen ion conductivity (mainly based on zirconia, ceria and gallates) and with protonic conductivity (mainly cerates and zirconates) included detailed studies of their ceramic and transport properties, contact resistance of grains in SOE, ageing processes depending on the mode of fabrication, temperature, composition of material, impurities, gas atmosphere, and other factors.

11

N. Sammes et al. (eds.), Full Cell Technologies: State and Perspectives, 11-18.
© 2005 Springer. Printed in the Netherlands.

Complex investigations on electrode materials and electrode kinetics were carried out with the aim to elaborate electrodes with high electrochemical activity. The route and mechanism of the electrode reactions, dependence of polarization on electrode materials and structure, gas atmosphere, and other factors were studied. As a result of this research the electrodes based on Ni, Co, Cu, manganite, and cobaltite having high working characteristics were elaborated.

Various alloys, cermets as well as lanthanum chromites with different dopants were used as interconnects in SOFC prototypes. The seals having different softening temperatures were elaborated and widely used for gas-tight assembling cells into stacks.

The mathematical models of such processes in SOFC as charge, heat and mass transfer, diffusion, viscous flow of gases in channels were elaborated. The models gave necessary information for optimizing the SOFC design.

The cell with cross-flow channels for fuel and oxidant was elaborated in the IHTE ("block cell"). The cell working surface is 30 cm2, wall thickness - 0.5 mm. 12 - 18 cells connected in series form the stack. The tests of single cells continued up to 10000h, of stacks 4000h. The characteristics of the cells and stacks changed insignificantly during the tests. Recently, a new concept of a block cell was elaborated. The modified design differs from the old one by higher specific parameters. It has no analogues anywhere.

A 1 kW SOFC prototype based on tubular cells was designed, manufactured and tested in 1989. The cell had a working surface of 63 cm2 and a mean power 12.5 W or 0.2 W/cm2 at 900^0C. 16 cells were assembled into a stack. The prototype consisted of 6 stacks. The prototype was fed with methane; internal partial oxidation was used for fuel processing. The prototype worked at fuel utilization of 0.9 and had efficiency about 40%.

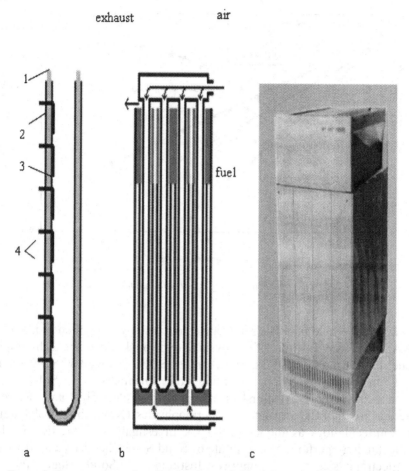

Figure 1. Cross-sections of a cell (a) and a stack (b) and a view of 1 kW prototype (c). 1 – electrolyte, 2 – anode, 3 – cathode, 4 – interconnect.

MOC CVD thin film technology as applied to YSZ electrolyte deposition on LSM porous substrate is developed in the IHTE. Electrolyte films of thickness up to 15 microns had acceptable gas-tightness for use in a SOFC. Several cells were manufactured and tested during last years. Specific characteristics of one of the cells are presented in Fig. 2.

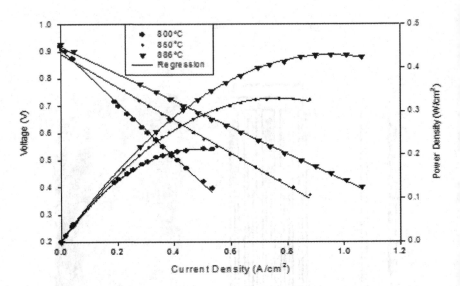

Figure 2. Specific characteristics of a cell with thin film YSZ electrolyte.

R&D in the field of SOFCs are carried out also by several other institutes of RAS. Boreskov Institute of Catalysis (Novosibirsk) develops technologies of fuel processing for SOFCs, one-chamber SOFCs as well as SOFCs with direct methane oxidation. Institute of Chemistry of Solids and Mechanochemistry (Novosibirsk) investigates new electrolyte and electrode materials for SOFCs. Institute of Thermophysics (Novosibirsk) elaborates thin film technology as applied to SOFCs. Investigations in the field of thin film technology perform also Institute of Solid State Chemistry and Institute of Electrophysics (Ekaterinburg), Institute of Solid State Physics (Chernogolovka). Institute of Electrophysics develops also nanomaterials and nanotechnology for SOFCs.

All these institutes participate in the Complex Program "Hydrogen Energy and Fuel Cells" announced by the Russian Academy of Sciences and the mining company Norilsk Nickel in November 2003. This program foresees carrying out fundamental research in the field of technologies related to all aspects of hydrogen energy and fuel cells. During first three years, the program will be focused to the following issues:

-   fuel cells (PEM and SOFC for stationary application including portable units);

-   fuel processors for PEM;

-   novel means for hydrogen storage.

Totally, more than 30 institutes of the RAS are involved in the Program.

Minatom institutions activity. All-Russian Research Institute of Technical Physics (VNIITF, Snezhinsk) and Institute of Physical and Power Engineering (IPPE, Obninsk) of Minatom develop SOFCs of tubular and planar design, respectively.

The VNIITF successfully tested 1.5 kW SOFC prototype in the end of 2003. Fig. 3 – 5 illustrate the concept and characteristics of the prototype and its units.

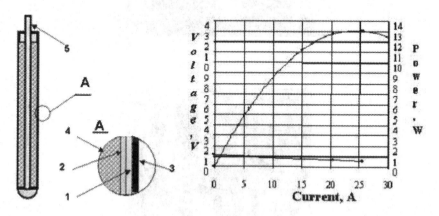

Figure 3. Cross-section of a single cell (left) and its current-voltage and current-power characteristics (right).

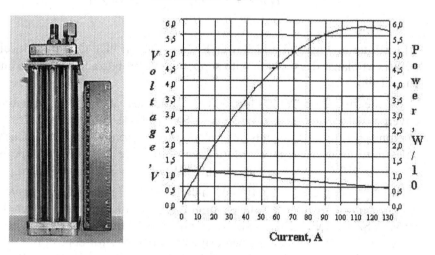

Figure 4. 8-cell bundle and its characteristics.

Figure 5. 14-bundle stack (top) and 4-stacks assembly.

Institute of Physical and Power Engineering (IPPE), Obninsk, develops SOFCs of planar design both with supported electrolyte and supported anode since 1996.

In the first case, YSZ electrolyte of thickness 250 – 400 microns is covered with Ni-cermet anode and LSM cathode of thickness 25 – 50 microns. IPPE has 4 patents for technology of porous catalytic interfaces "electrode-solid electrolyte". More than 200 single cells of this type were tested. At 950^0C, power density of 700 mW/cm^2 was achieved (Fig. 6).

In the second case, YSZ film of thickness 5 – 20 microns is deposited on a Ni-cermet disc of thickness around 1000 microns and covered by LSM+YSZ cathode. The anode has a multilayer structure: active layer and current collector. IPPE developed original technology for deposition of YSZ films in electrostatic field. At 800^0C, power density of 300 mW/cm^2 was achieved (Fig.6).

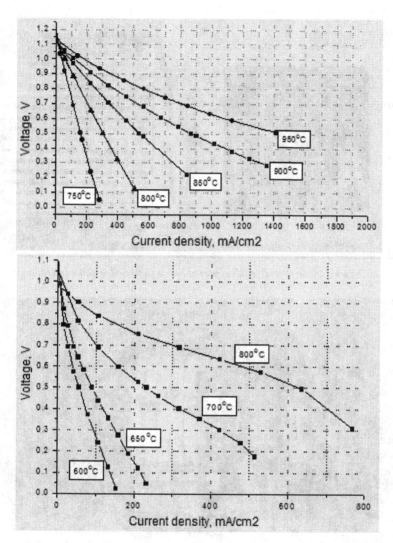

Figure 6. Current-voltage characteristics of a single cell: with supported YSZ electrolyte (top) and with supported Ni-cermet anode (bottom).

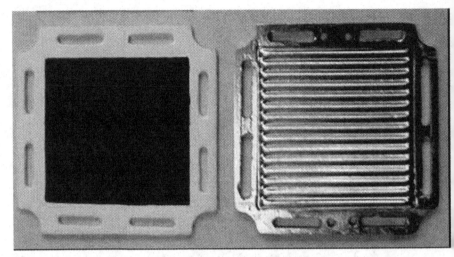

Figure 7. A single cell with supported anode (left) and metallic interconnect.

Figure 8. 5-cell stack of round cells with supported electrolyte before a test.

Round cells of 60 mm diameter with supported electrolyte were used for manufacturing 3- and 5-cell stacks (Fig. 8). Maximum duration of the tests was 200 hours. Maximum power at 900^0C and hydrogen as a fuel was ca. 30 W, i.e. specific power was about 260 mW/cm^2.

The level of R&D in the field of SOFC in Russia is rather high. Some technologies in this field developed by our institutions are very promising. The Program "Hydrogen Energy and Fuel Cells" gives good opportunity for rapid development of SOFCs and related technologies in Russia.

INTERMEDIATE-TEMPERATURE SOFC ELECTROLYTES

N. SAMMES[1], Y. DU[2]

[1]*Department of Mechanical Engineering, University of Connecticut*
[2]*Connecticut Global Fuel Cell Center, University of Connecticut, 44 Weaver Road, Storrs, CT 06269, USA*

Abstract: The electrolyte for solid oxide fuel cells (SOFC's) must be stable in both reducing and oxidizing environments and have sufficient ionic, as well as low electronic, conductivity at the operation temperature. Present SOFC's have extensively used stabilized zirconia, especially yttria stabilized zirconia, as the electrolyte. However, oxide ion conductors, such as doped ceria and perovskite-type oxides, have also been proposed as the electrolyte materials for SOFC's, especially for reduced-temperature of operation ($600^\circ C$ to $800^\circ C$), now known as intermediate-temperature solid oxide fuel cells (IT-SOFC).

Key words: Zirconia/Ceria/Lanthanum Gallate/Intermediate Temperature/SOFC

1. INTRODUCTION

During the last four decades, many oxide systems have been examined as potential electrolytes for intermediate temperature (IT) SOFC's [1]. An excellent review was presented by Etsell and Flengas in 1970 [2]; while more recent conductivity data are summarized by Minh and Takahashi [3]. The electrolyte for IT-SOFC's have to meet the following criteria: high ionic conductivity, low electronic conductivity, chemical and physical stability under reducing and oxidizing atmospheres at the operating temperature, ease of preparation and cost. In Fig. 1, the temperature dependence of the conductivity for several oxide ion conductors is shown. As shown in Fig. 1,

19

N. Sammes et al. (eds.), Full Cell Technologies: State and Perspectives, 19-34.
© 2005 *Springer. Printed in the Netherlands.*

bismuth oxide systems have the highest ionic conductivity, however, these oxides exhibit high electronic conductivity under reducing atmosphere, because of the reduction of Bi^{3+} to Bi^{2+} [3]. Certain perovskite-type oxides also exhibit high oxide ion conductivity; these are introduced later in the Chapter.

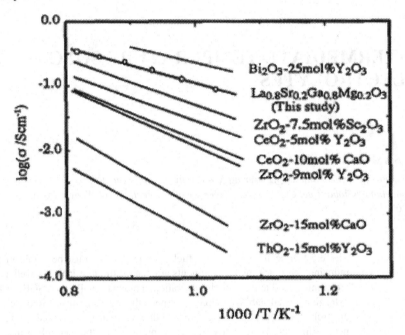

Figure 1. Effect of Temperature on Ionic Conductivity of Oxygen Ionic Conductors for SOFC Electrolytes (from 4).

Despite a tremendous amount of research activity over the past decade, criteria for the selection of new oxide ion conductors is poorly developed; stabilized zirconia, the first material to be discovered showing oxide ion conductivity, remains one of the best conductors for SOFC's, although doped ceria has been suggested as an alternative electrolyte for intermediate temperature SOFC's [5]. The contribution of electronic conductivity in the systems based on CeO_2 increases with increasing temperature under SOFC conditions. Both doped zirconia and ceria with high oxide ion conductivity show the fluorite-type structure, which is a face-centered cubic arrangement of cations with anions occupying all the tetrahedral sites.

The fluorite structure has a large number of octahedral interstitial voids. Thus this structure is rather open and rapid ion diffusion might be expected. A high temperature modification of zirconia has the fluorite-type structure, which is stabilized by addition of trivalent or divalent cations, such as Ca and Y, while pure ceria adopts the fluorite structure. Oxide ion conduction is provided by oxide ion vacancies and interstitial oxide ions. Intrinsic defects are fixed by thermodynamic equilibrium in pure compounds, while extrinsic defects are established by the presence of aliovalent impurities. To maintain electroneutrality, a soluble aliovalent ion in an ionic compound will be compensated by an increase in the concentration of an ionic defect. In the case of pure ZrO_2 and CeO_2, ionic conductivity is quite low, that is, the concentration of the oxide ion vacancies and interstitial oxide ions is low.

It was shown that the conductivity of doped zirconia and doped ceria vary as a function of dopant concentration, and has a maximum at a specific concentration. However, this maximum occurs at a much lower concentration than that expected. Attempts to explain this conductivity behavior have been made by Baker and Knop [6], and Hohnke [7], involving clusters in first and second coordination shells, and by Carter and Roth [8] based on structural effects. The dopants, Dy^{3+} and Gd^{3+}, with higher ion radius show a limited value of 8 mol %. The dopant Sc^{3+}, which has the closest ion radius to the host ion, Zr^{4+}, shows the highest conductivity and the highest dopant content. Similar conductivity dependence on the dopant was observed in the CeO_2 system. The highest conductivity was found at 10 mol % for Sm_2O_3 and at 4 mol % for Y_2O_3.

The diffusion of oxide ion vacancies is affected by the elastic strain energy, which is related to the size mismatch between the host and dopant cations [9]. Based on the earlier observations of Nowick [10], and Kilner and Steele [11], the importance of the defect pairs formed due to interaction between the oxide ion vacancies, $V_{\ddot{o}}$, and alivalent cations, M_{Ce}', in CeO_2 should be emphasized.

2. YTTRIA STABILIZED ZIRCONIA

The most advanced SOFC's employ oxide ion conducting zirconia-based electrolytes. The conductivity of the electrolyte determines their operation temperature. The temperature dependence of the electrical conductivity for zirconia-based oxides [12] is shown in Fig. 2.

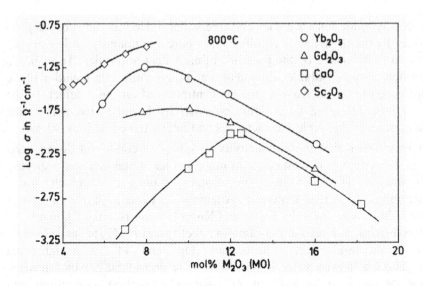

Figure 2. Effect of Dopants and Doping Level on Conductivity of Doped Zirconia at 800°C (from ref 13).

Yttria stabilized zirconia (YSZ), which has been used extensively in SOFC's, shows a conductivity of 0.14 S/cm at 1000 °C. Scandia doped zirconia (SSZ) has a higher conductivity and at 780 °C its value corresponds to that of YSZ at 1000 °C. In Fig. 3, the thickness of the electrolyte showing a resistance of 0.15 Ωcm^2 is shown on the right side of the vertical axis. The 0.15 Ωcm^2 resistance corresponds to 0.1V lost due to the electrolyte resistance at 0.5 mA/cm^2, where it is assumed that the cell voltage should be 0.7 V to maintain a total energy efficiency of greater than 50 %, and the cathode, the anode, and the electrolyte over-voltage contribute to equal parts of the cell voltage drop. The relationship between the temperature and the electrolyte thickness suggests that at an operation temperature of 1000 K, the thickness of the electrolyte should be less than 15 μm in YSZ based systems, while an electrolyte as thick as 150 μm could be used in the SSZ system, or alternatively the temperature could be lowered to approx. 715 K. A 15 μm thick electrolyte cannot be used in a self-supported configuration, because of the difficulty of handling such a thin electrolyte sheet.

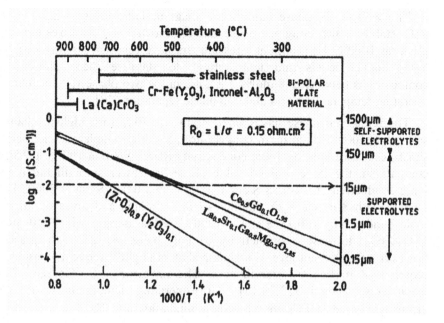

Figure 3. Comparison of the Specific Ionic Conductivities of the Most Commonly Interested YSZ, LSGM and GDC Electrolytes (from ref 14).

Room temperature monoclinic zirconia has little use as a SOFC electrolyte because it is predominantly an electronic conductor with low oxygen ion conductivity [15]. Cubic zirconia has high ionic conductivity but needs to be stabilized so that it retains its cubic structure at room temperature. Nernst discovered and reported in 1899 that mixtures of zirconia with other oxides such as magnesia showed high ionic conduction at high temperatures [16]. Two years later, he patented his further observation that the material composition (15% yttria and 85% zirconia) was suitable for electric-lamp glowers [17]. Westinghouse Electric Corporation has used a similar zirconia-based electrolyte in their SOFC development since 1962 [18].

MO and M_2O_3 oxides (Y_2O_3, Yb_2O_3, ScO_3, CaO, MgO) can stabilize the high-temperature polymorphic form of zirconia to lower temperatures by forming solid solutions with ZrO_2. The divalent and trivalent metal cations substitute for Zr^{4+} in the lattice sites, reducing their valence state and also creating vacancies in the oxygen sub-lattice. These oxygen vacancies enable doped zirconia to be an oxygen-ion conductor and therefore useful as a SOFC electrolyte. Of the materials mentioned above, yttria-stabilized zirconia (YSZ) offers the best combination of ionic conductivity and stability in the SOFC environment and is currently the material of choice for the electrolyte in commercial SOFC's.

The ZrO_2-Y_2O_3 phase equilibrium diagram indicates that 2.5-3 mol% Y_2O_3 stabilizes the tetragonal phase, and 8-9 mol% Y_2O_3 stabilizes the cubic phase at 1000°C [19], although researchers have reported slightly different Y_2O_3 dopant levels and temperatures to stabilize these phases. All compositions between 3-8 mol% Y_2O_3 are in the two-phase field at fuel cell operating temperatures, and undergo phase separation [20].

The tetragonal phase has a small grain size (0.5 μm), high mechanical strength (~1000 MPa at room temperature), and high toughness and thermal shock resistance, but low ionic conductivity (0.055 S/cm at 1000°C) [15]. In comparison, the large grained cubic phase has lower strength but higher ionic conductivity 0.14-0.18 S/cm at 1000°C and 0.052 S/cm at 800°C [21]. Doping with 8 mol% Y_2O_3 stabilizes cubic zirconia and gives the highest ionic conductivity (~0.18 S/cm at 1000°C) while doping with 9-10 mol% Y_2O_3-ZrO_2 decreases conductivity slightly (Fig. 4) [22]. In practice, however, yttria dopant level is maintained at, or slightly above, the minimum required (8 mol%) for stabilization. The properties of YSZ have been extensively studied [23,24,13]. The operating temperatures of YSZ electrolyte based SOFCs are typically around 800-1000°C.

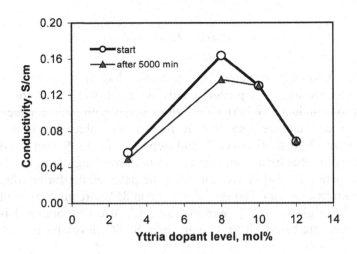

Figure 4. Effects of Y_2O_3 Concentration in Y_2O_3-ZrO_3 and Operating Time on Conductivity (after 15).

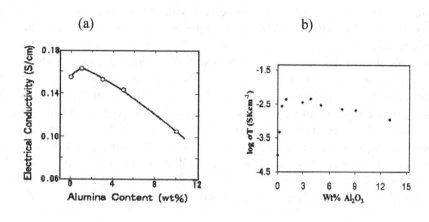

Grain Boundary Conductivity at 300°C (26).

Figure 5. Effect of Al₂O₃ Addition on Conductivity of 8YSZ (a) Bulk Conductivity at

1000°C (25); (b)

Adding a small amount of alumina affects the sintering behavior, electrical properties, and mechanical properties of YSZ [25-26]. 1- wt% Al_2O_3 results in dense sintered products, but above this level density decreases are observed with increasing alumina up to 20 wt%. Although maximum conductivity occurs at 1 wt%, the addition of Al_2O_3 causes an overall decrease in bulk conductivity (Fig. 5).

Another group of doped zirconia electrolyte materials is scandia-doped zirconia (SDZ). SDZ has high ionic conductivity relative to other doped zirconia compounds, as shown in Fig. 2. For example, 6-7 mol% and 8-10 mol% Sc_2O_3 doped ZrO_2 have ionic conductivities of 0.15-0.20 S/cm at 1000°C, and 0.11-0.12 S/cm at 800°C [22] respectively. Recent reports of cell performances with 6SDZ at 800°C are 0.6 W/cm² (140 μm 6SDZ support, LSM), 1.5 W/cm² (20 μm 6SDZ, LSM) and 2.4 W/cm² (20 μm 6SDZ, LSCF).

3. SCANDIA STABILIZED ZIRCONIA

Scandia doped zirconia (SDZ) is quite attractive as the electrolyte for SOFC's, especially for the intermediate temperature (600-800 °C) SOFC's. The long-term stability was not clarified before Yamamoto et al. reported the aging effect by annealing at higher temperatures [27]. ZrO_2 with 8 mol %

Sc$_2$O$_3$ exhibited a significant aging effect with annealing at 1000 °C (see Fig. 6). The conductivity of 0.3 S/cm at 1000 °C (as sintered) decreased to 0.12 S/cm after aging at 1000 °C for 1000 h. The conductivity value is comparable to that of ZrO$_2$ with 9 mol % Y$_2$O$_3$ after aging for 1000 h. On the other hand, ZrO$_2$ with 11 mol % Sc$_2$O$_3$ showed no aging effect by annealing at 1000°C for more than 6000 h. ZrO$_2$ with 11 mol % Sc$_2$O$_3$ shows a phase transition from the rhombohedral structure (low temperature phase) to the cubic structure (high temperature phase) at 600 °C, and a small volume change by the phase transition. The cubic phase was stabilized at room temperature by an addition of a small amount of CeO$_2$ and Al$_2$O$_3$. The conductivity of SDZ with CeO$_2$ and Al$_2$O$_3$ is slightly lower than the undoped SDZ. Similar aging effects to SDZ have also been observed in YSZ oxide ion conductors.

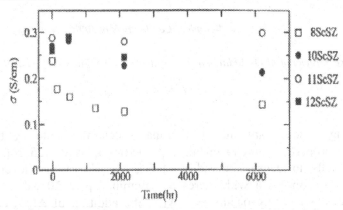

Figure 6. Conductivity of Scandia-Doped Zirconia (SDZ) at 1000°C (from ref 27).

4. DOPED CERIA

Doped ceria has been suggested as an alternative electrolyte for the low temperature SOFC's. Reviews for the electrical conductivity and conduction mechanism in ceria-based electrolytes have been presented by Mogensen et al. [28] and Steele [29]. Ceria possesses the fluorite structure as shown by stabilized zirconia. Mobile oxygen vacancies are introduced by substituting Ce^{4+} with trivalent rare earth ions. The conductivity of doped ceria systems depends upon the kind of dopant and its concentration. A typical concentration dependence of the electrical conductivity in the CeO$_2$-Sm$_2$O$_3$ system reported by Yahiro et al. [30] is shown in Fig.7. The maximum conductivity is observed at around 10 mol % of Sm$_2$O$_3$. The conductivity of the CeO$_2$-Ln$_2$O$_3$ system depends on the dopant ionic radius. The binding energy calculated by Butler et al. [31] shows a close relationship to the conductivity; the dopant with low binding energy exhibits high conductivity. Conductivity data for CeO_2-Gd_2O_3 and CeO_2-Sm_2O_3 possess ionic conductivities as high as 5×10^{-3} S/cm at 500°C; this conductivity value corresponds to 0.2 Ωcm^2 ohmic loss for an electrolyte of 10 μm thickness.

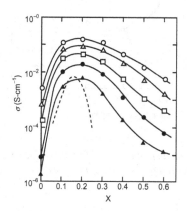

Figure 7. Concentration Dependence of the Electrical Conductivity in the CeO_2-Sm_2O_3 (30) ((○ 900°C) (△ 800°C) (□ 700°C) (● 600°C) (▲ 500°C) (----Ca stabilized zirconia at 900°C)).

These systems are quite attractive for electrolytes in low temperature SOFC's, and have been extensively examined. Ceria based oxide ion conductors are reported to have purely ionic conductivity at high oxygen partial pressures. At lower oxygen partial pressures, as prevalent on the anode side of the SOFC, the materials become partially reduced. This leads to electronic conductivity in a large volume fraction of the electrolyte extending from the anode side. When the fuel cell is constructed with such an electrolyte with electronic conduction, electronic current flows through the electrolyte even at open circuit, and the terminal voltage is somewhat lower than the theoretical value. In Fig. 8, total electrical conductivities (ionic and electronic) of $Ce_{0.8}Sm_{0.2}O_{1.9-\delta}$ are shown as a function of oxygen partial pressure Godickemeier and Gauckler [32-33] analyzed the efficiency of cells with $Ce_{0.8}Sm_{0.2}O_{1.9}$ by consideration of the electronic conduction. The maximum efficiency based on Gibbs free energy was 50 % at 800°C and 60 % at 600°C. Generally, the energy conversion in fuel cells should be discussed in terms of the efficiency based on the enthalpy at 298K (HHV). Therefore a factor of $\Delta G/\Delta H(298 \text{ K})$ should be used; $\Delta G(600°C)/\Delta H(298 \text{ K})$ is 0.84. Thus, the maximum efficiency of the cell with $Ce_{0.8}Sm_{0.2}O_{1.9}$ is 50 %. In practical operation, it is necessary to consider the fuel efficiency, which is generally less than 80 %. Therefore, the SOFC with $Ce_{0.8}Sm_{0.2}O_2$ should be operated at temperatures lower than 600 °C.

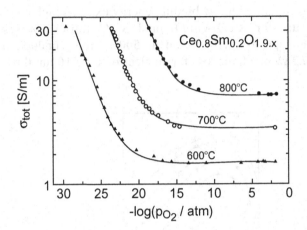

Figure 8. Total Electrical Conductivities (ionic and electronic) of Ce$_{0.8}$Sm$_{0.2}$O$_{1.9-\delta}$ as a Function of Oxygen Partial Pressure.

5. PEROVSKITE-BASED IONIC CONDUCTORS

In the search for new electrolyte materials, the perovskite based systems (ABO$_3$) have been considered as alternative options, particularly because ABO$_3$ can take on a number of different structures, and can be doped with aliovalent cations on both the A (for example Sr) and B (for example Mg)-sites. They can also accommodate very large concentrations of anion vacancies into their structures.

LaGaO$_3$-based perovskite type oxides, in particular, Sr- and Mg-doped LaGaO$_3$ (LSGM), exhibit high oxide ion conductivity [34-37]. The conductivity of La$_{0.9}$Sr$_{0.1}$Ga$_{0.8}$Mg$_{0.2}$O$_{3-x}$ is 0.12 S/cm at 800°C and 0.32 S/cm at 1000°C, which is similar to 9 mol% Sc$_2$O$_3$ doped ZrO$_2$ (0.31 S/cm) and higher by a factor of two compared to 8YSZ (0.16 S/cm) [38]. The major contributions to the high performance of LSGM electrolyte based fuel cells are the improvements in electrolyte conductivity and, more importantly, the exceptional structural and chemical compatibility with perovskite cathode materials, such as lanthanum cobaltite.

Sr- and Mg-doped lanthanum gallate is a complex system, which experiences phase changes and often co-exists with secondary phases. At room temperature, this material has an orthorhombic structure, which transforms to a rhombohedral structure at 445K [39]. Recent investigations [40-41] using high-resolution neutron diffraction have shown that the room temperature pseudo-orthorhombic changes to pseudo-rhombohedral between 250-500°C. The structure then transforms to the rhombohedral structure between 500-750°C.

Several phase diagrams of the La_2O_3-SrO-MgO-Ga_2O_3 system have been completed [40-41]. A solubility study [42] of Mg and Sr in the perovskite LSGM phases found that the perovskite phase containing Sr and Mg had a much better homogeneity range than if the perovskite phase contained only Sr or Mg. The Sr solubility in the perovskite with Sr and Mg could be as high as 20 mol% compared to only 2 mol% if only Sr was present. Fig. 9 illustrates the pure perovskite phase and impurity phase regions. LSGM must be as pure a phase as possible to optimize ionic conductivity.

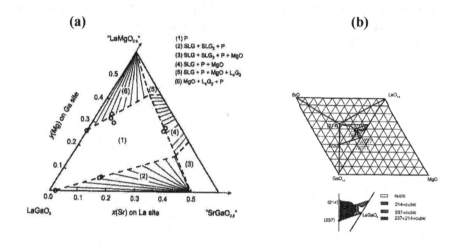

Figure 9. Phase Diagrams of (a) $LaGaO_3$-$LaMgO_{2.5}$-$SrGaO_{2.5}$ System and (b) $LaO_{1.5}$-SrO-$GaO_{1.5}$ Region (214=$LaSrGaO_4$, 237= $LaSrGa_3O_7$).

LSGM materials can be synthesized using a solid-state reaction [36,43,44], a sol-gel process [45,46], or the Pechini [47] process. Although LSGM has attractive properties, there are technical issues that must be resolved. For example secondary phases ($LaSrO_7$, $LaSrGaO_4$, $LaSrGa_3O_7$ or $LaSrGa_{0.9}Mg_{0.1}O_{3.95}$) are often formed [48]. The instability of LSGM at 1000°C [49] and 1500°C [50] has been reported. The unit cell volume for material sintered at 1500°C increased 0.27% (from 237.74 $Å^3$ to 238.39 $Å^3$) after being heated at 1000°C for 130 hours. It is believed that the volume increase is due to the segregation of the secondary phases $LaSrO_7$ and $SrLaGaO_4$ from the parent phase. This segregation alters the LSGM crystal structure thus decreasing the concentration of oxygen-ion vacancies, and also reducing the electrolyte oxygen ion conductivity. However, it is possible to prevent secondary phase segregation using appropriate processing conditions. For example, material kept in H_2 at 750°C for 132 hours did not exhibit phase separation. Therefore, LSGM should not be sintered at high temperatures for extended periods during cell fabrication. Slightly decreasing the A-site composition $(La_{0.9}Sr_{0.1})_{1-x}(Ga_{0.8}Mg_{0.20})O_{3-\delta}$, (x=0.02 and 0.05) has also been found to increase stability and conductivity [50-51].

Doping LSGM with, for example a second B-site dopant, such as Co or Cr, has been examined by a number of authors. $La_{0.8}Sr_{0.2}Ga_{0.8}Mg_{0.115}Co_{0.085}O_{2.8}$ has been examined as a potential electrolyte in a single cell SOFC, and was observed as having a maximum power density of 1.53 W/cm^2 at 800°C and 0.5 W/cm^2 at 600°C [52]. Another study [53] reports a similar fuel cell with a maximum power density of about 0.4 W/cm^2 at 650°C (Fig. 10).

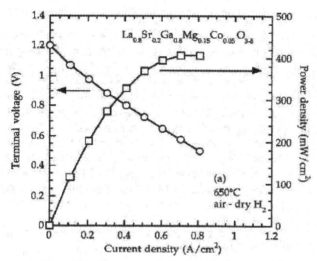

Figure 10. V/P Curves of a Fuel Cell: H_2+3 vol% H_2O, $Ni/La_{0.8}Sr_{0.2}Ga_{0.8}Mg_{0.115}Co_{0.085}O_{2.8}/Sm_{0.5}Sr_{0.5}CoO_3$, Dry Air.

The mechanical and physical properties of LSGM were examined by Sammes *et al.* [54]. $La_{0.9}Sr_{0.1}Ga_{0.8}Mg_{0.2}O_{3-\delta}$ was found to have a modulus of rupture (MOR) of 162 ± 14 MPa at room temperature and 55 ± 11 MPa at 900°C. MOR and fracture toughness of $La_{0.8}Sr_{0.2}Ga_{0.9}Mg_{0.1}O_{3-\delta}$ were observed to be 158 MPa and 1.63 MPam$^{1/2}$, respectively, which are much lower than the doped-ZrO_2 system. The TEC of 10^{-5} K^{-1} is similar to that of YSZ [55].

REFERENCES

[1] see, for example, Singhal, S., and Kendall, K, Editors, "High Temperature Solid Oxide Fuel Cells: Fundamentals, Design and Applications", Elsevier (2003).

[2] Etsell, T.H., and Flengas, *Chem. Rev.*, (1970) 70, 339.

[3] Minh, N.Q., Takahashi, T., in "Science and Technology of Ceramic Fuel Cells, Elsevier, Amsterdam (1995).

[4] Ishihara, T., Matsuda, H. and Takita, Y., *J. Am. Chem. Soc.*, (1994) 116, 3801.

[5] Steele, B.C.H., in *High Conductivity Solid Ionic Conductors*, ed: T. Takahashi, World Scientific Press, Singapore (1989).

[6] Baker, W.W., Knop, O., *Proc. Brit. Ceram. Soc.*, (1971) 19, 15.

[7] Hohnke, D.K., in *Fast Ion Transport in Solids,* eds: P. Vashista, J.N. Mundy, and G.K. Shenoy, North Holland, Amsterdam (1971).

[8] Carter, R.E., and Roth, W.L., in *Electromotive Force Measurements in High Temperature Systems,* ed: C.L. Alchok, I.M.M., london (1968).

[9] Kilner, J.A., and Brook, R.J., *Solid State Ionics,* (1982) 6, 237.

[10] Nowick, A.S., *Comments Solid State Phys.,* (1979) 9, 85

[11] Kilner, J.A., and Steele, B.C.H., in *Nonstoichiometric Oxides,* ed: O.T. Sorenson, Academic Press (1981).

[12] Yamamoto, O., *Electrochemica Acta,* (2000) 45, 2423

[13] Claussen, N., Ruhle, M. and Heuer, A.H., *Science and Technology of Zirconia II,* in *Advances in Ceramics.* 1984, American Ceramic Society: Columbus, OH. p. 555.

[14] Steele, B.C.H., *J. Mater. Sci.,* (2001) 36(5), 1053.

[15] Badwal, S.P.S., *Solid State Ionics,* (1992) 52, 23.

[16] Nernst, W., *Zeischrift fur Elektrochemie,* (1899) 6, 41.

[17] Nernst, W., *"Material for electric-lamp glowers,"* USA, US patent No.: U 00685730, 1901.

[18] Hooger, G., ed. *Fuel cell technology handbook.* CRC Press. 2003.

[19] Estell, T.H. and Flengas, N., *Chemical Reviews,* (1970) 70(3), 340.

[20] Scott, H.G., *J. Mater. Sci.,* (1975) 10, 1527.

[21] Badwal, S.P.S. and Foger, K., *Ceramics International,* (1996) 22, 257.

[22] Badwal, S.P.S., *Solid State Ionics,* (2001) 143, 39.

[23] Somiya, S., Yamamoto, N. and Yanagida, H., *Science and technology of zirconia III,* in *Advances in Ceramics.* 1988, American Ceramic Society: Columbus, OH.

[24] Heuer, A.H. and Hobbs, L.W., *Science and technology of circonia,* in *Advances in Ceramics.* 1981, American Ceramic Society: Columbus, OH.

[25] Bredikhin, S., Maeda, K. and Awano, M., *Solid State Ionics,* (2001) 144(1-2), 1.

[26] Feighery, A.J. and Irvine, J.T.S., *Solid State Ionics,* (1999) 121, 209.

[27] Yamamoto, O., Arachi, Y., Takeda, Y., Imanishi, N., Mizutani, Y., Kawai, M., and Nakamura, Y., *Solid State Ionics,* (1995) 79, 137.

[28] Mogensen, M., Sammes, N.M., Tompsett, G.A., *Solid State Ionics,* (2000) 129, 63

[29] Steele, B.C.H., *Solid State Ionics,* (2000) 129, 95

[30] Yahiro, H., Eguchi, Y., Eguchi, K., and Arai, H., *J. Appl. Electrochem.,* (1988) 18, 527.

[31] Butler, V., Catlow, C.R.A., Fender, B.E.F., and Harding, J.H., *Solid State Ionics,* (1983) 8, 109.

[32] Godickemeier, M., and Gauckler, L.J., *J. Electrochem. Soc.* (1998) 145, 414.

[33] Godickemeier, M., Sasaki, K., and Gauckler, L.J., *J. Electrochem. Soc.* (1997) 144, 1635.

[34] Huang, K., Tichy, R.S. and Goodenough, J.B., *J. Am. Ceram. Soc.,* (1998) 81(10), 2565.

[35] Huang, K., Tichy, R.S. and Goodenough, J.B., " *J. Am. Ceram. Soc.,* (1998) 81(10), 2576.

[36] Ishihara, T., Matsuda, H. and Takita, Y., *J. Am. Chem. Soc.,* (1994) 116, 3801.

[37] Feng, M. and Goodenough, J.B., *Eur. J. Solid State Inorg. Chem.,* (1994) t31(8-9), 663.

[38] Drennan, J., Zelizko, V., Hay, D., Ciacchi, F.T., Rajendran, S. and Badwal, S.P.S., *J. Mater. Chem.,* (1997) 7(1), 79.

[39] Marti, W., *J. Phys.: Condens. Matter,* (1994) 6, 127.

[40] Slater, P.R., Irvine, J.T.S., Ishihara, I. and Takita, Y., *Solid State Ionics,* (1998) 107, 319.

[41] Lerch, M., Boysen, H. and Hansen, T., *J. Phys. and Chem. Solids,* (2001) 62, 445.

[42] Matraszek, A., Kobertz, D., Singheiser, L. and Hilpert, K., "Thermodynamic studies of perovskietes on the basis of LaGaO3 and implications for SOFC." in *Seventh International Symposium on Solid Oxide Fuel Cells (SOFC VII).* 2001. Tsukuba, Japan: The Electrochemical Society, Inc., p. 319.

[43] Feng, M., Goodenough, J.B., Huang, K. and Milliken, C., *J. Power Sources,* (1996) 63, 47.

[44] Huang, P.N. and Petric, A., *J. Electrochem. Soc.,* (1996) 143(5), 1644.

[45] Polini, R., Pamio, A. and Traversa, E., "Sol-gel syntheses and phase purity of $La_{1-x}Sr_xGa_{1-y}Mg_yO_{3-z}$ solid electrolytes." in *Eighth International Symposium on Solid Oxide Fuel Cells (SOFC VIII).* 2003. Paris, France, p. 324.

[46] Huang, K., Feng, M. and Goodenough, J.B., *J. Am. Ceram. Soc.,* (1996) 79(4), 1100.

[47] Pechini, M.P., "*Method of preparing lead and alkaline earth titanaties and niobates and coating method using the same to form a capacitor,*" USA, US patent No.: 3330697, 1967.

[48] Djurado, E. and Labeau, M., *J. European Ceramic Soc.*, (1998) 18, 1397.

[49] Tao, S., Poulsen, F.W., Meng, G. and Orensen, O.T.S., *J. Mater. Chem.*, (2000) 10, 1829.

[50] Nakayama, T. and Suzuki, M., "Current status of SOFC R&D program at NEDO." in *Seventh International Symposium on Solid Oxide Fuel Cells (SOFC VII)*. 2001. Tsukuba, Japan: The Electrochemical Society, Inc., p. 8.

[51] Yamaji, K., Xiong, Y., Horita, T., Sakai, N. and Yokokawa, H., "Characterization of $(La_{0.9}Sr_{0.1})_{1+x}(Ga_{0.8}Mg_{0.2})O_{3-z}$ electrolytes with nonstoichiometric compositions." in *Seventh International Symposium on Solid Oxide Fuel Cells (SOFC VII)*. 2001. Tsukuba, Japan: The Electrochemical Society, Inc., p. 413-421

[52] Ishihara, T., Shibayama, T., Honda, M., Nishiguchi, H. and Takita, Y., *Chem. Commun.*, (1999): p. 1227.

[53] Ishihara, T., Shibayama, T., Honda, M., Nishiguchi, H. and Takita, Y., *J. Electrochem. Soc.*, (2000) 147, 1332.

[54] Sammes, N.M., Keppeler, F.M., Nafe, H. and Aldinger, F., *J. Am. Ceram. Soc.*, (1998) 81(12), 3104.

[55] Hayashi, H., Suzuki, M. and Inaba, H., *Solid State Ionics*, (2000) 128, 131.

SCANDIA-ZIRCONIA ELECTROLYTES AND ELECTRODES FOR SOFCS

J.T.S. IRVINE, T. POLITOVA, N. ZAKOWSKY, A. KRUTH, S. TAO, R. TRAVIS, O. ATTIA

School of Chemistry, University of St Andrews, St Andrews KY16 9ST, UK; Department of Mechanical Engineering, Imperial College London

Abstract: The aim of this study was to assess the potential of scandia stabilised zirconias for use in high temperature fuel cells. Such materials offer much better performance than the conventional yttria stabilised materials and now it seems that availability and cost may not be hugely problematic. This offers a very important way forward to fuel cell manufacturing. It has been shown that a small addition of 2 mol% yttria to scandia stabilised zirconia results in formation of cubic phase and so avoids major phase changes which we believe to be detrimental to long term electrolyte stability. This addition of yttria can be achieved without significant impairment of the electrical conductivity of the scandia-stabilised zirconia. We have shown that the mechanical properties of these scandia-stabilised zirconias are inferior to the yttria-stabilised zirconias, but the difference is not that great. Thus we have identified a new range of composition for use in solid oxide fuel cell electrolytes, which exhibits good compromise between phase stability and electrical properties. Some experiments were also performed to investigate the properties of titania doped scandia stabilised zirconia as a possible fuel cell electrode material. This has shown that the scandia containing titania zirconias were better than those of the yttria analogue. However, electronic conductivity was still not high enough to use such materials as electrodes without appropriate current collection.

Key words: SOFC/Scandia/Zirconia/Electrolyte/Electrode

N. Sammes et al. (eds.), Full Cell Technologies: State and Perspectives, 35-47.
© *2005 Springer. Printed in the Netherlands.*

1. INTRODUCTION

Solid oxide fuel cells offer clean generation of electricity at high efficiencies. Current developments are based upon an yttria-stabilised zirconia electrolyte which functions very well at temperatures in the 850-1000°C range for unsupported electrolytes and at temperatures as low as 700°C for supported thin films. Two design concepts predominate, the more expensive tubular design and the simpler planar design. The major weakness of the planar concept relates to interconnect and sealing problems. If a lower temperature electrolyte could be achieved, then much cheaper materials could be utilised for interconnects (steel, in fact) and the cost effectiveness of the planar design, in particular could be greatly enhanced. One other concern about the electrolyte that should be highlighted is stability; the tetragonal modification of yttria-stabilised zirconia undergoes a catastrophic transformation under hydrothermal conditions at about 300°C.[i]   Most commercial cubic zirconias are actually prepared with the composition $8mol\%Y_2O_3/92\%ZrO_2$, which strictly is just inside the two-phase cubic/tetragonal zirconia phase field[ii]. This means that on ageing these electrolytes at fuel conditions, tetragonal precipitates occur reducing conductivity[iii,iv]. Furthermore, on cycling this electrolyte between room temperature and operating temperature in the presence of water (a product of fuel cell operation) degradation and failure are highly likely due to the presence of the tetragonal form. Any compositional inhomogeneity, as we have observed in some of the "highest" quality commercial products, means that failure will occur. These problems can be reduced or even avoided if slightly higher yttria compositions are utilised[4].

Scandia-stabilised zirconia offers much higher conductivity than yttria-stabilised zirconia at 1000°C (x3) with a larger enhancement at lower temperature due to the lower activation energy (@0.65 eV vs. 0.95 eV)[v]; however, it has not been viewed as a serious alternative due to cost until recently. This has now changed with the increase in availability of scandia from Russia and China; presently, scandia is perhaps only 4 times as expensive per gram or approaching 2 times as costly per mole as yttria and it does seem that prices could decrease further. Although scandia-stabilised zirconias could be viewed as offering a route to even lower temperatures of operation than can be achieved using supported yttria-stabilised thin film electrolyte designs, we see the primary advantages of scandia systems as a dramatic decrease in internal cell resistance and offering more robust designs for low temperature operation. It is widely recognised in the field that once a low enough temperature of operation has been achieved to allow low cost steel interconnects to be utilised and sealing problems to be minimised, then there is no advantage for a further decrease in operating temperatures as most design concepts benefit from high temperatures, e.g. SOFC-gas turbine, and at these temperatures the internal reforming of hydrocarbons remains an attractive possibility.

2. OXIDE ION CONDUCTING ELECTROLYTES

Scandia-stabilised zirconia offers a route to even lower temperatures of SOFC operation than can be achieved using supported yttria-stabilised thin film electrolyte designs. Phase formation in the Sc_2O_3-ZrO_2 system with scandia contents between 5.0 and 15.0 mol% shows monoclinic, rhombohedral and cubic phases. Compositions with Sc_2O_3 content from 8 to 12 mol.% are members of a fluorite-type cubic solid solution. Impedance spectroscopy was applied to study the ionic conductivity. The best ionic conductivity was observed for the compositions with 10 and 11 mol.% Sc_2O_3. Long time annealing performed on samples with scandia content between 9 and 12 mol.% has been investigated at 800 ^{0}C for 1500 hours. Samples with 9 and 12 mol.% scandia showed significant decrease in conductivity. The 10 and 11 mol.% scandia samples exhibit high electrical conductivities and stabilities.

The effect of yttria co-doping of scandia zirconia on stabilisation and ionic conductivity has been investigated. Compositions in the ternary system $(Y_2O_3)_x(Sc_2O_3)_{(11-x)}(ZrO_2)_{89}$ (YxSc11-xZr89, $x = 0$-11) were prepared by solid state reaction and characterised by XRD, SEM and impedance spectroscopy. The electrical conductivity was studied as a function of temperature. The stability of the electrolyte materials was examined at the intermediate solid oxide fuel cell temperature of 800°C for up to 1500 hours. The contribution of the bulk and grain boundary resistivity of sintered and long term annealed compositions to total resistivity was estimated. Yttria additions were found to improve the phase stability of scandia stabilised zirconia. Even 1 mol. % Y_2O_3 addition eliminates the rhombohedral phase ($Sc_2Zr_7O_{17}$, the beta-phase) and stabilises the cubic structure at room temperature, Fig. 1. The best overall ionic conductivity was observed for compositions containing 2 mol. % Y_2O_3. The 1 and 2 mol. % Y_2O_3 compositions exhibit high electrical conductivities and good stabilities. Typically two linear regions were observed in the Arrhenius conductivity plot. The lower temperature region exhibited an activation energy of 1.3 eV and the high temperature region 0.75 eV. The difference between these activation energies decreases as yttria content increases and thus we attribute this change to short range ordering effect rather than simple near neighbour associations.

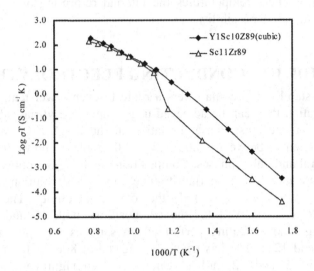

*Figure 1. Temperature dependences of conductivity of the Sc11Zr89 and Y1Sc10Zr89
compositions sintered at 1600 C for 16 hours*

Ionic conductivity and microstructure of the compositions in the $(Y_2O_3)_1$-$(Sc_2O_3)_x$-$(ZrO_2)_{99-x}$ and $(Y_2O_3)_2$-$(Sc_2O_3)_x$-$(ZrO_2)_{98-x}$ systems with total dopant content in the range 7-11 mol.% has been investigated as functions of temperature (300-1000 °C) with XRD, SEM and impedance spectroscopy. It has been demonstrated that the introduction of 8-11 mol.% additions of yttria and scandia fully stabilise the cubic phase to lower temperature and eliminate the cubic-rhombohedral phase transition. Introduction of 2 mol.% yttria into scandia stabilised zirconia results in an increased unit cell parameter and decreased grain size compared to the introduction of 1 mol.% yttria. Doping of scandia-zirconia with 2 mol.% yttria is preferable to doping with 1 mol.% yttria and leads to higher ionic conductivity at 800-1000 °C due to the lower activation energy for the ionic transport.

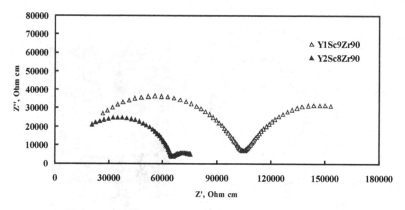

Figure 2. Impedance spectra plots of Y1Sc9Zr90 and Y2Sc8Zr90 samples, recorded at 353 ℃

All sintered samples of $(Y_2O_3)_2$-$(Sc_2O_3)_x$-$(ZrO_2)_{98-x}$ system showed high bulk conductivity with small contribution from the grain boundary resistivity. The grain boundary resistivity in $(Y_2O_3)_1$-$(Sc_2O_3)_x$-$(ZrO_2)_{99-x}$ system was more significant and comparable with the value of bulk resistivity at 400°C. The enhancement of electrical properties and the improving of the ceramic microstructure samples in the (Y_2O_3)-(Sc_2O_3)-(ZrO_2) system doped with 2 mol.% yttria compared with samples in the (Y_2O_3)-(Sc_2O_3)-(ZrO_2) system doped with 1 mol.% yttria is not simply due to microstructure and is indicative of improvements in the conduction of both the bulk and intergrain regions. This may reflect that there is still some small degree of secondary phase formation with only 1mol.% yttria co-dopant and so 1 mol.% Y_2O_3 doped samples show similar decreased conductivities to those of scandia zirconias that have not been rapidly quenched to retain complete cubic structure.

3. MECHANICAL TESTING

3.1. TEST PROCEDURES

A ceramic test rig for biaxial measurement (ball-on-ring) has been developed to accommodate a variety of different pellet sample dimensions from 21 to 28mm diameter inclusive. This was achieved by manufacturing a number of fittings as seen in Fig. 3 from a ceramic matrix composite material (Zircar RSLE 57), which is a low expansion high strength reinforced silica matrix composite product designed for use with temperatures up to 1100°C.

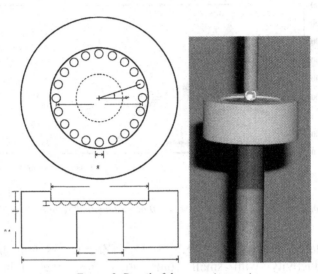

Figure 3. Detail of the ceramic test rig

The sample itself sits on a ring of sapphire balls – the number of balls being sufficient to give an approximately axisymmetric stress distribution during the test[vi], and the loading ball is also a sapphire ball. The fittings were set by means of alumina rods in an INSTRON 5584 testing machine. In order to permit measurements to be made at SOFC operating temperatures – taken to be up to 850°C – a tube furnace was adapted to fit the testing arrangement. The arrangement within the furnace includes independent temperature verification at the sample. The circular pellet arrangement in the test allows biaxial stress state to be established so that a greater proportion of the flaws will be subject to stress. This is thought to be particularly important for strength measurements.

During the test the cross head displacement is recorded as well as the compression load registered by the load cell. For temperature tests, the temperature is raised slowly (3°C per minute) and once the desired temperature is achieved and settled, the test is commenced. The properties are determined by use of standard equations which for elastic modulus[vii] and strength[6,viii]; the applicability of the equations were checked by use of finite element stress analysis in these geometries. Strength tests are clearly loaded until fracture occurs, and typical examples show initiation at the centre of the pellets and disintegration into two or three parts.

3.2. RESULTS

Four comparative materials of thickness 1.8mm, diameter 28mm were prepared as indicated in the above table. Elastic modulus measurements at room temperature varied somewhat from one specimen to another because consistency in manufacture is difficult and in particular, flaws can be introduced during the compaction of the materials. However, $(Y_2O_3)_8(ZrO_2)_{92}$ gave 80.0 GPa, $(Sc_2O_3)_{11}$ $(ZrO_2)_{89}$ showed 57.7 GPa, $(Y_2O_3)_2(Sc_2O_3)_9(ZrO_2)_{89}$ was 44.7 GPa, with $(Y_2O_3)_{11}(ZrO_2)_{89}$ 39.1 GPa, All the values here are low compared with literature values of 8mol.% yttria[ix]. Therefore it is not yet clear that the E values are reliable – there may be issues arising from the manufacture that mean consistently lower values. However, it is possible to deduce comparisons between the materials as they are produced in identical methods. It appears that the most favoured material, $(Y_2O_3)_2(ScO_3)_9(ZrO2)_{89}$ is more compliant than others, particularly 8 mol.% yttria, the most commonly used current SOFC electrolyte. Values of fracture strength and Weibull stress and modulus are shown in Fig. 4. Again these values make sensible comparisons, but are low compared with literature values for 8mol% yttria. This comparison does show that 11mol% scandia is slightly weaker, but not drastically so, than 8mol% scandia made by identical fabrication processes. 11mol% yttria is much less favourable.

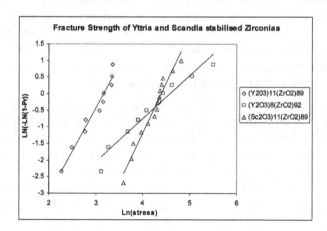

Figure 4. Weibull *Plots of Yttria and Scandia Zirconias*

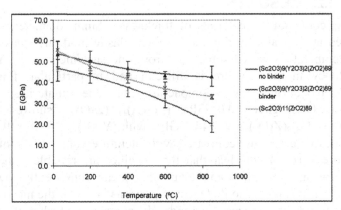

Figure 5. Elastic modulus measurements against temperature.

A further group of pellets tested were 2.0mm thick, 28mm diameter 11mol% scandia, and the preferred mixed dopant zirconia. Fig. 5 shows the results of many slow tests to determine the dependence of elastic modulus with temperature. This is similar to the behaviour of yttria zirconia under temperature, where a significant drop is evidenced from approximately 300°C upwards. For mechanical integrity the decrease in elastic modulus is good news because it means there is more flexibility to relieve any loads on the electrolyte material by strain. Note that the effect of the binder is to decrease the elastic modulus. There is a 7% reduction at room temperature – and this is very similar to that which would be expected by the increase of 5% porosity indicated by the difference in density[x].

These materials have also been assessed for strength using the biaxial flexure test. The comparison can be seen in Fig. 6, where an attempt has been made to derive Weibull statistics. There is less data than desirable, and the Weibull stress and modulus values are therefore not particularly reliable, although comparisons can be made. In this group of materials, it is clearly seen that the effect of the binder is to reduce the biaxial fracture strength. However, an increase in porosity is very significant on the effect of strength (mainly because of the increased number and size of defects that would be found), and here a 5% density difference leads to 30% reduction in strength. The strength of the 11 mol.% scandia is similar to the mixed doped zirconia, both including binder in manufacture.

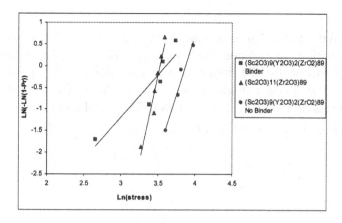

Figure 6. Approximate Weibull Plots for various Scandia Zirconias

3.3. CONCLUSIONS OF MECHANICAL EVALUATION

The testing has indicated that of all the investigated doped zirconias, the highest modulus and highest strength values belong to 8mol% yttria, that most traditional SOFC electrolyte. Compared with this, 11mol%scandia is 12% weaker and has significantly lower modulus (perhaps 20%). The favoured mixed dopant has similar strength to the scandia and a 7% lower elastic modulus. This mixed dopant is weaker than the traditional SOFC electrolyte material, but the difference is moderate, and the material has significantly lower elastic modulus allowing greater strain before mechanical integrity is affected. The conclusion is therefore that it would appear that there is not a bar mechanically to using $(Y_2O_3)_2(Sc_2O_3)_9(Zr_2O_3)_{89}$ as an SOFC electrolyte.

4. MIXED CONDUCTING ELECTRODES

The mixed conductor $Y_{0.2}Zr_{0.62}Ti_{0.18}O_{1.9}$ is a promising candidate as an SOFC anode material. In order to further improve the ionic conductivity, the conductivity of system Sc_2O_3-Y_2O_3-ZrO_2-TiO_2 was investigated when Y was substituted by Sc. Materials with single cubic fluorite structure were prepared by solid state reaction. The ionic and electronic conductivity were measured by 4-terminal d.c. method in 5%H_2 and, a.c. impedance spectroscopy in air and 5%H_2 respectively. It was found that TiO_2 may dissolve to about 18mol% in ternary system Y_2O_3-ZrO_2-TiO_2 and Sc_2O_3-ZrO_2-TiO_2, and 20mol% in the quaternary system Sc_2O_3-Y_2O_3-ZrO_2-TiO_2. It was also observed that the ionic conductivity is related to the oxygen vacancy concentration and the size of dopant ions and, electronic conductivity, to the lattice parameter, the sub-lattice ordering and degree of Ti-substitution. In addition, both ionic and electronic conductivities have been improved by the introduction of scandium into Y_2O_3-ZrO_2-TiO_2 system. The highest ionic conductivity (1.0×10^{-2}S/cm at 900°C) and electronic conductivity (0.14S/cm at 900°C) were observed for $Sc_{0.2}Zr_{0.62}Ti_{0.18}O_{1.9}$ and $Sc_{0.1}Y_{0.1}Zr_{0.6}Ti_{0.2}O_{1.9}$ respectively.

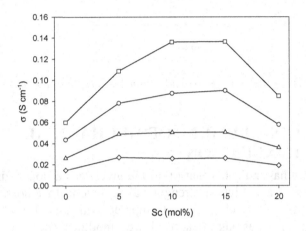

Figure 7. Electronic conductivity as a function of mol%Sc in $Sc_x Y_{0.2-x}Zr_{0.62}Ti_{0.18}O_{1.9}$. □ 900°C; ○ 800°C; △ 700°C; ◇ 600°C

Considering the required levels of both ionic and electronic conductivities for idea SOFC-anode materials, $Sc_{0.15}Y_{0.05}Zr_{0.62}Ti_{0.18}O_{1.9}$ is a promising candidate with ionic and electronic conductivities 7.8×10^{-3} and 1.4×10^{-1} S/cm respectively at 900°C. Electrochemical current-potential and impedance measurement techniques have been applied to a ceramic anode based on the mixed electronic and oxide ion conductor $Sc_{0.15}Y_{0.05}Zr_{0.62}Ti_{0.18}O_{1.9}$ (ScYZT). $Sc_{0.15}Y_{0.05}Zr_{0.62}Ti_{0.18}O_{1.9}$ had been previously identified as the best candidate for use as an SOFC anode material in the $Sc_x Y_{0.2-x}Zr_{0.62}Ti_{0.18}O_{1.9}$ series on the basis of both its ionic and electronic conductivity. Impedance spectroscopy measurements at 900°C typically showed high-, medium- and low-frequency arcs. These are interpreted as being related to the electronic current between anode and current collector, the O^{2-} ionic current exchange between anode and electrolyte and non-charge chemical process at the anode respectively. ScYZT significantly decreased the anode polarisation resistance thus decreasing the anode overpotential in comparison to when only a Pt electrode was used. It was found that interface and chemical impedance changed with changes in the electronic conductivity of ScYZT when a bias

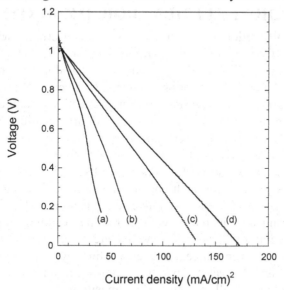

Figure 8. Fuel cell performance using Pt (a: 900℃, b: 950℃) and Sc0.15Y0.05Zr0.62Ti0.18O1.9 (c: 900℃, d: 950℃) as anode with Pt as cathode and YSZ as electrolyte.

potential was applied, however the overall polarisation impedance is not observed to change significantly. Where Pt was used as the anode on its own, the process is obviously different from that of ScYZT due to its lack of ionic conduction. The polarisation resistance for a Pt anode was 10 times greater than that with ScYZT at open circuit and was observed to change significantly with varying pO2. Similar phenomena to those with a Pt anode were observed when a porous YSZ interface was applied between YSZ electrolyte and a Pt current collector indicating that the significant decrease of polarisation resistance using ScYZT as anode is due to the mixed conducting or catalytic properties of the ScYZT. The changes in the anode overpotential as measured by d.c. and cyclic voltammetry methods may be attributed to the dependence of conductivity of ScYZT upon pO2. The three-probe fuel cell performance indicates that ScYZT is a promising anode material for use with an appropriate current collector.

5. NEUTRON DIFFRACTION STUDY OF SCANDIUM/YTTRIUM–DOPED ZIRCONIA

Neutron diffraction experiments were concerned with different compositions of the binary solid solution join, $Zr_{0.82}Y_{0.18-x}Sc_xO_{1.9}$, over the temperature range, 170-700 K; all with an overall dopant concentration of 18% but with different Sc:Y ratios. Two types of scattering were observed: narrow peaks arising from the fluorite Bragg reflections and broad peaks that were observed at some forbidden cubic fluorite lattice positions, i.e. hkl:h+k,h+l,k+l=2n; 0kl: k,l =2n; hhl: h+l=2n and h00:h=2n. The diffuse scattering is hence likely to origin mainly from localised tetragonal distortions. Diffuse scattering features were observed in all cases, however, they were less pronounced at higher temperatures. This suggests that the localised tetragonal distortion is reduced on heating. The magnitude of diffuse scattering also increases with increasing scandium dopant content. Isotropic temperature factors of oxygens are large, indicating significant distortion of the oxygen lattice, and increase linearly with temperature in the investigated range. At each temperature, temperature factors are of the same order of magnitude for all investigated compositions.

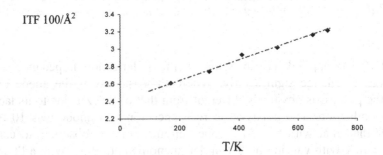

Figure 9. Oxygen temperature factors for composition
$Zr_{0.82}Sc_{0.09}Y_{0.09}O_{1.91}$ *as a function of temperature*

ACKNOWLEDGEMENTS

The authors would like to thank EPSRC (GR/R05772/01 and GR/R06052/0) for support.

REFERENCES

[1] Baumard and P. Abelard, Advances in Ceramics, Ed T. Clausen M. Ruhle and A.H. Meuer, 555.

[2] H.G. Scott, J. Mater. Sci., 1975, 10, 1527, I.R. Gibson PhD thesis, University of Aberdeen 1995.

[3] F.T. Ciacchi, K.M. Crane and S.P.S. Badwal, Solid State Ionics, 1994 74, 49.

[4] I.R. Gibson, G.P. Dransfield and J.T.S. Irvine, J. Europ. Ceram. Soc., 1998, 18, 661-667.

[5] S.P.S. Badwal, J. Mater. Sci, 1987, 22, 4125.

[6] T. Fett, Encyclopedia of Materials: Science & Technology, Elsevier Science Ltd, 2001, 823.

[7] D.R. Moore & S. Turner, Mechanical evaluation strategies for plastics, Cambridge University Press, 2001.

[8] K.D. Shetty, A.R. Rosenfield, P. McGuire, W.H. Duckworth, Am Cerm Soc Bull 59, 1193.

[9] A. Atkinson and A. Selcuk, Solid State Ionics, 2000, 134, 59.

[10] N. Ramakrishnan, V.S. Arunachalam, J Mater Sci, 1990, 25, 3930.

EXPLORATION OF COMBUSTION CVD METHOD FOR YSZ THIN FILM ELECTROLYTE OF SOLID OXIDE FUEL CELLS

Z. XU, S. YARMOLENKO, J. SANKAR

Center for Advanced Materials and Smart Structures, North Carolina A&T State University, Greensboro, NC 27411, USA

Abstract: Recently we developed a new method to deposit YSZ thin films based on combustion chemical vapor deposition (CCVD) technique. YSZ thin films have been processed on polished silicon and MgO substrates from combustion of an aerosol jet. Thin and uniform films have been obtained. The film growth rate was greatly enhanced by utilizing high reagent concentrations, proper substrate temperature, and application of DC bias.

Key words: Combustion Chemical Vapor Deposition/Thin Film Electrolyte/Solid Oxide Fuel Cells

1. INTRODUCTION

Yttria stabilized cubic phase zirconia (YSZ) has been used as electrolyte in solid oxide fuel cells (SOFCs) and oxygen sensors because of its high oxygen-ion conductivity over wide range of temperature and oxygen pressure [1]. In recent years, many research and development endeavors were placed in the thin film electrolyte fuel cells with the purpose to reduce the ohmic loss in the electrolyte and enhance the fuel efficiency of the fuel cells [2, 3]. Another important benefit of the thin film electrolyte is that fuel cells can run at reduced temperatures, which makes it possible to use less expensive materials for fuel cell construction. However, the porous surface of the cathode or anode gives rise to a great difficulty on the formation of a dense electrolyte thin film. In addition to the requirement of the film quality, cost is also a very important issue in the development of a coating technique.

N. Sammes et al. (eds.), Full Cell Technologies: State and Perspectives, 49-57.
© *2005 Springer. Printed in the Netherlands.*

The liquid fuel CCVD is a very promising technique to process metal oxide thin films. This method is based on combustion of fine aerosols of metalorganic solutions. Solutions were obtained by dissolving metal-organic reagents in organic solvent and were atomized into small sized aerosols by a nebulizer when it was mixed with oxygen. Thin films of Al_2O_3, SiO_2, CeO_2 and YSZ were deposited on various substrates [4, 5]. There is a potential to have conformal deposition of films on non-flat surface. Moreover, because there is no any vacuum chamber needed for the process, there is no size limitation, as well. Therefore, CCVD can be a low-cost and effective technique to produce YSZ thin films for industrial SOFC applications.

In this paper, we present the results of systematic study of YSZ thin films depositions with liquid fuel CCVD. The research of the YSZ deposition processes was focused on the enhancement of the film growth rate by exploring of the various CCVD parameters.

2. EXPERIMENTAL

A liquid fuel CCVD system developed for oxide film processing was introduced elsewhere [5]. Zirconium 2-ethylhexanoate (assay 96+%, Alfa Aesar) and Yttrium 2-ethylhexanoate (99.8%, Alfa Aesar) were chosen as reagents which were dissolved in toluene (HPLC grade, Fisher Scientific) separately. Mirror-polished MgO(100) and Si(100) were used as substrates. Three main parameters, such as reagent concentration, substrate temperature, and negative DC bias were investigated to maximize the film growth rate. The values of the parameters used in the experiments are shown in Table 1. The deposition time was 20 min. The characterizations of the samples included X-ray diffraction (XRD), scanning electron microscopy (SEM) and transmission electron microscopy (TEM).

Table 1. Parameters and their values used in the experiments

Variables	Values or Range
Reagent concentration (M)	5.0×10^{-4}, 1.25×10^{-3}, 2.0×10^{-3}, 2.75×10^{-3}, 3.5×10^{-3}, 4.25×10^{-3}, 5.0×10^{-3}
Substrate temperature (°C)	900-1300
DC bias (V)	0, 300, 700, 1100, 1500, 2000, 2500

3. EFFECT OF THE REAGENT CONCENTRATION

Fig. 1 shows the cross-sections of the films deposited at different reagent concentrations on Si(100) substrates. The dependence of the film growth rate on the reagent concentration is evident. Fig. 2 shows the linear relationship of the film growth rate with the reagent concentration within the range in this experiment. It does not mean that as high as possible a reagent concentrate should be used to obtain a high film growth rate. By observing the cross-sections of the samples, it is noticed that at the moderate high concentrations, e.g. 2.75×10^{-3}, 3.5×10^{-3}, and 4.25×10^{-3} M, a columnar film growth feature can be seen. The surface SEM observations also demonstrated polycrystalline structures of the films deposited at these concentrations. When the concentration reached 5.0×10^{-3} M, the columnar growth feature disappears in the cross-sectional image, cauliflower-like feature was seen on the surface of the film. Moreover, increase of the concentration below 4.25×10^{-3} M enlarged the crystal size in the films. Contrarily, the particle size of the film deposited at 5.00×10^{-3} M became very small and the film was of lower density. Therefore, the maximum reagent concentration of 4.25×10^{-3} M considered to be optimal from both the film growth rate and film quality points of view.

4. EFFECT OF THE SUBSTRATE TEMPERATURE

The substrate temperature controls the film growth modes and quality of the deposited films in chemical vapor deposition. The temperature affects the absorption/desorption, the reactivity and diffusivity of adatoms or clusters on the substrate surface, and hence affect the film growth rate and its morphology. Within the range of temperature employed in the experiments, two different film growth regimes were determined with a division point at about 1070 °C. In the lower temperature region that is referred to as surface reaction controlled regime, film growth rate was greatly affected by the substrate temperature with activation energy of 124 kJ/mol. The region of temperatures above 1070°C is referred to as gas phase diffusion controlled regime. In this regime film growth rate was still increased with the substrate temperature but with significantly lower activation energy of 20 kJ/mol (Fig. 3).

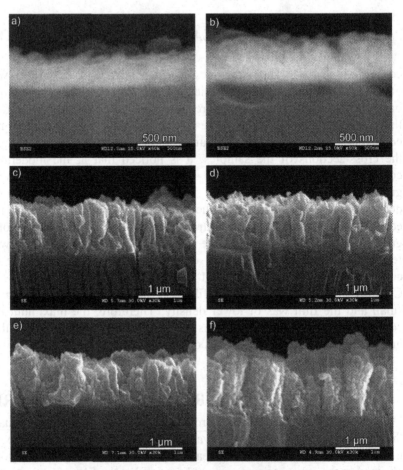

Figure 1. Cross-sectional micrographs of the YSZ films on Si(100) deposited at a substrate temperature of about 1180 °C, substrate to nozzle distance of 51mm, flow rates of oxygen and liquid fuel of 1600 and 2.0 cm^{-3}/min, respectively, and at reagent concentrations of a) 1.25×10^{-3}, b) 2.0×10^{-3}, c) 2.75×10^{-3}, d) 3.5×10^{-3}, e) 4.25×10^{-3}, f) 5.0×10^{-3} M, for 20 min.

Moreover, for the films deposited at the higher substrate temperatures, X-ray characteristic peaks were stronger (as shown in Fig. 4) and the particles were more faceted at one predominant orientation [6]. Our TEM study also revealed superstructure in the films deposited at higher substrate temperatures. Fig. 5 shows one of the examples. The superstructure shown in the micrograph is caused by Y_2O_3 doping and is responsible for O^{2-} conductivity. These results suggest that higher substrate temperatures give rise to higher film growth rates and better film quality.

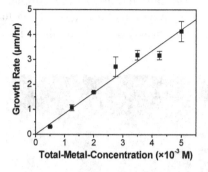

Figure 2. Film growth rate versus reagent concentration in toluene.

Figure 3. Film growth rate at different temperatures (total metal concentration 4.25×10^{-3} M)

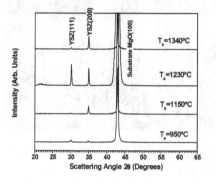

Figure 4. XRD spectra of the YSZ films deposited on MgO(100) substrates at different substrate temperatures.

Figure 5. TEM image shows the periodic superstructures of the Y_2O_3 (8 mol. %) doped YSZ film deposited at about 1200 °C.

5. EFFECT OF DC BIAS

DC bias was commonly used in CVD systems to enhance nucleation rate and/or deposition rate [7]. However, it was rarely applied in flame assisted vapor deposition because CCVD is usually conducted at the atmospheric pressure and in equilibrium state. Therefore, there must be a small amount of ions which can be accelerated by the applied electric field. In other applications, application of high DC voltage could lead to electrical discharge and enhance the decomposition of precursors in the reaction. Negative DC voltages from 0 to 2500 V were used in our experiment.

Fig. 6 shows the cross-sectional micrographs of some of the samples. In Fig. 6 (a), film is in columnar structure. When the applied DC voltage exceeded 700 V, forest-like microstructure was observed in the films as shown in Fig. 6 (b), (c) and (d).

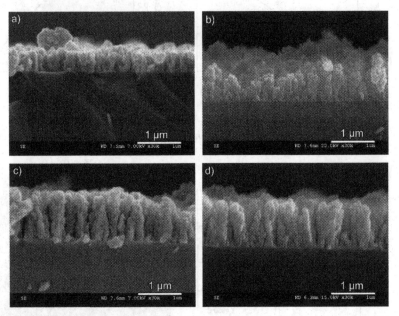

Figure 6. Cross-sectional micrographs of the films deposited on Si(100) substrates for 20 min at the condition of reagent concentration of 2.75×10^{-3}, flow rates of oxygen and liquid fuel of 1600 and 2.0 cm^{-3}/min, respectively, substrate to nozzle distance of 57 mm, substrate temperature of about 1170°C, and at different DC biases of (a) 0, (b) 700, (c) 1500, and (d) 2500 volts.

The "trees" seem to be formed by stacking of the crystallites in size ranging from 30 to 50 nm which is apparently much smaller than the particle size in the film deposited without DC bias. The mechanism to make the difference is not clear at present time. We assume that the application of the DC bias increased the concentration of species in the gas phase, i.e. the flux of the arriving atoms or clusters on the substrate, as well as the concentration of the ionized species, which results in more nuclei on the substrate. The increase of the concentration of species in the gas phase can also lead to less time for the adatoms to diffuse on the surface of the substrate or the deposited film, which can result in small particle in the deposited films. The similarity of the microstructures of the films deposited under DC bias and that deposited at high reagent concentration (see Fig. 1f) can support our assumption. The low density films in the forest-like microstructure can be densified by post thermal treatment. The images in Figure 7 indicate both the films deposited at 0 voltage bias and 2500 V bias become densified after thermal annealing.

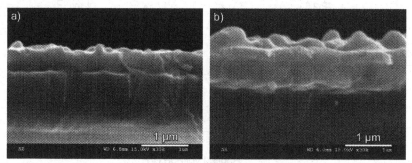

Figure 7. Cross-sectional micrographs of the annealed films at 1250 °C for 2 hrs. a) The same sample as shown in Figure 6 (a), b) the same sample as shown in Figure 6 (d).

It is worthwhile to identify the effect of the DC bias on the enhancement of the film growth rates. Fig. 8 shows the correlation between the film thickness and the applied DC voltage for the cases of the as-deposited films and thermally treated films. The correlation between the DC current and the applied voltage is also shown in the graph. A general trend of the film growth rate enhancement by the application of a DC bias can be distinguished. The thickness of the films deposited under DC bias of 1100 V is more than twice of that deposited without DC bias. The result supports the point of view that application DC bias increase the flux of the atoms, ions and/or clusters towards substrate. The general enhancement trend in the thermally treated and densified films is also apparent. The curves in the graph also reveal that 300 V bias was not enough to generate enhancement. On the other end of the curves, it is shown that a bias of above 1500 V was not necessary since the rate curves show saturation afterwards.

Our exploration of CCVD technique has shown that it can be used for development of dense YSZ films. By choosing properly high reagent concentration in the solution, high substrate temperature, and by applying negative DC bias, the coating rate and quality can be improved to meet the low-cost processing requirement for development of the SOFCs.

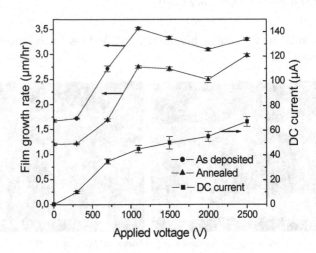

Figure 8. Correlation between the film thickness and the applied DC bias for the as-deposited and thermally treated films. Also the correlation between the DC current and the applied voltage in each experiment.

ACKNOWLEDGMENTS

This research was supported by NSF and DOE through the Center for Advanced Material and Smart Structures (CAMSS).

REFERENCES

[1] Sukhvinder P.S. Badwal, Ceramic Superionic Conductors, chapter 11 in Materials Science and Technology, A Comprehensive Treatment, Ed. R.W. Cahn, P.Haasen, E.J. Kramer, v. 11, 1994.

[2] B. Steele, Ceramic ion conducting membranes and their technological applications, Comptes Rendus de l'Academie des Sciences, Serie IIc: Chimie 1998, v. 1, N 9, p. 533-543.

[3] C. Bernay, A. Ringuede, P. Colomban, D. Lincot, M. Cassir, Yttria-doped zirconia thin films deposited by atomic layer deposition ALD: a structural, morphological and electrical characterisation. Journal of Physics and Chemistry of Solids, 2003, v. 64, N 9-10, p. 1761-1770.

[4]   A.T. Hunt, W.B. Carter, and J.K. Cochran, Jr., Combustion chemical vapor deposition: a novel thin-film deposition technique. Appl. Phys. Lett., 1993, v. 63, N 2, p. 266- 268.

[5] Z. Xu, J. Sankar, Q. Wei, Combustion chemical vapor deposition of YSZ thin films for fuel cell applications. MD (American Society of Mechanical Engineers), 2001, 95(Effects of Processing on Properties of Advanced Ceramics), p. 1-8.

[6] Z. Xu, S. Yarmolenko, and J. Sankar, Preparation and properties of YSZ electrolyte thin films via liquid fuel combustion chemical vapor deposition. Ceram. Eng. Sci. Proc., 2001, v.23, p. 711-718.

[7] S.D. Wolter, J.T. Glass, B.R. Stoner, Bias induced diamond nucleation studies on refractory metal substrates, Journal of Applied Physics, 1995, v.77, N 10, p. 5119-24.

DEVELOPMENT OF DMFC FOR PORTABLE APPLICATION

A. SMIRNOVA, D. KANG, N. SAMMES

Connecticut Global Fuel Cell Center, University of Connecticut, 44 Weaver Road, Storrs, CT 06269, USA

Abstract: This paper describes experimental results regarding manufacturing and electrochemical characterization of DMFCs with cathode catalyst layers based on supported and unsupported platinum catalysts. The cell performance in hydrogen and methanol solution, as different fuels, is compared and discussed in terms of electrochemical surface area (ESA) of the catalyst. The influence of airflow rate, humidification, and temperature on the DMFC performance is also presented. The cell based on a carbon supported cathode catalyst with 1.3 mg/cm^2 Pt loading shows better performance in comparison to the conventional cell with 3.6 mg/cm^2 of Pt-black.

Key words: PEMFC/DMFC/Carbon Supported Catalyst/Unsupported Catalyst

1. INTRODUCTION

Intense international academic and industrial research efforts have recently placed the direct methanol fuel cell (DMFC) on the brink of commercialization [xi,xii]. The major advantage of the DMFC relative to other fuel cells is the simplicity of the system, since it does not need a reformer, and start up is possible at ambient pressure. These advantages make it possible to miniaturize the DMFC as the power source for use in

59

N. Sammes et al. (eds.), Full Cell Technologies: State and Perspectives, 59-72.
© *2005 Springer. Printed in the Netherlands.*

stationary power generation applications, and make it especially suitable for portable power supplies.

The use of methanol as a fuel in comparison with hydrogen has several advantages. Methanol is a liquid fuel available at low cost, which can be easily handled, stored, and transported without an intermediate reformer. However, there are a few problems limiting the wide spread commercial application of the DMFC [xiii,14,15] and among them:

- Methanol Crossover through a proton conductive membrane
- Low activity of methanol oxidation catalysts
- Existing overvoltages on the cathode side due to reduction of oxygen and oxidation of methanol
- Carbon dioxide/water management

Methanol crossover is one of the major obstacles to prevent DMFC from commercialization. This effect, which is caused by diffusion of methanol through the membrane, reduces the cell efficiency. Competing with the oxygen reduction reaction on the cathode side, methanol adsorbs on the surface of Pt and is oxidized to CO_2. The main poisoning species formed during the chemisorption and oxidation of methanol is carbon monoxide [xiv], blocking the active sites on the surface of Pt [xv]. In order to increase the cell efficiency high Pt loadings are usually used approaching a few mg/cm^2 of precious metal [16] which substantially increases the cost of the fuel cell system.

Presently available carbon supported Pt catalysts have metal surface areas much higher than the surface area of pure Pt black catalysts. This suggests the possibility of using carbon supported Pt catalysts for the fabrication of highly performing MEAs with much lower precious metal loading and thus, lower cost.

The aim of this work was to study the influence of the cathode layer composition on the DMFC performance and compare the performance of the cells operating on different fuels, *viz.* H_2 and methanol.

The cell performances were estimated in correlation to the electrochemical surface area (ESA) of the catalyst and mass-transport limitations in the cathode catalyst layer. The experimental data are discussed in terms of temperature, methanol concentration, cathode gas humidity and flow rate.

2. EXPERIMENTAL

2.1. PREPARATION OF CATALYST PASTES AND MEMBRANE ELECTRODE ASSEMBLIES (MEA)

Catalyst powders, namely Pt-black and Pt/Ru-black, were purchased from Alfa Aesar and carbon supported catalyst (46.5% Pt) from Tanaka Inc.

Catalyst pastes with Pt and Pt/Ru-black unsupported catalysts were made by mixing the catalyst with DI water (1:2 wt % ratio). In addition, 5% Nafion 1100 solution (10 wt %) and 1,2-propanediol as a high boiling point solvent in N_2 flow were added under stirring conditions with further evaporation of iso-propanol as a volatile component of Nafion 1100 solution.

Preparation of pastes with carbon supported catalysts included mixing the catalyst powder with DI water (1:2 wt % ratio) and addition of 5% Nafion 1100 solution (25 wt %), iso-propanol, and 1,2-propanediol. Prepared mixtures were homogenized for 5 hours using an Ultra-Terrux homogenizer with further evaporation of volatile component.

In this work, three membrane electrode assemblies (MEAs) with the same composition of the anode Pt/Ru black layer and different cathode layers were prepared. The cathode layer of the first MEA (#1) was made using a Pt-black catalyst with 3.6 mg/cm^2 Pt loading on the cathode side. The cathode layer of the second MEA (#2) consisted of two layers: the external Pt-black layer in contact with the gas diffusion layer (2.8 mg/cm^2) and the internal Pt/C layer (0.3 mg/cm^2) in contact with the membrane. The cathode layer of the third MEA (#3) was made using carbon-supported Pt/C catalyst from Tanaka Inc. (1.3 mg/cm^2). The anode layers were made in a similar way using Pt/Ru-black with 5.0 ± 0.1 mg/cm^2 catalyst loading and 10 wt % of Nafion.

High Resolution TEM (HRTEM) was used for the estimation of the size of Pt particles in the carbon supported and unsupported Pt samples. The catalyst with Nafion was deposited on gold Quantifoil grids covered with carbon foil with ~1.2 µm circular holes. TEM analyses were performed in a Philips EM-420 TEM equipped with a Kevex-Noran Be-window EDS system, and operating at an accelerating voltage of 100 kV. High-resolution TEM (HRTEM) lattice images were obtained in a JEOL JEM-2010 TEM equipped with a UHR objective lens pole-piece ($C \cong 0.5$mm) and operated at an accelerating voltage of 200 kV. In this configuration, the point-to-point resolution at Scherzer defocus is less than 0.19 nm.

The MEAs were manufactured using a screen-printing decal technique. The electrode layers were printed on the surface of the inert Teflon film using the Systematic Automation Model 810 Series Screen Printer and the screen with polyester thread (110 mesh) from SAATI Print. After drying, MEAs were hot pressed with Nafion 117 membrane (EW 1100) as received from DuPont at 70 kg/cm^2 and 130°C for 5 min followed by peeling off Teflon films from both sides of the MEAs. Further experiments were provided in the Electrochem hardware with 6.25 cm^2 active area and included two Teflon (0.25 mm thick) gaskets to maintain a constant 0.1 mm pinch from each side of the MEA. Carbon paper SIGRACET 10BB from SGL Technologies GmbH was used as a gas diffusion layer (GDL).

Cyclic voltammetry (CV) was used for estimation of the electrochemical surface area of the supported and unsupported catalysts. The corresponding CV plots for cathode and anode catalyst layers were obtained using a Princeton potentiostat. The measurements were made in the range of 0.01-0.8 V with a 100 cc/min gas flow rates and a 20 mV/sec scan rate at room temperature. In the case of the cathode catalyst layer, pure nitrogen was supplied to the cathode as a working electrode and hydrogen to the anode. CV measurements for the anode as a working electrode (WE) were made with nitrogen on the anode side and hydrogen gas on the cathode.

The cell performance was evaluated using the Teledyne test Station equipped with the humidification chambers, mass flow and temperature controllers, and 50 Amp load box from Scribner, which allowed simultaneous correction of the cell potential for resistive losses in membrane (iR-drop).

3. RESULTS AND DISCUSSION

3.1. CYCLIC VOLTAMMETRY MEASUREMENTS

The total active area of the catalyst is typically much higher than the electrochemical surface area determined from cyclic voltammetry measurements. That means that not all catalyst particles are in contact with the ionic conductor Nafion and only a certain amount of it participates in electrochemical reaction.

Fig. 1 represents the *in-situ* electrochemical characterization of the cathode catalyst layers based on supported and unsupported catalysts. At two times lower loading of carbon supported catalyst (MEA #3), its limiting current density and thus, electrochemical surface area (ESA) is 3 times higher than the corresponding surface area of unsupported catalyst (MEA #1). This is due to the smaller size of the Pt particles in carbon supported catalyst and thus, higher surface area in contact with Nafion (Table 1).

The ratio between the total and electrochemical surface area of unsupported catalyst (approximately 2:1) indicates that only about 50 % of

Pt particles are in contact with Nafion. The other part is disconnected from the polymer electrolyte and therefore, is inactive. This effect is due to the relatively large size of Pt-black particles (Fig. 2a,b). Additional agglomeration of the Pt-black particles aggravates it even further (Fig. 2c). However, in the case of carbon-supported catalyst the interfacial area, where polymer electrolyte, catalyst, and hydrogen as a reactant meet within the electrode approaches 90% (Table 1), which correlates with catalyst structure possessing evenly distributed Pt particles on the surface of spherical particles of carbon (Fig. 2d).

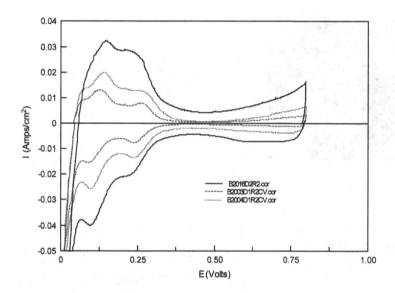

Figure 1. Cyclic voltammetry plots for the DMFCs (WE – cathode):
MEA #1- Pt/black; Loading 3.6 mg/cm²;
MEA #2 - Pt/black + Pt/C; Total loading 3.1 mg/cm²;
MEA #3 - Pt/C; Loading 1.3 mg/cm²

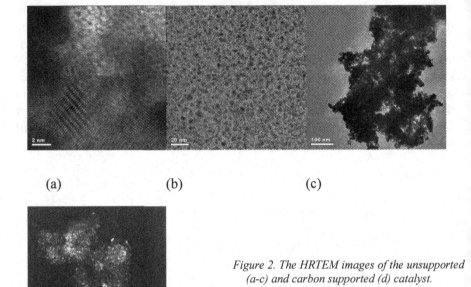

(a)   (b)   (c)

Figure 2. The HRTEM images of the unsupported
(a-c) and carbon supported (d) catalyst.

(d)

Table 1. Comparison of the total Pt area and Electrochemical Surface Area (ESA) estimated
from cyclic voltammetry.

MEA #	Catalyst in the cathode layer	Pt surface area (CO adsorption)	ECA, m^2/g (CV)
1	Pt	28	15
2	Pt and Pt/C	-	30
3	Pt/C	121	110

In all three MEAs the rate of methanol oxidation was facilitated by the platinum-ruthenium unsupported catalyst, which in the presence of CO as a byproduct of the reaction, exhibit an electrochemical activity higher than pure Pt. However, compared to Pt supported and unsupported catalysts, the electrochemically active surface area of PtRu alloys cannot be determined by hydrogen adsorption using cyclic voltammetry due to the overlap of hydrogen and oxygen adsorption potentials, and the tendency for hydrogen to absorb in the ruthenium lattice [xvi]. However, under the same operation conditions, cyclic voltammetry can be used for qualitative estimation of the similarity in the PtRu anode layer properties.

The cyclic voltammetry plots obtained for the anode as a working electrode (Fig. 4) indicate that the anode layers for the three MEAs made with screen printing technique have similar loadings, which correlates using the weights of Pt/Ru decals estimated prior to the hot-pressing procedure.

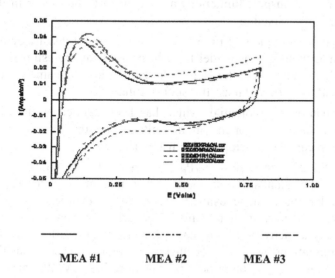

MEA #1   MEA #2   MEA #3

Figure 4. Cyclic Voltammetry plots for the MEA anodes in (WE - anode).

3.2. CELL PERFORMANCE IN HYDROGEN AS A FUEL

It can be assumed, that membrane resistance, cathode overpotential, and diffusion resistance in the GDL, mainly determine the cell performance in the presence of hydrogen as a fuel. Consequently, the cathode overpotential, being usually higher than the anode one, consists of the diffusion resistance and the potential drop due to the ohmic resistance of the cathode.

In order to estimate the input of the cathode catalyst layer on the MEA cell performance, each cell before testing in methanol solution has been evaluated in PEMFC operation conditions, namely with hydrogen on the anode side and oxygen or air on the cathode. Fig. 6 shows the dependencies of cell voltage and cell voltage compensated for membrane resistance vs. current density.

All three cells demonstrated similar performance in oxygen, however in air the difference in cell performances is much more pronounced. Taking into account that all three MEAs were manufactured under the same conditions, namely using the same membrane, gas diffusion, and anode catalyst layer, this difference could be attributed to mass transport limitations in the cathode catalyst layer. This somewhat expected result can be explained by 2.5 times higher Nafion loading in carbon supported catalyst in comparison to Pt/blacks since, for the catalyst layers with the same thickness, mass transport limitations are expected to be higher in the layer with higher Nafion loading.

The experimental plots of iR-free voltage vs. current density obtained for O_2 or air and hydrogen as a fuel have been used for the estimation of the factors, which determine the cell polarization losses, namely activation potential, Tafel slope, and mass transport limitations.

The activation overpotential, which has been defined as the difference between the theoretical open circuit voltage and the cell voltage at 10 mA/cm^2, was calculated using the Nernst equation:

$E=E_o+(RT/2F)$ $ln(P_{H2}P_{O2}^{1/2}$ $/P_{H2O})$, where $E_o= -\Delta Go(P,T)/2F$, and $\Delta Go(P,T)$ is the standard free energy of formation of water vapor. This calculation has been made using the assumption that in the low current density region, namely at 10 mA/cm^2, the polarization data is controlled only by the catalytic activity of the catalyst, and is not influenced by mass transport limitations generally occurring at higher current densities. The calculated values of activation polarization were 282 mV, 266 mV, and 272 mV for the cells #1, 2, and 3 correspondingly.

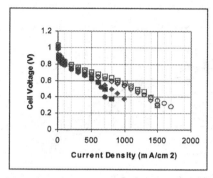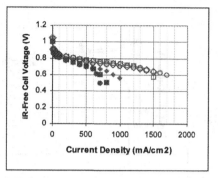

Figures 5 and 6. Cell voltage (a) and compensated cell voltage (b) as a function of current density in O₂/air and hydrogen as a fuel. Open symbols - performance in oxygen; symbols with background - performance in air.

O- MEA #1; ☐- MEA #2; ◊- MEA # 3

The GDL layer resistance has been estimated using a well-known approach for limiting currents (I_{lim}), according to which for non-reacting diffusion the current density in the mass transport limited region can be normalized by the factor $1-I/I_{lim}$. The plot of the iR-corrected cell voltages vs. normalized current densities gives the voltage losses due to the mass transport limitations in the GDL. Gradually increasing with current density, these losses at 500 mA/cm² were found to be 5mV for MEA #1 and 12 mV for MEA #2 and #3, which is much lower than the corresponding activation and membrane resistive losses. Furthermore, in the case of DMFC these losses could be even less pronounced because of the lower current density operation region.

3.3. CELL PERFORMANCE IN METHANOL AS A FUEL

The comparative data of the cell performances obtained in methanol/air (Fig. 7) shows that the best performance was achieved for cell #3 with carbon supported catalyst. As it is shown in Fig. 7, the performance of the cell # 3 improved with time, which was not observed for the cells #1 and #2 tested under similar conditions. This effect can be explained by the presence of a higher concentration of Nafion in the Pt/C cathode catalyst layer, which in this case probably needs time to approach equilibrium and humidification.

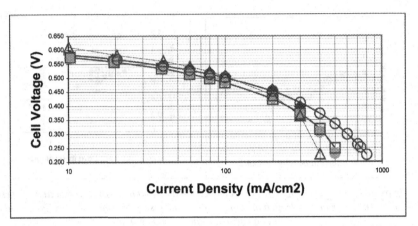

Figure 7. Performance of three MEAs measured in air and 1 mol/L methanol solution as a fuel

◆- MEA #1 after 2 weeks in methanol solution;

●- MEA #2 after 2 weeks in methanol solution;

☐-MEA #3 after 5 hours in methanol solution; ◇- MEA #3 after 12 hours in methanol solution; ○- MEA #3 after 2 weeks in methanol solution.

The temperature dependence of the cell voltages vs. temperature has been measured at 60, 70, 80, and 90°C for the cells in 1M methanol/air at ambient pressure. In the range 70-80°C the cell voltages increased with temperature, however at 90°C all three cells demonstrated an insignificant drop in performance, probably due to the drying of Nafion. Thus, for all three cells maximum performance was observed at 80°C. Typical dependence of the cell voltage vs. current density at different temperatures is given in Fig. 8. Increase in voltage (50 mV) was observed with increasing temperature from 60 up to 80°C, however the further increase in temperature, up to 90°C, did not seem to improve the performance further.

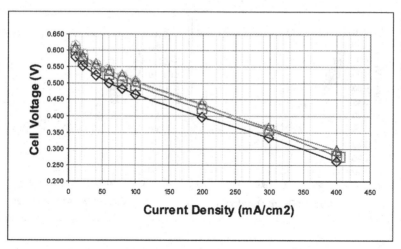

Figure 8. Cell voltage (a) vs. current density for MEA #3 in methanol/air at:
◇- 60°C; ▢- 70°C; △-80°C; ○- 90°C;
Flow rate of methanol solution: 5 mL/min

The effect of air flow rate on the DMFC performance (Fig. 9) has been estimated at 80°C and a constant flow of 1.0 mol/L methanol solution. A significant effect of the flow rate of humidified cathode gas was postulated as being due to several reasons. One of them is the reduced effect of methanol crossover on catalyst 'poisoning' due to partial oxidation of methanol and correspondingly the decrease of the cathode overpotential, which is confirmed by a continuous increase in the OCV from 785 mV at 200 cc/min to 830 mV at 5000 cc/min of the air flow rate. This effect could also be due to increasing the flux of oxygen through the membrane, which decreases the anode overpotential due to oxidation of carbon monoxide as a byproduct of methanol reaction of oxidation. This data is in correlation with the results of the earlier work by Scott *et al* [xvii] showing improvement of the DMFC performance in oxygen under pressurized conditions.

A particularly significant increase in the DMFC performance can be seen at high current densities, which is probably related to the significant decrease of voltage losses due to the less pronounced mass transport limitations.

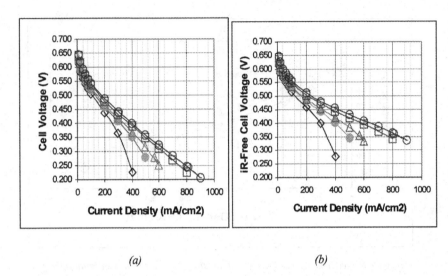

(a)   (b)

Figure 9. Cell voltage (a) and compensated cell voltage (b) vs. current density for MEA #3 at different flow rates of air: ◇- 200 cc/min; ● -500 cc/min; △- 1000 cc/min; □- 3000 cc/min; ○-5000 cc/min; Methanol flow rate: 5 ml/min (OCV is not shown)

The cell voltage in air at 1.0 M concentration of methanol solution was stable during 300 hours of operation (510 mV at 100 mA/cm^2). That is 50 mV higher than the performance of commercially produced DMFCs at the same current density.

The performance of the cells has been tested in different relative humidity (RH) conditions, namely at 100% RH and in dry air (0% RH). Fig. 10 represents typical cell performance in air/1 mol/L methanol solution for one of three cells. Under humidified conditions, the cell performance was gradually improving by increasing the air flow rate, which was typical for all cells and could be explained by the combination of a few factors, such as better oxygen reduction kinetics and less concentration of carbon monoxide produced on the cathode as a result of methanol oxidation. This effect, which was more pronounced for thin membranes, is still noticeable for thick Nafion 117 membrane and increases the cell performance by 50 mV. However, the cell performance in dry air decreased while increasing the air flow-rate (Fig. 10). In dry air, at flow rates higher than 300 cc/min, the cell voltage dropped because of the increased cathode layer resistance as a result of its continuous drying.

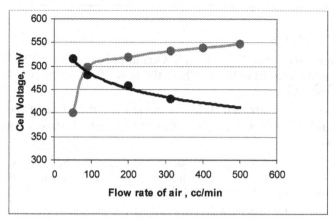

Figure 10. DMFC voltage vs. flow rate of air at 100 mA/cm² current density in 1.0 M methanol solution. Upper curve - humidified air; lower curve- dry air.

This effect could be also attributed to the insufficient membrane water management in relatively thick Nafion 117 membrane and thus, to higher cathode layer resistance.

4. CONCLUSIONS

The DMFC based on carbon supported catalyst with low catalyst loading (1.3 mg/cm²) has been successfully tested in a methanol/air environment. The cell shows better performance in comparison to the cell based on unsupported catalyst with twice the Pt-black loading. These results are explained by the higher surface area of Pt carbon supported catalyst and are in good correlation with CV and BET data. The results show that carbon supported catalyst can be successfully used as the electrode material for the fabrication of relatively cheap cathode catalyst layers in DMFC. Further work is needed to estimate the lower concentration limit of the catalyst, which is sufficient to maintain stable performance and long-term endurance.

ACKNOWLEDGEMENTS

This work was carried out under the auspices of Battelle, Prime Contract # DAADOS-99-D-7014/DO 0007 subcontract 172791 for the US Army Communications and Electronics Comman

REFERENCES

[xi] Costamagna P. and Srinivasan S. Quantum jumps in the PEMFC science and technology from the 1960s to the year 2000. J Power Sources 2001; 102:242-252

[xii] Heinzel A., Barragan V.M. A review of the state-of-the-art of the methanol crossover in direct methanol fuel cells. J. Power Sources 1999; 84:70-74

[xiii] Scott K., Taama W., Argyropoulos, K. Sundmacher, The impact of mass transport and methanol crossover on the direct methanol fuel cell. J. Power Sources 1999; 83:204-216

[xv] PtRuRhNi nanoparticle electrocatalyst for methanol electrooxidation in direct methanol fuel cell. J Power Sources 2004; 224:236-242

[xvi] Lamy C., Lima A., LeRhun V., Delime F., Coutanceau C.H., Leger J. M., Recent advances in the development of direct alcohol fuel cells. J Power Sources 2002; 105:283-296

[xvii] Antolini E., Formation of carbon-supported PtM alloys for low temperature fuel cells: a review. Materials chemistry and physics 2003; 78: 563-573

[xviii] Scott K., Taama W., J. Cruickshank, Performance of a direct methanol fuel cell, J Appl Electrochemistry 1998; 28: 289-297

SYNTHESIS OF POROUS AND DENSE ELEMENTS OF SOFC BY ELECTRON BEAM PHYSICAL VAPOR DEPOSITION (EBPVD)

F.D. LEMKEY, B.A. MOVCHAN

International Center for Electron Beam Technologies of E.O.Paton Electric Welding Institute, National Academy of Science of Ukraine, 68 Gorky St. Kiev-150, 03150, Ukraine

Abstract: EBPVD synthesis of certain elements that constitute the functional components of SOFCs i.e. the porous electrodes and dense electrolyte were examined. Partially stabilized zirconia (YSZ) plus 10 wt % Ni porous electrode elements were produced in the form of thick layers. The microporous structure was formed during non-equilibrium deposition of the vapor phase condensates whose two phases do not appreciably interact in the solid state and resulted from a so-called "shadowing" effect. During initiation and subsequent growth of various crystallographic faces of the nuclei at different rates, certain microrelief forms and the faces and microprotrusions screen the adjacent regions of the surface from the vapor flow. This results in microporosity aligned normal to the electrolyte layer and permits fuel molecules to reach the electrolyte interface. Subsequent deposition of a thin dense crack free YSZ electrolyte layer resulted in the synthesis of two of the four operating components, i.e. anode, electrolyte, cathode and interconnects. This high rate vapor condensation synthesis of SOFC elements is in sharp contrast to the current powder technology route that uses tape casting or sintering of expensive colloidal powder. The potential value effectiveness of EBPVD processing appears to make it an attractive alternative to the high cost of powder technology.

Key words: Solid Oxide Fuel Cells/Electron Beam Deposition/Porosity/Zirconia Ceramics

N. Sammes et al. (eds.), Full Cell Technologies: State and Perspectives, 73-80.
© 2005 *Springer. Printed in the Netherlands.*

1. INTRODUCTION

The selection of materials and processing of the four elements, porous anode, solid electrolyte membrane, mixed ion-electron conducting cathode and interconnecting elements presents enormous challenges to the construction of a solid oxide fuel cell. Each material must have the electrical properties required to perform its function in the cell at sufficiently high temperature and have enough chemical and structural stability to endure each fabrication operation. To demonstrate the applicability of electron beam physical vapor deposition (EBPVD) to the partial construction of a planar cell the porous anode supported approach was examined. The direct method of high rate EBPVD opens up the possibility of producing inorganic materials with either fully dense or micro-porous structures. Thick films, foils or free standing sheets up to 1-3mm thick can be prepared as well as dense/porous coatings on finished articles [1]. Average size of pores in condensate layers can be varied from several nanometers (nm) to several microns (μm). Total porosity may be produced up to 50-55 %.

YSZ is currently deposited on all last stage aircraft gas turbine components by EBPVD and there exists a rich history [2,3] in the development of these porous YSZ over-layers as thermal barrier coatings (TBC). Because these TBCs operate at temperatures exceeding 1100 °C and are strain compliant with higher expanding superalloy substrates due to its columnar porosity after thousands of engine thermal cycles; it was thought that a similar approach to the formation of a porous anode would be useful. The nickel phase would be distributed within the YSZ columnar and/or dendrite crystallites and thus provide for greater mechanical stability during thermal cycling at the lower temperature interval of both 'high and intermediate' temperature SOFCs. Dense thin YSZ layers required for electrolyte can be produced by reducing the substrate temperature (<500 °C) or overheating the condensate to temperatures >1400 °C. Therefore possibilities are open to produce two-layer condensates having a porous base and a dense top layer in one technological cycle of deposition.

2. METAL-CERAMIC POROSITY

Microporous structures can be derived from condensates with phases that do not interact in the solid state e.g. Ni-ZrO_2, Al_2O_3-ZrO_2 and Ni15Cr5Al-Al_2O_3 [4]. The microporous structure is formed during non-equilibrium condensation of the vapor phase. One of the main mechanisms of microporous structure formation is based on the so-called "shadowing" effect. During initiation and subsequent growth of various crystallographic faces of the nuclei at differing rates, certain microrelief develops on the condensation surface. The faces and microprotrusions, growing at maximum speed, screen the adjacent slower growing regions on the surface from the vapor flow. This results in the formation of inner microvoids. The shadowing effect is enhanced if the angle of the vapor flow incident on the condensate surface is less than 90° or the second phase particles form and grow on the condensation surface as shown in Fig. 1.

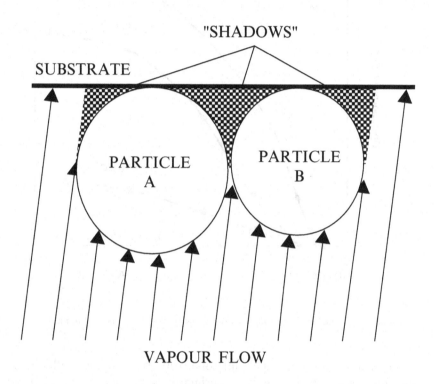

Figure 1. Scheme of the "Shadowed" areas created by spherical Particles A and B

The technology of producing microporous condensates consists of simultaneous electron beam evaporation in vacuum of the matrix material and the addition material from two adjacent copper water-cooled crucibles up to 70mm diameter. Alternatively the evaporation from one crucible of a composite ingot (matrix + additive) and subsequent condensation of the mixed vapor flow on the substrate can be employed. The distance from the evaporators to the substrate was equal to 300-320 mm. Substrate preheating in the range of 700-1000 °C influences the degree of open porosity as shown for a representative metal oxide system in Fig. 2.

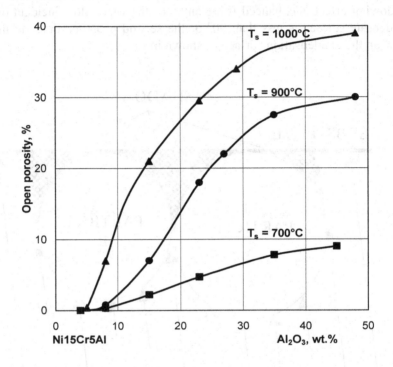

Figure 2. Dependence of open porosity on Al₂O₃ content and deposition temperature T_s

Greater amounts of open microporosity were found with the rise in substrate temperature. This is due primarily to particles coarsening with the increase in condensation temperature and of the respective sizes of "shadow" sections. Coalescence of "shadowed" areas during the condensate formation creates a system of open microporosity. Over the range of 20-30 wt % Al_2O_3 the average pore sizes are 0.32 µm, 0.55 µm, and 0.80 µm at deposition temperatures of 700, 900, and 1000 °C respectively. Microporous structure within the examined metal–ceramic condensates changes insignificantly after long-term annealing at 1100 and 1200 °C. A measurable increase (12-14 %) in the open porosity observed is probably a result of coalescence of isolated micropores and their merging with the open pore system.

3. ZrO_2+Ni POROUS ANODE

Microporous structures form over the entire range of concentrations, i.e. both in the metal (Ni) matrix with ceramic (ZrO_2) additive and a ceramic matrix (ZrO_2) with an additive of metal (Ni). Fig. 3 a, b illustrates examples of the microstructure of condensates of Ni-17 wt % $ZrO_2(7Y_2O_3)$ and $ZrO_2(7Y_2O_3)$-10wt % Ni, respectively, of 0.4-0.5mm thickness, deposited on a stationary substrate at Ts=1000 °C. Note a certain difference in the shape and size of pores at practically the same porosity level, i.e. ~30vol.%. The

Figure 3a. Cross Section of Ni +17wt %YSZ

microporosity of freestanding condensates was determined using the methods of hydrostatic weighing and mercury porometry with a 'Porosizer-9300' apparatus.

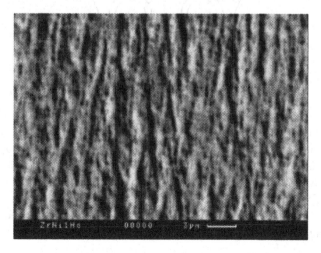

Figure 3b. Cross Section of YSZ + 10wt%Ni

4. YSZ ELECTROLYTE LAYER ON POROUS YSZ+Ni

A diagram of UE-204 electron beam physical vapor deposition which was adapted to produce two-layer discs of $ZrO_2(Y_2O_3)$ + Ni or Ni + $ZrO_2(Y_2O_3)$ of 0.5m diameter and different thickness of the microporous layer (0.5-1.5 mm) and dense layer (5-20 μm) is shown in Figure 4.

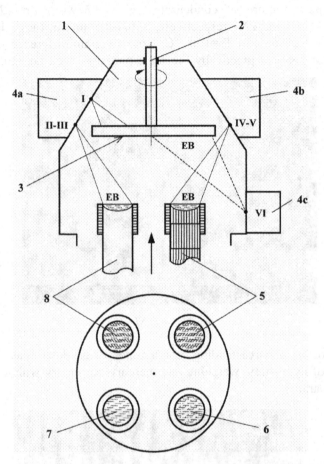

Figure 4. UE-204 EBPVP Unit for Synthesis of Two Layer Discs

This unit has a working chamber (1), vertical shaft (2) and rotating substrate in the form of a disc (3). Chambers (4a), (4b), and (4c) accommodate six electron beam guns: I – VI. EB guns II and III are designed to evaporate $ZrO_2(Y_2O_3)$ and Ni ingots; EB guns IV and V are designed for evaporation of reserve ingots (5) and (6). EB gun I is designed for preheating the rotating substrate (disc), EB gun VI is used for heating (densifying) the condensate surface.

The process for deposition of a two-layer composition is conducted as follows. First the EB gun I is used to preheat a rotating substrate (disk) up to the condensation temperature T_s. Then a small amount of CaF_2 is evaporated from the surface of reserve ingot (5) or (6), and a thin (5-10μm) separation layer of CaF_2 is deposited on the disk surface. After that guns II and III are switched on. Ingots (7) and (8) $ZrO_2(Y_2O_3)$ and Ni respectively are evaporated and a microporous layer of the required thickness is deposited. At the final stage of deposition of the top layer nickel evaporation is interrupted (by switching off EB guns II and III) and gun VI is switched on to overheat (1400 °C) and compact the surface layer to the required thickness. After completion of the deposition process and substrate cooling the two-layer condensate is separated from the substrate. Examples of two layer (anode and electrolyte) elements are shown in Figs. 5 and 6 for varying amounts of YSZ porosity levels. It will be noted that the sharpness of the anode/electrolyte interface may provide a model for the better understanding of the electrochemical processes at the metal-electrolyte-gas three-phase boundary.

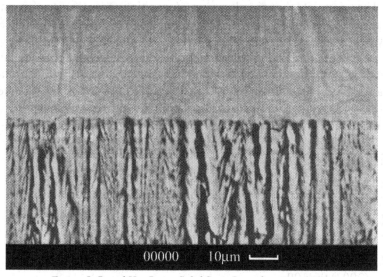

Figure 5. Rapid Hot Beam Solid State Densification of YSZ Layer

There is a high probability that the above continuous technological process may be extended and three-layer anode/electrolyte/cathode disks maybe deposited by further evaporation of the cathode components from ZrO_2 + Ag, or La_2O_3 or Mn_2O_3, for example. A preliminary cost estimate demonstrates the high effectiveness of the EBPVD processing route. For EBPVD the cost of 1Kg of YSZ ingot is ~50 $ plus the cost of other components in the form of low cost ingots, e.g. Ni.

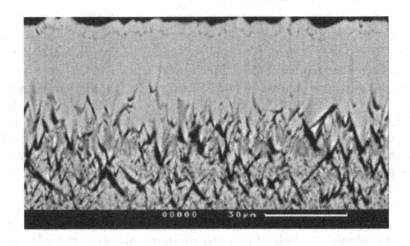

Figure 6. Microstructure of a section of as-deposited two-layer ZrO_2 $(7Y_2O_3)$ –
$Ni/ZrO_2(7Y_2O_3)$ condensate (polished andnon-etched surface.

The cost of the simultaneous deposition process is also moderate although the capital cost of the equipment can exceed 500,000 $. With powder ceramic processing the powder costs can exceed 2000 $/Kg plus the cost of other components and the multi-step processes to produce laminated systems from fine powders which are moderate or high in value. The potential cost effectiveness of direct processing of SOFC elements by EBPVD appears to offer an attractive alternative to other fine powder consolidation technologies.

REFERENCES

[1] Movchan B.A. Modern Science of Materials in the 21st Century, Naukova Dumka Kiev (1998) 318-332

[2] Parker D.W. Mater. Design 14, (1993) 345

[3] Meir S.M.,Gupta D.K., J. Eng. Gas Turbine Power 116 (1994) 250

[4] Movchan B.A., Lemkey F.D. Surface and Coatings Technology 165, 1, (2003) 90-100

FULLERENE STRUCTURES SUITABLE FOR FUEL CELLS ELECTRODES

N. KOPRINAROV, M. KONSTANTINOVA, G. PCHELAROV,
M. MARINOV

Central Laboratory of Solar Energy and New Energy Sources, Tzarigradsko chaussee blvd.72, Sofia, Bulgaria, koprin@pronto.phys.bas.bg

Abstract:　　　Objects of the presentation are fullerenes, carbon nanotubes, layered pyrographite surfaces and nanohorns. The different methods for obtaining these at present are discussed together with their advantages and shortfalls. The characteristic specifics of the separate types of structures and similarities and differences between them are outlined. Emphasis is placed on those parameters that should be of interest in view of the application as electrodes. Special attention is paid to their mechanical and electrical properties as well as the active surface yield. Examples are shown presenting the current tendencies regarding the investigation of these types of materials. Their prospective applications are discussed in view of their qualities and high prices at present.

Key words:　　Fuel Cell/Fuel Cell Electrodes/Carbon Nanostructures/Fullerene/Nanotube/

Nanohorns/Nanotechnology

N. Sammes et al. (eds.), Full Cell Technologies: State and Perspectives, 81-95.
© 2005 *Springer. Printed in the Netherlands.*

1. INTRODUCTION

Characteristic for present day fuel cell constructive solutions is that the whole range of processes is located in three relatively well segregated spatial zones. The hydrogen decomposes into protons and electrons in the first zone. Protons are transported within the second zone. Finally   protons, electrons and oxygen atoms combine and yield water within the third zone. All necessary functions of the first zone can be achieved if it is filled with a porous conducting material to play the role of an electrode. A range of requirements is sought of this material, for the cell to function well these have to assure the optimum conditions for all processes. The first requirement is directed towards the porosity. Along the route that the hydrogen or other work substance follows the resistance must be at a minimum, such that the access of gas flow to the cell is not limited at the maximum work mode. Each limitation in the quantity of gas inflow leads to limiting power ratings of the cell. As hydrogen decomposes at the electrode surface, this surface must be big enough so the over all quantity of gas required in a given work mode may decompose. The area and work temperature are extremely important in view of the activation polarization parameter. To facilitate proton uptake, together with the gas the electrolyte must also have assess to the over all electrode work area. In order that the cell is not large in size, it is necessary to obtain a large electrode size for a minimum in volume. Using a porous structure most easily does this. The effective decomposition of hydrogen into a proton and an electron is different for different materials. This process occurs best on a platinum surface, or when platinum is utilized as a catalyst deposited on other metals. The free electrons obtained after hydrogen decomposition are collected and transported to the external connections of the cell at a minimum resistance. The value of the resistance determines the size of the internal losses in the cell when in use. This problem is known as Ohmic polarization. Together with these conditions, it is desirable for the electrode material to be mechanically stiff, light, cheep, non degradable during exploitation, not to be damaged during eventual work gas contamination and to withstand the optimum work temperature of the cell. The combination of electrons with protons and oxygen occurs in the third zone, this in terms of functional requirements is to certain extent analogous to the first zone. This is the electrode with a positive polarity in the external circuit. Here the additional problem Concentration polarization appears. This results from restrictions to the transport of the fuel components to the reaction location. This usually becomes significant at a high current because the formation of product water and excess humidification block the reaction sites.

Carbon was one of the first electrode materials to be utilized in electrochemistry. During many years of utilization of carbon as an electrode it was able to assert its position due to its qualities. Its good electrical conductivity, favorable chemical stability throughout the whole fuel cell work temperature range, its mechanical stability, high thermal conductivity, low weight, low price as a chemically pure material, ease of manipulation, manifestation in a large diversity of physical structures with different properties, allow for the synthesis of composite materials optimized to a maximum degree with the electrode requirements.

Carbon is one of the most common elements in the universe. From a chemical point of view one of its most unique qualities is the ability of its atoms to combine with one another in quite diverse ways. This for example allows for stable rings of 3, 4, 5, 6, 7, 8 and more atoms to be built. The atomic bonds can have different spatial orientations that allow for the spatial in sequence ordering of the carbon atoms to be very diverse: straight line atomic order, in-plane atomic order and three-dimensional atomic order. The three dimensional arrangement where each atom forms a spatial angle of 120 degrees with its neighbors represents the most dense packing – the diamond crystal. Graphite is the most stable modification of carbon, however this should not be assumed a-priori. In practice chemically pure carbon can be obtained from many kinds of source products by employing different technologies. The final carbon structure is dependent on the source material and method of synthesis. The reason for this is that for most methods of carbon synthesis part of the bonds between the carbon atoms in the source product remain intact until the final build-up of the end structure and all the time are an obstacle to the near perfect arrangement of the atoms.

The carbon structure noted for the highest disorder is the carbon cluster. The bonds are determined from the spatial position of the atoms at the time of bond formation and the temperature facilitating bonding. Such an amorphous carbon for example is the powder obtained during acetylene burning known as acetylene black. Carbon black and furnace blacks that are obtained from partially oxidative combustion of furnace oils are conducting. They have a large surface area (>500 m^2/g) and porosity (pore size <2nm), these are stable and make good contact with the electrolyte.

If carbon black is heated to a very high temperature, for example 2800 °C, the energy, gained by the atoms is high enough so part of the bonds within the clusters is broken and the atoms reform in graphite planes. The method is known as graphitization. The graphite planes are neither ideal nor are they in perfect arrangement with respect to each other because graphitization takes place simultaneously within the whole space and the defects that arise are unable to leave the bulk of the material.

A disordered carbon material with a structure such as that illustrated in Fig.1 is Carbon foam. This can be obtained from different precursors, including pitches by graphitization. For example, the company MER Corporation (1) markets Carbon foam with the following characteristics:

density g/cm3 0.016- 0.62; pores per inch 20-600; approx. pore size (μ) 1270 – 40; thermal conductivity W/mk 0,05 – 210.

In crystal graphite, the graphite planes are arranged at a precisely repeated distance from each other also sliding with respect to one another, so that the atoms from one of the planes can fall within the energetically most favorable position with respect to the atoms of neighboring planes. Due to the highly dense packing of the atoms in graphite the intercalation of the electrolyte and gas molecules in the bulk is highly limited. In view of the fuel cell needs, this insufficiency is far more significant than its advantages due to stability and good electrical conductivity.

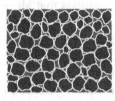   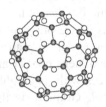

Figure 1. Carbon foam   Figure 2. C_{60}

During 1985 the land mark discovery by Kroto, Curl and Smally (2), (3), (4), (5), (6) clearly showed that carbon can build-up spatially closed structures. The first stable particle of this new carbon modification Fig.2 was found to be built of 60 carbon atoms and was named buckminsterfullerene after the famous architect who designed the Montreal dome structure. Later it became apparent that such particles built of 5 and 6 atom rings can be larger in diameter than C_{60} and may be built of greater numbers of atoms, theoretically tending to infinity. Their practical synthesis is dependent on the conditions of atomic bond formation and the stability of the structure. If during spatial structure build-up there is an incorporation of carbon rings with 3, 4, 7 and 8 atoms then the number of theoretically possible spatially closed structures would become formidable (7), (8). However, the probability of building very complex stable structures is very small and it is highly unlikely, that such structures will arise spontaneously or will be easy to fabricate.

The unique qualities of fullerene type carbon structures are numerous. For example, the energy levels for electrons in C_{60} are located respectively at three levels 3.75, 4.6 and 5.8 eV (9). C_{60} has the behavior of an isolator, but if doped with elements from the halogen group its conductivity in accordance with the level of doping may convert it to a conductor while at low temperatures it can be a superconductor (10), (11), (12), (13), (14). The fact that C_{60} becomes conducting only after doping and that the doping elements can migrate between the separate molecules does not allow for the utilization of C_{60} when employed as an electrolyte. The diameter of C_{60} is 7.2 Å, which is sufficiently large so that most chemical elements may be enclosed within the molecule interior. It is expected that the C_{60} electrons will interact with the electrons of the element enclosed within the C_{60} and in this way will lead to the desired level of doping (15), (16). In this case the doping element atoms will be stationary and if these are close to an electrolyte will not be able to pass through.

During 1990 Iijima (17), (18), (19), (20) described new kinds of fullerene structures. It appeared that these could be obtained by warping the graphite plane to form a cylindrical structure Fig.3.

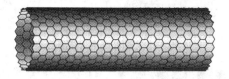

Figure .3 Carbon nanotube

To facilitate the theoretical calculations so as to explain the specific properties of the nanotubes it is suitable to introduce a wedge coordinate system Fig.4a. The angle between the two axes is 60 degrees while the axes orientation is perpendicular to two neighboring sides of the 6-atom ring. In this instance, the condition for in-plane tube synthesis is to cut a rectangle, that has a base beginning at the start of the coordinate system and ends at the same point of the hexagon with coordinates (i, j) (21). In order to distinguish the tubes from each other, it is acceptable for them to be named in accordance with the coordinates of the ring where their base will be cut. Shown in Fig. 4a are 3 such examples respectively (6,1), (6,2), (6,3). In this case the length of the tube coincides with the height of the rectangle. As the lengths of the bases of the three rectangles are different, after warping the tubes will have different diameters – Fig. 4b. It can also be seen that if the base is obtained following a cut along the (0, j) coordinates, the 6 atom ring walls are in parallel to the edge of the cylindrical surface. These tubes are called parallel or zigzag tubes. If the cut is along the hexagon with coordinates (j, j), the tops of the hexagons are perpendicular to the edge after warping. These tubes are known as perpendicular or armchair. When during graphite plane warping the bonding of the rings is achieved by spiral shift a chiral NT is created (Fig. 4c) (22). As shown in Fig.4c the step change is equivalent to one ring, but this can be different – 1, 2 or more rings. If electrical current is allowed to flow through different tubes, (for example (6,1), (6,2), (6,3) from Fig. 4a) it will flow in parallel with to the height of the rectangles cut and the charge carriers will meet the carbon atom network at different angles. As the resistance of the 6 atom ring is different in each direction, the electrical resistance of the tubes in each case will likewise be different. Also depending on the diameter, there is a spatial carbon bond deformation – Fig.4d.

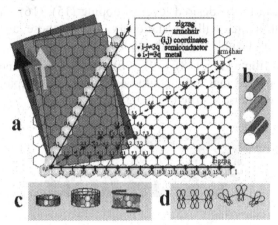

Figure 4. a- coordinate system and parts of the carbon nanotube network (6,1), (6,2), (6,3), в-the carbon network after warping into to form tubes, c- examples of parallel, perpendicular or chiral NTs, d-electron cloud deformation following network warping.

The theoretical calculations carried out show, that due to the reasons considered, the tubes for which the relation i - j = 3q is valid, exhibit a conductivity of a metallic nature, while those with the relation i - j ≠ 3q were found to exhibit semiconductor conductivity. In accordance with the data presented in (21), the inverse proportionality of the bonding between the tube diameters and the width of the forbidden gap can be observed in Fig. 5. There are NT with a diverse range of band gap widths, including NT with a metallic conductivity, that for very low temperatures can be superconducting (23). Given certain conditions the NT can conduct current in a ballistic way and not dissipate heat while the current densities are stable and extremely high, $J > 10^7 A/cm^2$ (24).

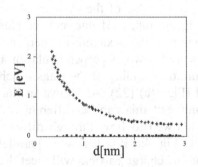

Figure 5. The forbidden band gap width of SWNT as a function of diameter.

The only problem at present, which is not insignificant, is that currently there are no methods in existence to obtain a specific previously chosen type of NT. Usually NT have a wide range of possible diameters each exhibiting a different conductivity. The segregation of the nanotubes by properties can only be done very ineffectively within wide limits, this makes them extremely expensive (25).

After single wall nanotubes (SWNT) arise and since there is a large supply of carbon atoms on their surfaces, they change to multiwall nanotubes (MWNT). These are numerous tubes concentrically grown within one another (26), (27), (28). Their diameters are different and as discussed above each of these will possess different electrical properties. This was confirmed by scientists at IBM (29) who by the introduction of a strong electric current can build the consecutive component tubes of a MWNT structure, then starting from the layer on the outside can burn down to a tube with the required band gap. The diameter of a SWNT and the diameter of the innermost tube of a MWNT structure can be very different. Most often this is from a few nanometers to several tens of nanometers. Recently there was an announcement that the narrowest possible nanotube has been obtained. In this case the carbon atoms are arranged in a row.

SWNT are of greater interest than MWNT due to their simpler construction and because all the carbon atoms are arranged on their surface. There exist many methods and method modifications aimed at securing the maximum yield of SWNT. The most important factors facilitating SWNT growth are: the maximum possible temperature for C atoms introduced within the growth region; the presence of large quantities of catalysts Fe, Ni and Co (30), (31) in an atomic state and a high speed of drift of the NT away from the zone of synthesis. In obtaining these growth conditions, SWNT formed in parallel are often observed, these resemble ropes. Their sizes can be between 10 and 20 nm across and up to 100 μm in length (32).

Researchers at Cambridge University, UK, have spun fibers and ribbons of carbon nanotubes directly from the reaction zone of a furnace. They were able to wind-up continuous fiber "without apparent limit to length" (33). Usually the length of the NT is from a few hundred to several tens of thousands of times larger than the diameter of the tube. This allows for the use of different techniques to grow nanotube ropes or layers of tubes obtained by deposition. For example, researchers from Rice University, US, found that quenching a 4 wt.% nanotube solution into ether and filtering it produced a "bucky-paper" of entangled "super-ropes" about 300-1000 nm thick. (34).

During synthesis NT seldom remain with their ends open. Usually a hemispherical carbon network closes them. The atoms from this part of the tube are not as strongly bonded together as the atoms forming the tube walls and may be easily removed by etching. The NT opened by etching, can be filled with different chemical elements (for example, Pb, Bi, Cs, S, Se; oxides of Ni, Co, Fe, V, Mo, Sn, Nd, Sm, Eu, La, Ce, Pr, Y, Zr, Pd, Ag, Au; Au_2S_3, AuCl, CdS). When SWNT are modified in this way they gain new mechanical and electrical properties (35).

Of special interest in the case considered by us is the possibility to store gases and especially H_2 (36) on or within the NT. The introduction of H_2 within the tubes can be done through openings obtained as defects in the walls of the tubes or by intercalation between the atoms of the walls of the tube (37). In the first case the introduction is achieved without the presence of additional energy while in the second case it is necessary to surmount the energy barrier of 1,5 eV. In both cases if a gradient in the concentration of H_2 is along the tube, this will not only lead to the accumulation of H_2 in the tube, but also to the transportation of H_2 in the direction of the concentration gradient.

Based on the two properties that NT have - a large conductivity and the possibility for transporting H_2 not only interstitially between the NT, but also through their bulk, it can be expected, that they may replace the carbon fibers within the fuel cell electrodes. Fore example, a fuel cell with a construction such as that shown in Fig. 6 when utilizing an electrode made of a rope of NT or a NT mat would have a significantly smaller resistance for H_2 transport within the SWNT and a much larger active surface.

Figure 6. The cross section of PEMFC with carbon fiber electrodes.

Currently there are 3 basic methods utilized in NT synthesis. The first is by AC or DC arc discharge between carbon electrodes, most often in a rarified inert gas ambient. The NT obtained are predominantly multiwall and have a few defects, but a large percentage of other types of nanostructures and amorphous carbon are also obtained. The second method is Chemical Vapor Deposition (CVD). This method is highly productive, however the NT have many defects. When this method is employed to deposit NT on a surface covered with a catalyst, the NT grow perpendicular to the surface and are predominantly multiwall structures. The largest percentage of pure material is obtained by sublimation of C with the aid of an impulse laser. The inadequacy of this method is its low yield. NT, obtained by any of these methods can be purified by etching away the undesirable carbon structures with hydrochloric acid or oxidation in O_2 or air (38).

The MWNT end caps usually have quite varied forms Fig. 7, including the form of a cone with an angle of 20 degrees. In our investigations of carbon nanostructures obtained by the arc discharge method (39) we came across conical structures on one side, as well as structures conical on both sides (Fig.8). As these were not found to have any cylindrical segments they represent a new type of structure that essentially should be labeled as conical. Due to the specifics characteristic for the arc discharge method these structures are multilayer.

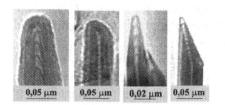

Figure 7. MWHT end caps structures   *Figure 8. Conical carbon*

During 1999 the research group of S. Iijima (40), (41) by laser ablation of graphite without a catalyst, obtained single wall conical structures with a product yield of 50 g/h and a purity of 70-80%, sufficiently pure and not requiring further purification. The spatial angle of the conical surface is 20 degrees. In the literature these structures are termed nanohorns (SWNH). Sketches imitated closed (a) and open (b) SWNH are shown in Fig.9.

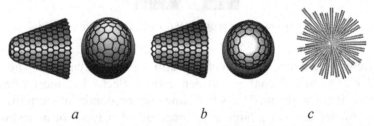

a   *b*   *c*

Figure 9. Sketches imitated closed (a), open (b) SWNH and dahlia nanohorns (c)

At synthesis these are grouped in spatial deposits with sizes of around 100 nm, similar to *Dahlia,* Fig. 9c. This comes from the fact that the quantity of carbon atoms released by the laser pulse is large and the energy of the separate atoms is likewise significant, as a result these are dispersed in all directions and traverse a long path. The large work pressure obstructs the speedy dispersal of the atoms and facilitates deposit formation. The form of the SWNH deposit is also dependent on the nature of the atmospheric ambient. When the work gas is He or N_2 the spherical assembly with a budlike shape (*bud*like nanohorns) is predominant 70-80 %, while for an Ar work gas ambient 95 % of the formation resembles the *dahlia* flower (42), (43). Their structure has fewer defects than that of the *bud*like nanohorns.

Parts of the SWNH are closed. These can be opened by etching away the caps in an atmospheric ambient of O_2 at 693 °C or in H_2O_2. Hence their active surface is increased from 300 $(m^2.g^{-1})$ to 1175 $(m^2.g^{-1})$, as access to their interior is assured (44). Heating part of the SWNH above 550 °C leads to their breakdown. When catalysts such as Fe, Ni, Co are utilizing in SWNH synthesis part of these remain in the deposit. Chemical methods are used to remove the catalyst. For example, Fe is removed by oxidation in air and washing with hydrochloric acid (HCl). Some parts of the iron are found to be joined with C while others are covered with a layer of carbon. During etching the quantity of covered iron is reduced, forming empty spaces at some locations that increase the material absorption capabilities (45).

Gas storage on a particular substance occurs in accordance with four different mechanisms namely, physical adsorption, chemisorption, absorption, and occlusion. The first two mechanisms take place on the surface while the other two proceed within the cracks and empty spaces in the bulk. In the last case the gas molecules are broken down and the composing atoms combine with the atoms of the absorbing substance. In view of the processes taking place at the FC, the principal interest is focused on the physical adsorption as in this case the gas and absorbing substance are not changed and the adsorption – desorption process is fully reversible. The main factor affecting this process is the surface cleanliness. Theoretically the chemisorption can be quite high, but the desorption at normal temperatures is strictly limited and this leads to drastic deterioration of process reversibility. The theoretical and experimental investigations carried out by (46) for the case of SWNH show, that the specific surface area of the internal space is 318 m^2 g^{-1}, while the specific surface area of the external pores (interstitial) is 200 m^2 g^{-1}. The pore volume in the internal space is 0.11 mL g^{-1}, while the pore volume of the external pores is 0.36 mL.g^{-1}. In accordance with this model the infiltration of the gas within the bulk of the separate tubes takes place through the defects in the carbon network. The gas molecules first accumulate on the sharp edges of the tubes, then around the openings and eventually they distribute along the whole surface. Accumulation along the external surfaces is uniform and the density of the gas molecules per unit surface gradually increases.

The way in which one chemical element deposits on the surface of another is dependant on the strength of the interaction between the two elements and the state of the surface. When noble elements such as Au, Pt, Ag, deposit on a plane surface, for most elements they are found to interact weakly with the surface of the substrate and aggregate as large island structures. If some kind of chemical bonding to the surface atoms is found to exist the noble element is deposited in the form of a thin layer or as very small islands that as they grow thicker combine to form a continuous layer. On the other hand the surface activity of an element is not only dependent on the activity of the element itself but also on the state of element bonding of the atoms across its surface. When the surface has defects or a small radius of curvature the bonding between the atoms in the bulk and the surface is reduced and this leads to an increase in the chemical activity of the surface. For these reasons it is reasonable to expect that the deposition of Pt with the aim of fulfilling the functions of a catalyst on a graphite surface and across the surface of carbon nanostructures will be different. This is confirmed by the technology for the deposition of Pt on a SWNH surface described in (47). Platinum is deposited by utilizing the colloidal method. NaHSO3 and H_2O_2 were added to an H_2PtCl_6 solution to form Pt oxide colloids. The SWNH powder was put into the Pt oxide colloid dispersion, and the Pt oxide colloids were adsorbed on the SWNH surface. After eliminating the Cl, Na, and S ions, the samples were dried and reduced in H_2.

The comparison between Pt deposited on SWNH and on carbon black demonstrates, that Pt can be deposited as a catalyst on SWNH too. The size of the Pt clusters on SWNH is about 2 nm and these are two times smaller than the clusters that deposit on carbon black. This affords the opportunity for them to have the same active area for a smaller quantity of platinum used, also the finer in size the better in performance. When platinum is added to the carbon target it can be evaporated together with the sublimation of C and may be deposited on SWNH. In this fashion the synthesis of SWNH and their coverage with a catalyst may be done simultaneously.

The qualities: ease of type fabrication in a pure and repeatable fashion; a very large surface area and possibility for the gasses and liquids to easily infiltrate, have been utilized by NEC, Japan Science and Technology Corporation and the Institute of Research and Innovation to create methanol-fueled polymer electrolyte cells (48). In the year 2003, NEC demonstrated a FC used as the power supply of a PC with a maximum power output of 18 watts, this was half the size of the notebook PC itself. Also a FC with a maximum output of 3 W that powered a function-rich cell phone with a consumption of about 1W while in operation. The power source was almost the same size as a cellular phone (49). According to NEC the energy density of the fuel cells demonstrated is about the same as for comparable lithium-ion batteries, but it is expected that the fuel cells produced by 2005 will yield an energy density that is three times higher. The experts at NEC claim that the critical issues for fuel cells are cost and size. At present, carbon nanohorns cost $500 per gram. It is expected however that by 2005 their price will be lower than $1. The use of platinum will also be cut to one-tenth the current level. The target energy density by 2005 is 50 to 100 mW/cm^2 (50).

At the present moment the prices of the various carbon products that are an object of this study differ in very wide boundaries depending on their purity, the producer and the quantity ordered, however in each instance these are fairly high. For example, there are different offers for buckyballs (SES ResearcH USA 2002) per gram in $: C_{60} 99.9+ %, purified - 110.00, C_{70} 99.0 % - 525.00, C_{76} 95.0% - 500.00, C_{84} 95.0% - 420.00; MWNT - 110.00, SWNT - 375.00; Purified Open treated MWNT - 170.00, Purified Open end SWNT - 950.00. We shall refrain from listing more prices as the tendency is for the prices to fall drastically with the introduction of new production and purification technologies and the constant rise in production volume. In accordance with the studies of the scientists from the Center for Solar Energy and Hydrogen Research (ZSW) in Ulm (50) the price of the electrodes of a single FC represents 5 % of the total cost of the FC. This calculation takes account of all the production expenses including all mounting operations. It follows, that the high carbon product prices affect the price of the FC to a limited degree due to the small percentage of carbon utilized in the end product. In counter weight to the high cost of these carbon materials for the FC, in order to assess the viability of introducing these new products we must also take into account all their advantages. For example, their reliability, weight, size and so forth, these also play a key role in assuring the optimum operation of the FC.

REFERENCES:

[1] MER Corporation, 7960 South Kolb Road, Tucson, Arizona 85706.

[2] Kroto, H., Science, (1988)242, 1339.

[3] Curl, R., Smalley, R., Science, (1988) 242, 1017

[4] Curl, R., Smalley, R., Scientific American (1991), 54.

[5] Curl, R., Carbon, (1992),30 No8, 1149.

[6] Kretschmer, W., Lamb, L., Fostiropoulos, K., Huffman, D., Nature, (1990) 347, 354

[7] Terrones, H., Mackay, A., Carbon, (1992) 30 No 8, 1251.

[8] Fowler, P., Manolopoulos, D., Ryan, R., Carbon, (1992) 30 No 8, 1235.

[9] Weaver, J., J. Phys. Chem. Solids, (1992) 53, 1433.

[10] Iwasa, Y., Watanabe, S., Kaneyasu, T., Yasuda, T. , Koda, T., Nagata, M., Mizutani, N., J. Phys. Chem. Solids, (1993) 54 No 12, 1795.

[11] Haddon, R., Hebard, A., Rosseinsky, M., Murphy, D., Glarum, S., Palstra, T., Ramirez, A., Duclos, S., Fleming, R., Siegrrist, T., Tycko, R., American Chem. Soc. 5, (1992), 71.

[12] Haddon, R., Hebard, A., Rosseinsky, M., Murphy, D., Duclos, S., Lyons, K., Miller, B., Rosamilia, J., Fleming, R., Kortan, A., Glarum, S., Makhija, A., Muller, A., Eick, R., Zahurak, S., Tycko, R., Dabbagh, G., Thiel, F., Nature, (1991) 350, 320.

[13] Wang, H., Schlueter, J., Cooper, A., Smart, J., Whitten, M., Geiser, U., Carlson, K., Williams, J., Welp, U., Dudek, J., Caleca, M., J. Phys. Chem. Solids,(1993) 54, No 12, 1655.

[14] Saito, S., Oshiyama, A., J. Phys. Chem. Solids, (1993) 54 No 12, 1759.

[15] Chai, Y., Guo, T., Ch. Jin, Haufler, R., Chibante, L., Fure, J., Wang, L., Alford, J., Smalley, R., J. Phys. Chem., (1991) 95 , 7564.

[16] Ostling, D., Rosen, A., Z. Phys., (1993) 26, 279.

[17] Iijima, S., Nature, (1991) 354, 56.

[18] Iijima, S., Ichihashi, T., Ando, Y., Nature, (1992) 356, 776.

[19] Iijima, S., Ajayan, P. , Ichihashi, T., Phys. Rev. Lett., (1992) 69, 3100.

[20] Ajayan, P., Iijima, S., Nature, (1992) 358, 23.

[21] Dresselhaus, M., Carbon filaments and nanotubes, L. Biro et al. (eds), (2001),11.

[22] Fonseca, A., Hernadi, K., Nagy, J., Lambin, Ph., Lukas, A., Syntethic Metals, (1996) 77, 235.

[23] Kasumov, A., Deblock, R., Kociak, M., Reulet, B., Bouchiat, H.,Khodos, I., Gorbatov, Yu., Volkov, V., Joumet, C., Burghard, M., Science, (1999) 284.

[24] Frank, St., Poncharal, Ph., Wang, Z., Heer, W., Science, (1988) 280, 1774.

[25] Konya, Z., Carbon filaments and nanotubes, L. Biro et al. (eds), (2001), 85.

[26] Ebbesen, T., Ajayan, P., Nature, (1992) 358, 220.

[27] Withers, J., Loutfy, R., Carbon, (1993) 31 No 5, 685.

[28] Ebbesen, T., Physics Today, (1996), 26.

[29] www.research.ibm.com/resources/news/20010425_Carbon_Nanotubes

[30] Hernadi, K., Fonseca, A., Nagy, J., Bernaerts, D., Lukas, A., Carbon, (1996) 34, No 10, 1249.

[31] Hernadi, K., Fonseca, A., Nagy, J., Bernaerts, D., Riga, J., Lukas, A., Syntethic Metals, (1996) 77, 31.

[32] www.sesres.com/Nanotubes.asp /

[33] www.nanotechweb.org/articles/news/3/3/7/1, 11 March 2004/.

[34] http://nanotechweb.org/articles/news/2/12/7/1, 12 December 2003 /

[35] Fonseca, A., Nagy, J., Carbon filaments and nanotubes, L. Biro et al. (eds), (2001), 75.

[36] Dillom, A., Jones, K., Bekkedahl, T., Klang, C., Bethune, D., Heben, M., Nature,(1997) 386, 377.

[37] Frauenheim, T., Seifert, G., Koehler, T., Elstner, M., Lee, S., Lee, Y., Nanostructured carbon for advansed Applications, G. Benedek et.al. (eds), (2001), 347.

[38] Yang, Ch., Kaneko, K., Yudasaka, M., Iijima, S., Physika B, (2002) 323, 140.

[39] Koprinarov, N., Marinov, M., Pchelarov, G., Konstantinova, M., Stefanov, R., J. Phys.Chem., (1995) 99, 2042.

[40] Iijima, S., Yudasaka, M., Yamada, R., Bandow, S., Suenaga, K., Kokai, F., Takahashi, K., Chem. Phys. Lett., (1999)165, 309.

[41] Iijima, S., Physica B, (2002), 323,1.

[42] Bekyarova, E., Kaneko, K., Kasuya,, D., Murata, K., Yudasaka, M., Iijima, S., Langmuir, (2002) 18, 4138.

[43] Kasuya, D., Yudasaka, M., Takahashi, K., Kokai, F., Iijima, S., J. Phys. Chem. B, (2002) No106, 4947.

[44] Beuyarova, E., Kaneko, K., Yudasaka, M., Murata, K., Kasuya, D., Iijima, S., Adv. Mater., (2002) 14, 973.

[45] Yang, C., Kaneko, K., Yudasaka, M., Iijima, S., Nano Letters ,(2002) 2 No. 4, 385.

[46] Murata, K., Kaneko, K., Kanoh, H., Kasuya, D., Takahashi, K., Kokai, F., Yudasaka M., Iijima, S., J. Phys. Chem., (2002) B 106, 11132.

[47] Yoshitakea,T., Shimakawaa,Y., Kuroshimaa, S., Kimuraa, H., Ichihashia, T., Kuboa,Y., Kasuyad, D., Takahashic, K., Kokaic, F., Yudasakab, M., Iijima, S., Physica B, (2002) 323, 124.

[48] www.asiabiztech.com /Nikkei , August 31, 2001

[49] www.nec.co.jp/press/en

[50] www.nec.co.jp/press/en

[51] www.initiative- rennstoffzelle.de/en/ibz/live/nachrichten/detail/109

METALLIC MATERIALS IN SOLID OXIDE FUEL CELLS

V. SHEMET, J. PIRON-ABELLAN, W.J. QUADAKKERS,
L. SINGHEISER

Forschungszentrum Jülich, Institute for Materials and Processes in Energy Systems,
52425 Jülich, Germany, (t) +49 2461615560, (f) +49 2461613699, v.shemet@fz-juelich.de

Abstract: Fe-Cr alloys with variations in chromium content and additions of different elements were studied for potential application in intermediate temperature Solid Oxide Fuel Cell (SOFC). Recently, a new type of FeCrMn(Ti/La) based ferritic steels has been developed to be used as construction material for SOFC interconnects. In the present paper, the long-term oxidation resistance of these class of steels in both air and simulated anode gas will be discussed and compared with the behaviour of a number of commercial available ferritic steels. Besides, in-situ studies were carried out to characterize the high temperature conductivity of the oxide scales formed under these conditions. Main emphasis will be put on the growth and adherence of the oxide scales formed during exposure, their contact resistance at service temperature as well as their interaction with various perovskite type contact materials. Additionally, parameters and protection methods in respect to the volatilization of chromia based oxide scales will be illustrated.

Key words: Solid Oxide Fuel Cell/Interconnect/Ferritic Steel/High Temperature Conductivity

97

N. Sammes et al. (eds.), Full Cell Technologies: State and Perspectives, 97-106.
© *2005 Springer. Printed in the Netherlands.*

1. INTRODUCTION

Metallic materials for be used as interconnects in SOFCs should fulfil a number of specific requirements [1, 2]. Crucial properties of the materials are high oxidation resistance in both air and anode environment, low electrical resistance of the oxide scales formed on the alloy surface as well as good compatibility with the contact materials. Additionally, the value of the coefficient of thermal expansion (CTE) should match with those of the other cell components [3]. These requirements can potentially be achieved with high chromium ferritic steels [4], however, previous studies [5] have shown that none of the commercially available ferritic steels seems to possess the suitable combination of properties required for long term reliable cell performance.

One of the most important problems found during stack operation using metallic interconnect materials is the formation of volatile chromium oxides and/or oxy-hydroxides [6, 7] at the cathode side of the cell leading to serious deterioration of the cell performance [7]. Several authors proposed various protective coating types to prevent the deleterious effect of volatile Cr-species [8].

Recently, a new class of FeCrMn(La/Ti) ferritic steels (see Table 1) has been developed to be used as construction materials for SOFC interconnects [4,9]. In the present study, the long-term oxidation resistance of some of these FeCrMn(La/Ti) steels in both air and simulated anode gas has been studied and compared with the behaviour of a number of commercially available ferritic steels. Main emphasis was put on the growth and adherence of the oxide scales formed during exposure, their contact resistance at service temperature as well as their interaction with various perovskite type contact materials.

Table 1. Studied model and semi-commercial FeCrMn(La/Ti) alloys.

Steels designation	Batches	Major features
JS-1	HNA, HMZ	high Mn, Ti, La
JS-2	HXV	high Mn, low Ti, La
JS-3	HUF, JEW, JEX	low Mn, Ti, La

2. EXPERIMENTAL DETAILS

The model ferritic Cr steels with variation in Mn, Ti and La content were manufactured by Krupp/Thyssen NIROSTA (KTN). The main features of the various studied alloy types are given in Table 1. The studied commercial alloys, supplied by Hitachi metals, Krupp/Thyssen VDM and Rolled Alloys, respectively are listed in Table 2. For oxidation studies, samples of dimensions 20x10x2 mm were cut from the prevailing semi-finished products, ground to 1200 grit surface finish and finally cleaned in acetone. Discontinuous oxidation tests were carried out at 800 °C in air and in Ar-4%H_2-2%H_2O. For weight measurements the exposures were interrupted every 250h. For more detailed analysis of the oxidation kinetics, isothermal oxidation tests using a SETARAM thermobalance were carried out. The oxide scales formed during oxidation in the various atmospheres were studied by light optical microscopy, scanning electron microscopy (SEM) with energy dispersive X-ray analysis (EDX) and X-ray diffraction (XRD). The contact resistances of the oxide scales were measured using a conventional four point method. For these studies, samples of 10x10x2 mm were ground to 1200 grit surface finish and finally pre-oxidized for 100h at 800°C in air. Subsequently, a layer of Pt-paste was applied to both oxidized surfaces. For the electrical connection a Pt-mesh was used. The contact resistance was monitored in-situ during 500h exposure at 800°C in air.

For studies concerning the compatibility of the steels with contact layers, samples of 20x10x2 mm were ground to 1200 grit surface finish and pre-oxidized for 100h at 800°C in air. Subsequently, various La-based perovskite type contact pastes [10] were applied on top of one side of the samples which were then exposed for 1000h at 800°C in air.

Table 2. *Chemical compositions of the studied commercial ferritic steels (Mass.- %).*

Alloy designation	Fe	Cr	Mn	Ti	Al	Ni	Si
1.4509	Bal.	18	0.38	0.12	0.03	0.12	0.7
446	Bal.	25	0.5	0.05	-	-	0.29
1.4742	Bal.	17	0.31	0.01	1.04	0.18	0.93
ZMG232	Bal.	22	0.45	-	0.19	-	0.35
Crofer 22APU	Bal	23	0.4	0.05	< 0.1	0.16	< 0.1

3. RESULTS AND DISCUSSIONS

Fig. 1 shows the oxidation behaviour under cyclic oxidation conditions for several FeCrMn model alloys with and without Ti and La additions compared with the behaviour of some of the most promising commercial ferritic steels. The JS-1 alloy (batch HNA) and the commercial alloy ZGM232 show the highest weight changes mainly due to the extensive internal oxides formed beneath the scale. The other studied alloys show similar oxide scale growth rates because either no internal oxides are present at all or they appear in form of very fine internal precipitates. The two commercial alloys 1.4509 and 446 exhibited substantial scale spallation during the oxidation test but the chromium concentration is still sufficiently high to re-form the chromia based scale.

Fig. 2 shows the oxidation behaviour of alloys JS-1 and JS-3 (batches HNA and JEX) oxidized in both air and simulated anode gas. A general tendency is that the oxide scales formed in anode gas are slightly thinner than those formed in air. As has previously been reported for Cr-based alloys [11], for all studied alloys the scales formed in the anode gas generally possess better adherence than those formed in air or oxygen.

Fig. 3 shows the instantaneous K_p-values at 50h determined during isothermal oxidation in air for a number of ferritic steels with and without Mn, Ti and/or La additions in comparison with a number of other SOFC relevant alloys as function of the reciprocal temperature. The addition of Ti to the FeCr/Y alloy leads to a substantial increase in the oxidation rate. The enhancement increases with increasing Ti content, in agreement with previous findings related to the behaviour of NiCr-based alloys in steam reforming gas [12]. Comparing the alloys FeCrMn and JS-1 (batch HNA) it is clear that in the entire range of temperatures studied, only minor changes in the Mn, La and Ti contents can substantially affect the oxidation rate. Based on these findings, the suitable oxidation resistance of the ferritic steels requires adequate addition and careful control of these elements in the alloy. By optimum additions of the oxygen active elements, oxidation rates can be achieved which are similar to those of chromium based ODS alloys, which have frequently been proposed as interconnect material for high temperature SOFCs [13, 14].

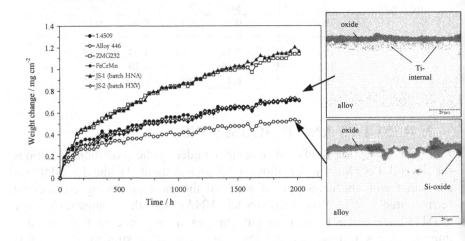

Figure 1. Oxidation behaviour under cyclic conditions of several commercial and model ferritic steels at 800°C in air.

The electrical resistances of the oxides scales formed on different FeCrMn(La/Ti) alloys during exposure at 800°C are shown in Fig. 4a. The results are compared with data obtained for a number of commercial alloys, i.e., 1.4742 (X10CrAl18), 1.4509 (X2CrTiNb18), alloy 446 and alloy ZGM232. The values obtained for alloys 1.4509, 446 and ZMG232 showed a very wide scatter range, although on all three materials, the oxides formed mainly consist of a duplex $MnCr_2O_4/Cr_2O_3$ scale.

The highest values of approximately 20 $\Omega\,cm^2$ were obtained for the commercial alloy 1.4742 which has in some cases been considered as a potential candidate to be used as interconnect in intermediate temperature SOFCs. These high values can be explained by the fact that this alloy, depending on the exact alloy composition and surface treatment, in some cases tends to form a very protective alumina scale [5], which, however, possesses a very poor electrical conductivity. In contrast, the new JS-3 alloys (batches JDA, JEW, and JEX) show very low contact resistance values of approximately 10 $m\Omega\,cm^2$, i.e., values which are two to three orders of magnitude smaller than those of most commercial alloys.

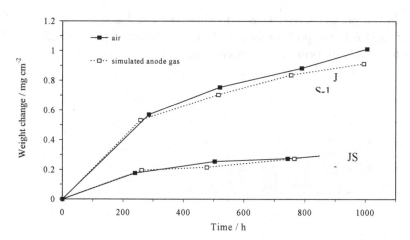

Figure 2. Oxidation behaviour of selected model ferritic steels at 800°C in air and Ar-4%H₂-2%H₂O.

In Fig. 4b the contact resistances as function of time for various JS-3 alloys compared with the commercial alloy 1.4016-3C are plotted. The contact resistance values for 1.4016-3C are higher than those of newly developed JS-3 alloys and increase with decreasing temperature. The temperature dependence was more pronounced for the commercial alloy than for the JS-3 alloys. At lower temperatures the JS-3 alloys show relative higher conductivity compared to the commercial 1.4016-3C alloy.

In a planar cell configuration the connection between the interconnect and the cathode material is frequently achieved by using a contact layer which compensates geometrical irregularities of the cathode and/or interconnect surface [5]. Because the contact layers have to possess a high electronic conductivity, they commonly consist of La-based perovskites[15].

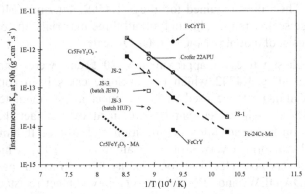

Figure 3. Instantaneous K_p values at 50h as function of temperature during isothermal oxidation in air compared with values for the ODS alloy Cr5Fe-Y$_2$O$_3$.

In the present study, selected contact pastes, La(Sr)MnO$_3$ and LaCoO$_3$ were applied on top of pre-oxidized sample surfaces of a selected JS-1 alloy (batch HMZ) in form of a slurry coating.

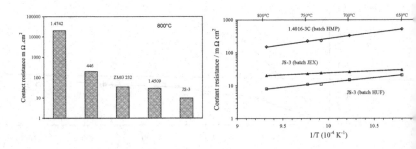

Figure 4. Contact resistance of the oxide scales formed on various ferritic steels after 500 h oxidation at different temperatures in air.

The coated samples were then oxidized for 1000h at 800°C in air. Subsequently, the alloy/coating interaction was studied by SEM/EDX. The results in Fig. 5 illustrate that the alloy/perovskite interaction strongly depends on the contact layer composition.

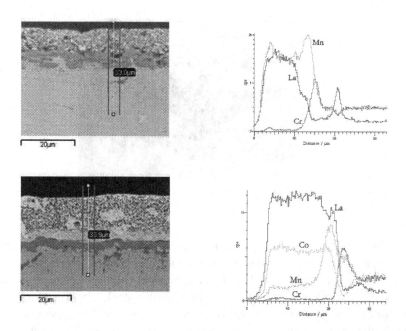

Figure 5. Interface layer formed between ferritic steel (batch HMZ) with a) La(Sr)MnO₃,
and b) LaCoO₃ contact layer during 1000h exposure in air at 800°C.

In case of La(Sr)MnO$_3$ the formation of spinel MnCr$_2$O$_4$ takes place at the interface with the contact paste. Some chromium was detected on the surface of the contact layer. The use of LaCoO$_3$ as contacting material might reduce the chromium diffusion through the contact paste because no enrichment of this element was observed. This effect may be due to the formation of a dense interaction layer which consists of a mixture of spinel and perovskite compounds which are virtually free of Cr.

The BaO-CaO-Al$_2$O$_3$-SiO$_2$ (BCAS) glass ceramic seal is often used for joining dissimilar materials, i.e. ceramic cells, metallic manifolds and metallic interconnects [2]. The ceramic seal should possess a satisfactory matching of the thermal expansion coefficient with the cells and the chosen alloy and it should also exhibit a long-term stability under the operation conditions.

In the present study, selected BCAS, were applied on top of the sample surfaces of ferritic steels with/or without Al and Si additions. The coated samples were then oxidized in air and Ar-H$_2$-H$_2$O.

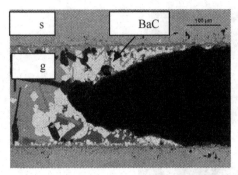

Figure 6. SEM cross section of FeCrMn(Ti/La) steel with glass after 300 h oxidation in air at 800 °C.

The results in Fig. 6,7 show that the alloy/glass/gas interaction strongly depends on the gas and steel composition. During air oxidation a barium chromate (white phase in Fig. 6) is formed at the three-phase boundary of glass in contact with a high-chromium ferritic steel. The mismatch of CTE between $BaCrO_4$ and steel leads to high stresses in the glass-ceramic seal with the consequence that the glass-ceramic is separated from the steel surface after joining.

If the atmosphere is changed to $Ar-H_2-H_2O$, a different interaction pattern between steel and glass-ceramic seals can be found. Fig. 7 presents the cross sections of the steel plates covered with glass.

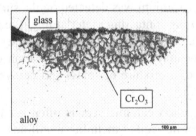 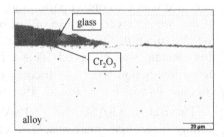

Figure 7. Metallographic cross sections of ferritic FeCrMn(Ti/La) steels with glass sealant after 400 h of exposure in $Ar-H_2-H_2O$ at 800 °C. a) Si and Al containing steel, b) low Si and Al containing steel.

In the case of Al/Si containing steel, a non protective scale had been formed and the inner oxidation of chromium was observed. During the exposure of 100 h the reaction front of the inner oxidation reached a depth of about 120 μm. Due to the chromium internal oxide formation, the increased volume results in bulging of the steel surface. It is needless to say that this expansion leads to stresses in the joint, and as a consequence the glass-ceramic is disrupted and partially spalls. This effect may be due to the formation of low melting phases at the gas/alloy interface, which accelerate the grain boundary corrosion. In contrast, the low Si/Al containing steel (JS-3) shows long term stability of the alloy/glass/gas interface.

4. CONCLUSIONS

Commercial high Cr ferritic steels as interconnect material have large drawbacks because the materials tend to exhibit substantial scale spallation during long term oxidation, especially under cyclic oxidation conditions. Besides, the formed oxide scales frequently exhibit high electrical resistances. These effects are mainly related to a poor control of the oxygen active minor alloy constituents. For these reasons, new alloys of the type FeCrMn(La/Ti) were developed to be used as interconnect materials for SOFCs. These alloys form oxide layers with low growth rates being constituted by a duplex $MnCr_2O_4/Cr_2O_3$ scale. The formation of the external spinel layer considerably reduces the chromium evaporation [8]. Also, it has been demonstrated that this type of scale has low contact resistance values, i.e., two to three orders of magnitude lower than that of alumina-rich scales frequently formed on commercial ferritic steels. Besides, the combination of these new steels with suitable perovskite-type contact layers allows formation of interaction layers with low electronic resistance [10], which further reduce the formation of volatile Cr-species. The chemical interaction of metallic interconnects with $BaO-CaO-Al_2O_3-SiO_2$ glass-ceramic seal significantly varies for different steels. Small changes in the chemical composition of the steel influences the corrosion behaviour, especially in Ar-H_2-H_2O.

ACKNOWLEDGMENT

The authors are grateful to their colleagues Mr. Mahnke and Mr. Cosler for carrying out the oxidation experiments, Mr. Schmidt for preparing the contact layers, Mr. Gutzeit for the metallographic investigations and Mr. Wessel for the SEM/EDX analyses.

REFERENCES

[1] W.J. Quadakkers, J. Pirón-Abellán, V. Shemet and L. Singheiser, Materials at High Temperatures 20 (2)115 –127.

[2] S.P.S. Badwal, R. Deller, K. Foger, Y. Ramprakash, J.P. Zhang, Solid State Ionics 99, 297-310, (1997).

[3] Th. Malkow, U.v.d. Crone, A.M. Laptev, T. Koppitz, U. Breuer, W.J. Quadakkers, Proc. Of the 5th International Symposium on Solid Oxide Fuel Cells (SOFC-V), Aachen, Germany, 2-5 June 1997, Ed. U. Stimming, S.C. Singhal, H. Tagawa, W. Lehert, Electrochemical Society Inc., Penn., 1244-1252, (1997).

[4] W.J. Quadakkers, Th. Malkow, J. Pirón Abellán, U. Flesch, V. Shemet, L. Singheiser, 4th European SOFC Forum, 10-14 July 2000, Lucerne, CH, Proceedings, Edt. A. McEvoy, 827-836, (2000).

[5] J. Pirón-Abellán, V. Shemet, T. Malkow*, L. Singheiser, W.J. Quadakkers, paper submitted to Fuel Cells

[6] W.J. Quadakkers, H. Greiner, W. Köck, H.P. Buchkermer, K. Hilpert, D. Stöver, 2nd European SOFC Forum, 6-10 May 1996, Oslo, Norway, Ed. B. Thorstensen, 297-306, (1996).

[7] S.P. Jiang, J.P. Zhang, L. Apateanu, K. Foger, Journal of the Electrochemical Society 147, 4013-4022, (2000).

[8] C. Gindorf, L. Singheiser, K. Hilpert, M. Schroeder, M. Martin, H. Greiner, F. Richter, Proc. 6th International Symposium on Solid Oxide Fuel Cells (SOFC-VI), Honolulu, Hawaii, 17-22 October 1999, Edt. S.C. Singhal, M. Dokiya, The Electrochem. Soc., Pennington NJ, USA, 812-821, (1999).

[9] J. Pirón Abellán, V. Shemet, F. Tietz, L. Singheiser, W.J. Quadakkers, 7th International Symposium on Solid Oxide Fuel Cell (SOFC-VII), EPOCHAL, Tsukuba, Japan, 3-8 June 2001, Proceedings The Electrochem. Soc., Pennington, NJ, USA, 2001, 811-819, (2001).

[10] O. Teller, W.A. Meulenberg, F. Tietz, E. Wessel, W.J. Quadakkers, Proc. 7th International Symposium on Solid Oxide Fuel Cells (SOFC-VII), EPOCHAL, Tsukuba, Japan, 3-8 June 2001, Proceedings The Electrochem. Soc., Pennington, NJ, USA, 895-903, (2001).

[11] Cambrigde W.J. Quadakkers, J. F. Norton, S. Canetoli, K. Schuster, A. Gil, Proc 3rd Int. Conf. Microscopy Oxid., 16-18 Sept. 1996, Cambridge (Ed. by S.B. Newcomb, J.A. Little), Institute of Materials, London, 599-609, (1997).

[12] P.J. Ennis, W.J. Quadakkers, Conference High Temperature Alloys - Their Exploitable Potential, 15-17 Oct. 1985, Petten, NL, Proceedings, Elsevier London, 465-474, (1988).

[13] U.v.d. Crone, M. Hänsel, W.J. Quadakkers, R. Vaßen, Fresenius Journal of Analytical Chemistry 358, 230-232 (1997).

[14] W.J. Quadakkers, M. Hänsel, T. Rieck, Materials and Corrosion 49, 252-257 (1998).

[15] W.J. Quadakkers, H. Greiner, M. Hänsel, A. Pattanaik, A.S. Khanna, W. Malléner, Solid State Ionics 91, 55-67 (1996).

SOFC WORLDWIDE – TECHNOLOGY DEVELOPMENT STATUS AND EARLY APPLICATIONS

L. BLUM, R. STEINBERGER-WILCKENS, W.A. MEULENBERG, H. NABIELEK

Forschungszentrum Jülich, D-52425 Jülich, Germany

Abstract: Solid Oxide Fuel Cells (SOFC) of various types and designs have been developed world wide through the last two decades. They offer interesting advantages over other fuel cell types, but also have inherent materials problems that have caused a slower development pace as, for instance, compared to the low temperature Polymer Electrolyte Fuel Cell (PEFC). Due to their high operating temperature in the range of 700 to 1000°C, SOFC can be used with a variety of fuels from hydrogen to hydrocarbons with a minimum of fuel processing, can be coupled with gas turbines for the highest electrical system efficiency known in power generation, deliver process heat in industrial applications or supply on-board electricity for vehicles, to name but some typical applications. This report summarizes the more prominent SOFC development strands and gives an overview of the achievements of the various R&D groups. The analysis includes a benchmark that attempts to compare cell and stack characteristics on a standardized basis.

Key words: Solid Oxide Fuel Cells/Design Concepts/Cell Data/Stack Data

1. INTRODUCTION

Within the last ten years SOFC development has made big progress which can clearly be seen from the tenfold increase in power density. A declining interest in SOFC could be observed towards the end of the last century, when several of the leading companies terminated their activities, amongst them Dornier in Germany and Fuji Electric in Japan. Nevertheless a tremendous increase in activities occurred during the last years with

107

N. Sammes et al. (eds.), Full Cell Technologies: State and Perspectives, 107-122.
© 2005 Springer. Printed in the Netherlands.

companies re-starting their activities and new industry and research institutions starting SOFC-related work.

This report tries to give an overview on the main development lines and summarizes the development status, reached at the end of 2003, by presenting a comparison of obtained results in cell and stack development. The authors concentrate on the published results of industry and the larger research centres. The numerous activities at universities are not taken into account to facilitate the overview.

2. DESIGN CONCEPTS

There are mainly two different concepts under development – the tubular and the planar design. As far as proof of long term stability and demonstration of plant technology are concerned the tubular concept is far more advanced. In comparison, the planar development offers higher power density. In this chapter the various design variants are presented, followed by a description of the main companies involved and the status of cell and stack technology.

2.1. TUBULAR CONCEPTS

The most advanced tubular SOFC has been developed by Westinghouse, since 1998 Siemens Westinghouse Power Corporation (SWPC). Their concept is based on a porous cathode tube, manufactured by extrusion and sintering. The tube length is 1.8 m with a wall thickness of 2 mm and an outer diameter of 22 mm (see Fig. 1). The active length is 1.5 m, which is coated with a ceramic interconnect by atmospheric plasma spraying with zirconia electrolyte deposited by EVD (Electrochemical Vapour Deposition) and with an anode made by slurry dip coating and dual atmosphere sintering [1]. Attempts are being made at replacing the very expensive EVD process by other techniques. The cells are connected to bundles via nickel felts. The high ohmic resistance of this concept requires an operating temperature between 900 and 1000°C to reach power densities of about 200 mW/cm^2. To overcome this problem SWPC is working on a modified concept, using flattened tubes with internal ribs for reduced resistance (see Fig. 1).

The Japanese company TOTO uses the same design, but started earlier with the implementation of cheaper manufacturing technologies. They chose shorter tubes of 0.5 m length with an outer diameter of 16 mm [2].

A different tubular design is pursued by Mitsubishi Heavy Industries (MHI/Japan, Fig. 2). The single cells are positioned on a central porous support tube and connected electrically in series via ceramic interconnector rings, which leads to an increased voltage at the terminals of a single tube. In contrast to the first two types fuel is supplied to the inside of the tube and air to the outside [2, 3].

The university of Birmingham with their spin-off company Adelan and the US company Acumentrix are developing anode supported micro tubes with a length of 50 mm.

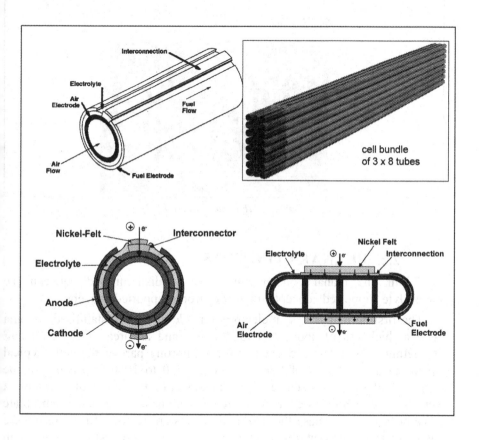

Figure 1. SWPC – tube design, cell bundle and flattened tube

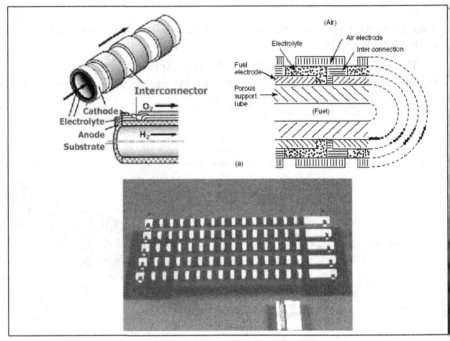

Figure 2. MHI – tube design [2, 3]

2.2. PLANAR CONCEPTS

Within the planar concepts, it is to be distinguished between the electrolyte supported concept and the electrode supported concept.

The first normally uses an electrolyte made of yttrium stabilised zirconia with a thickness of about 100 to 200 μm and an area of 10 x 10 cm^2 (sometimes up to 15 x 15 cm^2) as the supporting part of the cell. Typical operating temperatures of this concept are 850 to 1000°C because of the relative high ohmic resistance of the thick electrolyte. In case of operation at very high temperatures ceramic interconnects made of lanthanum-chromate have been used. Because these ceramic plates are restricted in size, require high sintering temperatures, have different thermal expansion behaviour in oxidising and in reducing atmosphere and have comparatively bad electrical and thermal conductivity there is an obvious trend to metallic interconnect plates. The advantage of ceramic plates is the negligible corrosion and therefore low degradation which sustains the interest in this material. The metallic interconnect plates on one hand allow the reduction of operating temperature and on the other hand an increase in size. Especially the good thermal conductivity reduces the temperature gradients in the stack and allows bigger temperature differences between gas inlet and outlet, which reduces the necessary air flow for cooling. Because conventional high temperature alloys have a thermal expansion coefficient too high when compared to zirconia a special alloy was developed (chromium with 5% iron and 1% yttria), which was used by Siemens and is still used by Sulzer Hexis.

In the Siemens concept (see **fig. 3**) the small cells of 5 x 5 cm² were arranged in parallel on a big bipolar plate in a 4 x 4 configuration and a cross flow arrangement. Stacks up to 80 layers were assembled and tested. The biggest stack delivered a power of about 5.4 kW operating with hydrogen/air at 950°C. The further development was done with cells 10 x 10 cm² and an increased plate size, suited for assembling 9 cells. When the Westinghouse tubular activities were taken over in 1999, the planar development was stopped. The Fraunhofer Gesellschaft (FHG) department IKTS in Dresden based its work on the development at Siemens.

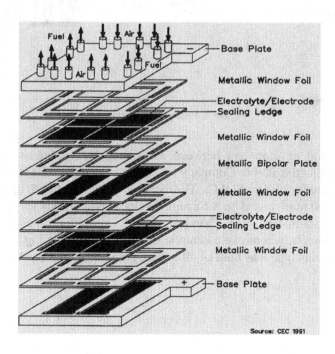

Figure 3. Siemens Window Frame design

The special steel developed by Siemens together with the Austrian company Plansee is still used by Sulzer Hexis within their electrolyte supported stack design. Fuel is supplied to the centre of circular cells with a diameter of 120 mm and flows in parallel with the air to the outer rim of the cell, where the fuel gas that has not reacted within the cell, is burned. Air is supplied from the outside and heats up, while flowing towards the centre (see fig. 4). The stack is typically operated at 950°C. Up to 70 cells are stacked together, delivering 1.1 kW.

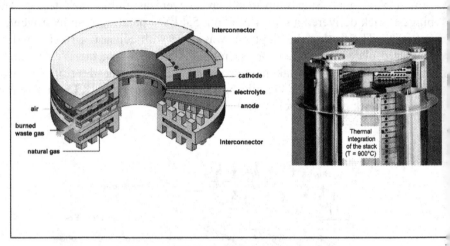

Figure 4. Sulzer Hexis Cell and Stack [4]

A joint development between Mitsubishi Heavy Industries (MHI) and Chubu Electric Power Company is the so called MOLB-Type (Mono-block Layer Built) planar SOFC. The cells are manufactured up to a size of 200 x 200 mm², based on a corrugated electrolyte. In this way the electrolyte also contains the gas channels, which simplifies the design of the interconnects, where plane ceramic plates are used (see Fig. 5). The biggest stack of this type was built of 40 layers, delivering 2.5 kW at 1000°C.

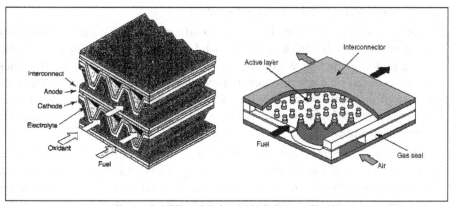

Figure 5. MHI and Cubu: MOLB Design [3, 5]

Since the electrolyte resistance is the most important obstacle on the way to further reduce the operating temperature the reduction of the thickness of the electrolyte is a main challenge. This can be done by shifting the mechanical stabilising function from the electrolyte to one of the electrodes. Mostly, the anode is favoured, because it has a good electrical conductivity. Therefore no increase in ohmic resistance is incurred by increasing the electrode thickness. Also, the nickel cermet has a good mechanical stability, which allows the manufacturing of larger components than with ceramic electrolyte substrates (see Fig. 6).

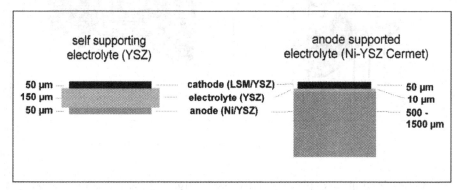

Figure 6. Anode supported cell concept

As one of the first institutions, this development was started in 1993 at Forschungszentrum Jülich and in the meantime is favoured by many developers throughout the world as the 'next generation' of SOFC. This concept allows reducing the operating temperature down to the range of 700 to 800°C whilst retaining the same power density as electrolyte supported cells at 950°C. At the same time this design allows the use of ferritic chromium alloys for interconnects, because its thermal expansion coefficient corresponds to that of the anode substrate.

At Forschungszentrum Jülich an anode substrate is manufactured by warm pressing with a thickness of 1 to 1.5 mm on which an electrolyte made by vacuum slip casting with a thickness of 5 to 10 µm is applied. The stack design is based on a co- or counter- flow arrangement. The latter is favoured in case of methane operation with internal reforming. Fig. 7 shows the stack design and a 40 layer stack delivering 5.4 kW at 800°C with methane and internal reforming.

Similar concepts are pursued e.g. by Global Thermoelectric, PNNL/Delphi and Risø National Laboratory. ECN and its spin-off company InDEC (Innovative Dutch Electrochemical Cells) manufacture electrolyte supported cells as well as anode supported cells.

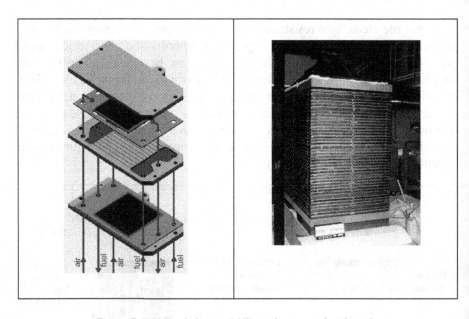

Figure 7. FZJ Stack design, 5 kW stack, operated with methane

Other institutions, like the DLR in Stuttgart, have developed concepts using pure metal substrates instead of the anode cermet to improve mechanical and redox stability. Up to now they have realised stacks with three to four layers.

A completely different design has been developed by Rolls Royce. Short electrode and electrolyte stripes are applied onto a porous ceramic substrate, which functions as mechanical supporting element. The single cells are connected electrically in series using short stripes of ceramic interconnects

and are operated at about 950°C. The concept is shown in Fig. 8. Rolls
Royce is currently working on the realisation of a 1 kW stack

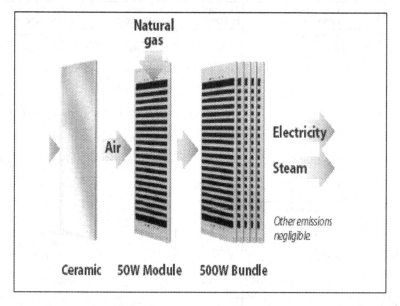

Figure 8. Rolls Royce Stack design concept

3. IMPORTANT DEVELOPERS

End of the nineties some of the most important developers in Europe,
Daimler-Benz/Dornier and Siemens, terminated their activities in planar
SOFC. After an interim phase the number of engaged companies has again
increased, though.

In the last two decades of the last century Westinghouse (now Siemens
Westinghouse Power Corporation SWPC) dominated the development in the
US. Since the "SECA" – programme was started, the situation has
completely changed and several consortia have been formed and activities
restated in the field of planar SOFC.

In Japan in the nineties more than 10 companies were involved in planar
SOFC development. After the goals of the "Shunshine" project of Nedo
could not be achieved, however, a re-orientation took place.

In the following tables the most important developers are listed.

Table 1: Developers in Europe

Country	Institution/ company	Employees in SOFC (ca.)	Own stack development	Main focus in development
Denmark	Haldor Topsoe			system, reformer
	Risoe	>70	planar: until 1999: ceramic IC, electrolyte substrate since 2000: metallic IC, anode substrate	materials, cells, stack
Germany	FZJ	70	planar, metallic IC, anode substrate	materials, cells, stack, manufacturing, system, modelling
	Siemens	18	tubular, "flat tube"	materials, manufacturing
	DLR-Stuttgart	10	planar, metallic IC, metallic substrate	materials, cells, stack, manufacturing
	BMW	9	planar, metallic IC, metallic substrate	stack, manufacturing, system
	IKTS-Dresden	6	planar, metallic IC, electrolyte substrate	stack
	HC Stark/Indec (NL)			powders, cell manufacturing
France	EDF/GDF	10	--	fuel conditioning, cell testing
	CEA; Fuel Cell Network (universities)	planned: 250	--	materials
United Kingdom	Ceres Power (spin-off from Imperial College)	16	planar, CGO electrolyte for 550°C, metallic IC, metallic substrate	sealing, cell, stack, manufacturing, system
	Rolls Royce	60	planar, on porous ceramic substrate	materials, cells, stack, manufacturing, system
Netherlands	ECN	15	--	cell development, manufacturing
Switzerland	Sulzer Hexis	>60	metallic IC, electrolyte substrate; anode substrate	materials, cells, stack, manufacturing, system

Table 2: Developers in North America

Country	Institution/ company	Employees in SOFC	Own stack development	Main focus in development
USA	SWPC	> 120	tubular (cathode substrate) "flat tube"	materials, cells, manufacturing, system
	Delphi Automotive Systems (collaboration with PNNL)	65	planar, metallic IC, anode substrate	cells, stack, system/APU
	McDermott Technology (former SOFCo) together with Ceramatec	30	planar, ceramic IC, electrolyte substrate	materials, cells, stack, reformer, manufacturing
	GE (former Honeywell - former Allied Signal)	15	planar, metallic IC, anode substrate	materials, cells, stack, reformer,
	PNNL (Pacific Northwest National Laboratory)	goal: 60 (2002)	planar, metallic IC, anode substrate	materials, cells, modelling
	ANL (Argonne National Laboratory)	5	planar, metallic IC, electrolyte substrate	materials, cells, modelling
	LLNL (Laurence Livermore National Laboratory)	?	planar, anode substrate	cells, stack,
	Acumentrics	?	tubular (anode substrate)	cells, stack, system
	ZTek	?	planar, metallic IC	cells, stack, system
Canada	Global Thermoelectric (now part of FCE)	80	planar, metallic IC, anode substrate	materials, cells, stack, manufacturing, system
	FCT (Fuel Cell Technology) together with SWPC	?	--	system

Table 3: Developers in Asia and Australia

Country	Institution/ company	Employes in SOFC	Own stack development	Main focus in development
Japan	Chubu EPCo (CEPCo) together with MHI	together	MOLB Design: planar, ceramic IC, electrolyte substrate	materials, cells, stack, manufacturing, system
	MHI (Mitsubishi Heavy Industries) with EPDC	120	tubular (porous support tube, serial connection")	materials, cells, stack, manufacturing, system
	Mitsubishi Materials	?	Gallate electrolyte, 800°C	materials, cells, stack
	TOTO together with Kyushu EPCo	?	tubular (cathode substrate)	materials, cells, stack, manufacturing
	Kyocera	?	unique cylindrical planar	materials, cells, stack, manufacturing
	Tokyo Gas	?	planar, metallic IC, anode substrate	cells, stack
Korea	KIER (Korean Institute of Energy research	?	anode supported flat tube	Stack, System (pressurised)
Australia	CFCL	>100	electrolyte substrate, since 2001 shift to ceramic IC	materials, cells, stack, manufacturing, system

4. DEVELOPMENT STATUS

4.1. CELLS

In the field of cell development many activities are ongoing, especially at various universities. Therefore it is quite difficult to compile comparison data, especially if they are supposed to be based on comparable operating conditions. In Fig. 9 this has been attempted for anode supported cells at 750°C operating temperature, comparing the most common cathode materials.

Besides the power density the manufacturable cell size is an important feature in characterizing the potential of the technology. A comparison of the achieved values is given in Fig. 10.

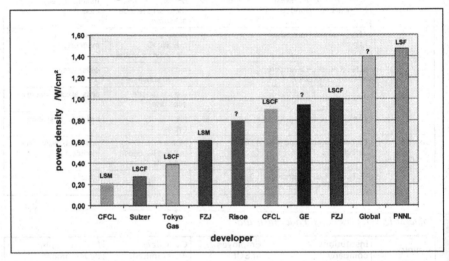

Figure 9. Anode supported cells- power density at 0.7 V and 750°C, using $H_2 + 3\%H_2O$

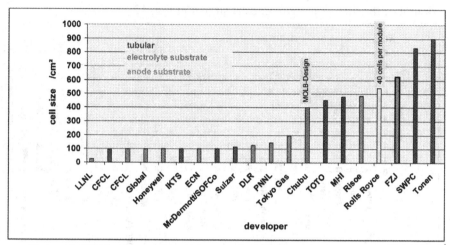

Figure 10. SOFC cells - maximum size manufactured by various companies [6 – 13]

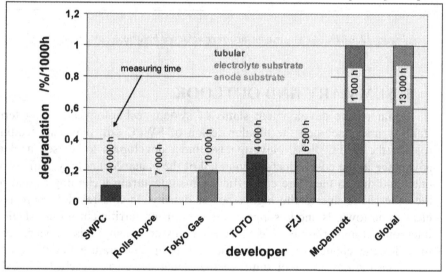

Figure 11. SOFC cells - degradation rates in inert environment (except for Global, who
measured in metal housings) {1, 8, 10, 12]

Meanwhile the degradation rates of planar cells are in the same range as
the tubular cells of SWPC (see Fig. 11), whereas the demonstrated operation
time is still much shorter for the planar cells.

4.2. STACKS

Compared to cell development there are much less developers with
proprietary stack technology. Some of them have changed the design in

recent years, restarting developments at lower power. The achieved power output of the different companies is shown in Fig. 12.

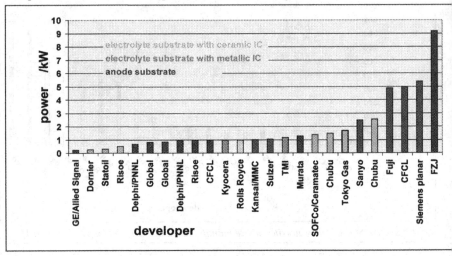

Figure 12. SOFC stacks - maximum power achieved byvarious companies {6 - 13]

5. SUMMARY AND OUTLOOK

As far as the development status of system technology and long term stability are concerned the tubular design of SWPC still plays the leading role in the SOFC field. Nevertheless, most developers today see a clear advantage in the cost reduction potential of the planar technology. This is on one hand due to the more cost efficient manufacturing technologies and on the other hand due to the higher power density. In this relation there is a clear trend towards anode supported design, using ferritic chromium steel as interconnect material. Besides increased power density this concept also provides the chance of reducing the operating temperature below 800°C. Although the development status of the planar design is clearly behind the tubular, considerable progress could be achieved during the recent years. A consolidation of activities can be observed, especially in the US, driven by the SECA programme and also in Japan by increased engagement of industry. A great push was created by the envisaged application as APU (auxiliary power unit), especially in Germany and in the US. In the stationary field the main focus is on small units in the kW range for residential energy supply and up to several 10 kW for small to medium sized CHP applications.

REFERENCES

[1] Kabs, H.;"Advanced SOFC Technology and its Realization at Siemens Westinghouse", Bilateral Seminars 33, Materials and Processes for Advanced Technology: Materials for Energy Systems, Egyptian-German Workshop, Cairo 7.-9. April 2002, ISBN 3-89336-320-3, edited by D.Stöver, M.Bram, (2002), 91-101

[2] Fujii, H.;"Status of National Project for SOFC Development in Japan", Solid Oxide Fuel Cells Meeting, 18. November 2002, Palm Springs, USA, (2002)

[3] Fujii, H.; Ninomiya, T; "Status of National Project for SOFC Development in Japan", European Solid Oxide Fuel Cell Forum, Vol. 2, Lucerne, Switzerland 1.-5. Juli 2002, edited by J. Huijsmans, (2002), 700-707

[4] Schmidt, M.; "The Hexis Project: Decentralised electricity generation with waste heat utilisation in the household", Fuel Cells Bulletin 1, No 1, (1998) 9-11

[5] Nakanishi, A.; Hattori, M.; Sakaki, Y.; Miyamoto, H.; Aiki, H.; Takenobu, K.; Nishiura, M.; "Development of MOLB Type SOFC", European Solid Oxide Fuel Cell Forum, Vol. 2, Lucerne, Switzerland 1.-5. Juli 2002, edited by J. Huijsmans, (2002), 708-715

[6] Proceedings 6th Int. Symposium Solid Oxide Fuel Cells (SOFC VI), The Electrochemical Society, Pennington, NJ, USA, (1999)

[7] Proceedings 7th Int. Symposium Solid Oxide Fuel Cells (SOFC VII), The Electrochemical Society, Pennington, NJ, USA, (2001)

[8] Proceedings 8th Int. Symposium Solid Oxide Fuel Cells (SOFC VIII), The Electrochemical Society, Pennington, NJ, USA, (2003)

[9] Proceedings Fuel Cell Seminar, Portland 2000

[10]Proceedings Fuel Cell Seminar, San Diego 2003

[11]Proceedings Fourth European SOFC Forum, Lucerne, 2000

[12]Proceedings Fifth European SOFC Forum, Lucerne, 2002

[13]Gardner, F. J. et all, SOFC Technology Development at Rolls Royce, Journal of Power Sources 86 (2000) 122 – 129

RECENT RESULTS OF STACK DEVELOPMENT AT FORSCHUNGSZENTRUM JÜLICH

R. STEINBERGER-WILCKENS, L.G.J. DE HAART, I.C. VINKE,
L. BLUM, A. CRAMER, J. REMMEL, G. BLASS, F. TIETZ,
W.J. QUADAKKERS

Forschungszentrum Jülich, D-52425 Jülich, Germany

Abstract: Since the mid-nineties several generations of SOFC stacks have been designed
              and tested incorporating the anode substrate-type cells developed in Jülich.
              The 6th generation, the so-called F-design stacks, with metallic interconnect
              has been the 'work horse' used for testing materials, cells and manufacturing
              processes in cell and stack development since its introduction in the year 2001.
              Stacks with up to 60 layers have been operated in recent years, delivering up to
              13 kW of electric power. The ferritic parts are made of the commercial steel
              type CroFer22APU.
              This paper summarises the current developments at FZJ and presents some
              recent results obtained with new materials.

Key words: SOFC/Solid Oxide Fuel Cells/Stacks/Cells/Systems/Testing

1. INTRODUCTION

The Forschungszentrum Jülich (FZJ) in Germany has been active in the field of SOFC technology since 1990. Research work includes all aspects from materials development and characterisation, over cell and stack design, modelling and manufacturing, to testing of cells, stacks and systems.

Since the mid-nineties several generations of SOFC stacks have been designed and tested based on the anode substrate type cells developed in Jülich. The main topics addressed in the development of the cell and stack technology are high performance, low degradation and the potential for low-cost production.

N. Sammes et al. (eds.), Full Cell Technologies: State and Perspectives, 123-134.
© 2005 *Springer. Printed in the Netherlands.*

2. STACK DESIGN

The so-called E- and F-designs for stacks with planar anode substrate type cells and metallic interconnects constitute the 'work horses' at Forschungszentrum Jülich used for testing SOFC materials, cells and manufacturing processes in cell and stack development since its introduction in the year 2000 [1]. Ferritic interconnects were chosen since they offer a high electric conductivity and thus the potential for high power density in the stacks. On the other hand the ferritic material gives rise to chromium evaporation (which can poison the cathodes) and is prone to massive corrosion at temperatures above approximately 900°C [2].

The E-design stack, shown in Fig. 1A, consists of a relatively thick metallic interconnect with machined gas channels on both sides and a metallic 'picture frame' for the cell, brazed to the interconnect by a metal solder. Sealing is obtained by glass ceramic sealants applied to the planar cell and frame surfaces using an automated dispenser. A fine Ni-mesh is spot-welded to the fuel side of the interconnect in order to improve the electrical contact between interconnect and anode substrate. On the air side usually a lanthanum cobaltite (LC) contact layer is sprayed onto the ribs of the interconnect.

The F-design, shown in Fig. 1B, is basically similar to its predecessor, with the same counter-flow configuration and distribution manifolds. The new stack design incorporates features developed towards a reduction in manufacturing steps, though offering a greatly simplified processing, thus reducing costs. Materials investment is reduced by using thinner interconnects. Moreover, the gas channels have been omitted on the fuel side resulting in a considerable reduction of manufacturing time. The fuel gas distribution over the anode is supplied by a coarse Ni-mesh, which simultaneously provides the electrical contact between interconnect and anode substrate. The Ni-mesh displays a pressure drop which is only twice that of the gas channels used in the E-design and therefore within tolerable limits.

In the new F-design the metal frame holding the cell is no longer brazed to the interconnect. This saves the time consuming and costly high temperature vacuum metal brazing step. Instead the metal frame is 'glazed' to the interconnect with the same glass-ceramic, which is used for sealing the repeating units in the stack. This 'glazing' and sealing occurs now in a one-step process during the first heating of the assembled stack.

Recent activities have been aimed at developing a light-weight stack design suitable for automobile applications, for instance for on-board electricity generation ('auxiliary power unit', APU). For this purpose the F-design was adopted to the manufacturing of the interconnects from sheet metal by stamping. Interconnect and frame are welded together and form a 'cassette' which offers the desired mechanical stability (cf. G-design in Fig. 2).

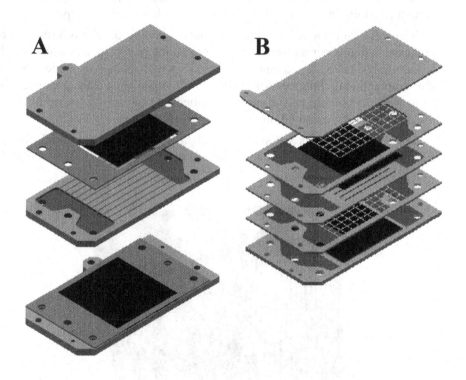

Figure 1. Schematic views of the E- (left) and F-design (right) for stacks with planar anode substrate type cells and metallic interconnects

Due to the reduced thickness of the interconnect and frame the weight can be reduced by up to 90% which again, together with the much simplified manufacturing process, results in a considerable reduction of costs.

3. STACK DEVELOPMENT FOR THE 20 KW SOFC SYSTEM DEMONSTRATOR

In the context of setting up a fully operational 20 kW SOFC system demonstrator at FZJ a new concept for packaging the 'hot' parts of the system (meaning parts operating above 400 °C) has been developed.

The resulting module design is shown in Fig. 3. Pre-reformer, air-preheater, afterburner and stack are integrated into a common housing which is thermally insulated and can be operated in self-sustaining mode without the conventional furnace for balance of heat-flow. The pressure for contacting and sealing of the stack is applied through spring loaded bolts on the outer side of the insulation.

This design reduces volume and weight by avoiding additional piping. Because all components are manufactured out of the same ferritic chromium

Figure 2. Stamped sheet metal 'repeating unit' of the Jülich G-design light-weight stack. Interconnect, cell and frame are welded into single units that are then joined into a stack by a glass ceramic sealant.

steel as used for the interconnect plates of the stack, thermal mismatch is avoided and no additional effort for compensation of expansion coefficients is required. The high integration of components in the housing results in a high efficiency of heat transfer, but also in a close thermal linkage of the processes in the system. The components are designed for combination with a stack with a nominal power of 5 kW, based on a cell size of 20 x 20 cm².

First operating results from the system, expected in the first half of 2005, need to prove whether this concept performs satisfactorily in transitional modes of operation.

4. RESULTS

4.1. TESTING IN THE 5- TO 10-KW RANGE

The first multi-kilowatt, 40-layer stack (40 cells of 20 x 20 cm²) was put into operation in mid-April 2002. It was manufactured according to the E-design and all ferritic parts were made of the commercial steel type 1.4742. The stack delivered 9.2 kW$_{el}$ in operation with hydrogen and 5.4 kW$_{el}$ with methane as fuel gas [1]. Whilst a small fraction of hydrogen was added to

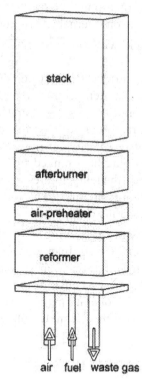

Figure 3. Integrated Stack Concept – "Integrated Module"

simulate pre-reformed natural gas, the stack nevertheless operated at nearly 100% internal reforming. It was run with a constant current density of 0.5 A/cm² for the first 1000 h of continuous operation. At these conditions the fuel utilisation was around 59 %. After a reduction of the furnace temperature the stack continued to operate in thermal self-sustaining mode, still producing between 4.0 and 4.5 kW_{el} with methane as fuel. Current density was lowered to 0.3 A/cm² after 1000 h, until it was switched off after a total of approx. 1250 h. Average degradation was in the range of 10%/1000 h at an average stack temperature below 850 °C, typical for the unprotected ferritic steel 1.4742.

This stack design at FZJ is intended to deliver insight into the interaction of materials (esp. steel, cell materials and sealants). Therefore neither the optimisation of the operating parameters, a maximisation of the power output, nor the optimisation of stack design parameters towards any given application were topics of development work at this stage. It should be noted that the total power output of this stack was at the time the highest ever reached by a planar stack with ferritic interconnects. Further noteworthy data are the high power density of around 630 mW/cm² and the operation in thermally self-sustaining mode on internal reforming.

April 2004 a power-stack with 60 cells (20 x 20 cm²) was put into operation (Fig. 4). It delivered 13.3 kW_{el} running on hydrogen (with 10% steam) as fuel at 0.74 A/cm² current density (0,6 W/cm² power density at an average of 0.83 V per cell). The temperature ranged from 709 °C (min) to 808 °C (max) at this operating point. Operated on methane (again, simulating partially pre-reformed natural gas) the stack rendered 11.9 kW_{el} at 0.74 A/cm² current density (0.55 W/cm² power density at an average of 0.74 V per cell). The temperature here ranged from 598 °C (min) to 809 °C (max). At no point in the stack the temperature was allowed to exceed 800°C in order to guarantee a long lifetime. The maximum possible power obtainable from this stack was thus not reached.

The stack was equipped with cells with LSCF cathode and the contacting and chromium retention layer LCC10. Interconnects and frames were made of the steel CroFer22APU (from the first commercial batch) and sealed with Ba-Ca-Al-Si glass. Degradation was in the order of 3% per 1000 hours within the first 1000 hours of operation (Fig. 5). This reduction in total power was predominately due to the untimely voltage loss of very few single cells.

4.2. LONG TERM DURABILITY TESTS

FZJ has adopted the design with stainless steel interconnect plates due to the steel's high electrical conductivity und the ensuing high power densities and compact stack designs. At the same time this material has a promising potential for low cost production. Still, the problems of chromium species evaporation from the steel and the corrosion behaviour of steel in SOFC relevant atmospheres need to be carefully addressed in order to achieve an operating lifetime relevant for stationary SOFC applications. One first step was the development of the steel JS-3 (now commercially available as CroFer22APU) [2]. This steel shows the formation of a slowly growing, thermodynamically stable oxide scale which exhibits reduced formation of volatile chromium species and at the same time possesses a high electrical conductivity.

In the year 2002, E-design short-stacks (2 cells 10 x 10 cm²) made of JS-3 coated with the cathode contact layer LCC2 [3] were operated. Long term testing over 4000 hours resulted in an improved durability compared to earlier stacks with the ferritic steel 1.4742 [1]. Stack voltage degradation was in the range of 2 to 3% per 1000 hours of operation. More recent tests using JS-3 together with improved contact and retention layers (LCC10) have shown substantially lower degradation rates of below 0.5% per 1000 hours over more than 2500 hours of operation (one example is shown in Fig. 6).

Figure 4. The 60-cell stack operated at FZJ from April 2004. 13.3 kWel power production with hydrogen fuel, 11.9 kWel with methane.

A parallel route in reducing corrosion and interaction phenomena during stack operation is to lower the stack temperature. Since this also considerably reduces the attainable power density more efficient cathode materials need to be employed. The LSCF cathodes currently used deliver similar results as LSM cathodes but at 50 K lower temperatures. This corresponds to results reported in the past years (for instance [4], [5], [6]).

The combination of new steel materials, new cathodes and protective coatings could well extend the operation time of stacks towards 40 000 hours. (assuming 0.5% degradation per 1000 hours a total power loss of 20% would have to be foreseen) which would be necessary in order to address the market of stationary combined heat and power (CHP).

Fig. 7 shows the voltage versus time plot of a short stack with LSCF cells over a testing time of 2500 hours.

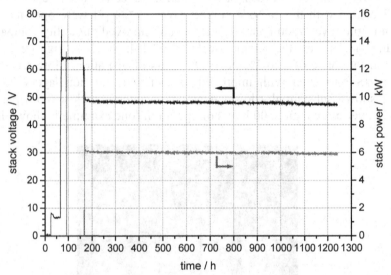

Figure 5. Time dependence of the voltage and power of the 60 cell F-design stack.

The stack displayed a degradation of around 3.5% on both cells. This is remarkably high in comparison with the results shown in Fig. 6, considering the much lower temperature. Keeping in mind that this stack was made from the first batch of commercial steel CroFer22APU we assume this degradation is mainly due to the slightly less protective properties of the oxide scales on the commercial steel as compared to those on JS-3. This results in a stronger interaction with our standard Ba-Ca-Al-Si-glass sealings.

4.3. THERMAL CYCLING TESTS

The satisfactory operation of several E-design short-stacks with JS-3 interconnects over some 2000 hours gave rise to a first targeted test of thermal cycling capabilities (Fig. 8). After an initial period of 1800 h of galvanostatic operation, a short-stack was subjected to regular heating up (2K/min, approx. 5 to 7 h in total) and cooling down cycles (natural cooling down in the furnace, approx. 20 h to 220°C). Weekends were used to cool down to 75°C in a 48 hour cycle (see insert in Fig. 8).

The stack was cycled 20 times, operated for 500 hours in order to determine the steady-state operation and again cycled 20 times. The initial degradation of the cells was also around 2-3%/1000 h. As can be seen from the figure, the ageing did not depend on the number of cycles and did not remarkably change with the number of cycles.

Post-operation analysis showed that the glass sealant used was of high quality and had hardly suffered in the testing. This would explain that apparently no major leakages occurred and the stack remained in a stable operational mode.

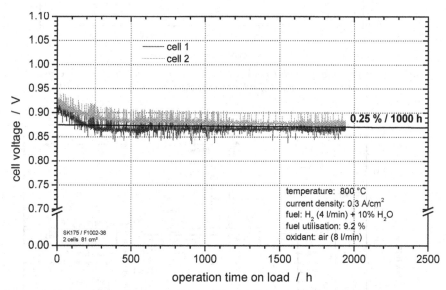

Figure 6. Operation of a short stack (2 layers) with LSM cathodes at 800°C and 300 mA/cm². The average degradation over 1500 hours was less than 0.5% per 1000 h. The interconnects were made of JS-3 steel with a coating of LCC10.

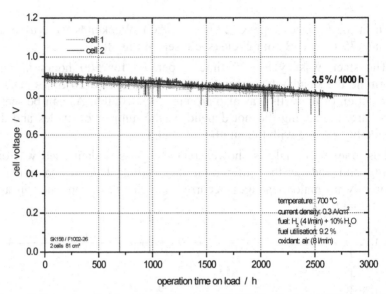

Figure 7. Operation of a short stack (2 layers) with LSCF cathodes at 700°C and at 300
mA/cm². The average degradation over 2500 hours was 3.5% per 1000 h

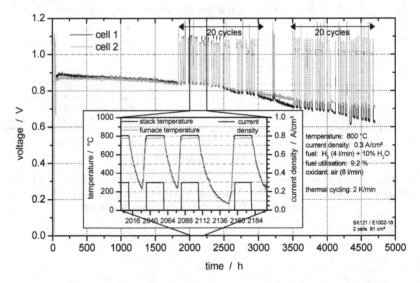

Figure 8. Time dependence of the cell voltages in an E-design short-stack during
galvanostatic operation at 0.3 A/cm² and 800 °C with H2, followed by two periods of 20
thermal cycles each; insert shows the time-temperature (down to 220 °C and 75 °C, resp.)
profile of the cycles.

4.4. I- V-CHARACTERISTICS

After a first series of encouraging tests with short-stacks manufactured according to the new F-design, a number of larger stacks with cells 20 x 20 cm² were manufactured, assembled and operated. Fig. 9 shows the current-voltage and current-power characteristics of a 10 cell F-design stack operated both on hydrogen and (simulated partially pre-reformed) methane.

With hydrogen a maximum power of nearly 2.5 kW was reached, with methane operation 2.0 kW. Fuel utilisation is in the range 60 to 70%. With similar 10-cell E-design stacks operated under the same conditions the maximum power reached was substantially less; 1.6 kW with hydrogen and 1.0 kW with methane, respectively [1]. Since both metallic parts of the F-design repeating unit, interconnect and cell frame, only have to be machined at one side, the flatness of both parts is improved, which in combination with the coarse Ni-mesh results in a far better and more reproducible contacting of cell and interconnect. The single layers in the stack all show power densities close to the best values obtained for single cells measured in ceramic housing with low fuel utilisation, i.e. 0.63 W/cm² .

The flattening of the current-voltage curves at higher current densities (see Fig. 9) is caused by the temperature increase in the stack during operation under load. This is more pronounced in the curve for methane operation, because at lower current densities the cooling due to the methane reforming reaction at the anode still dominates the temperature profile.

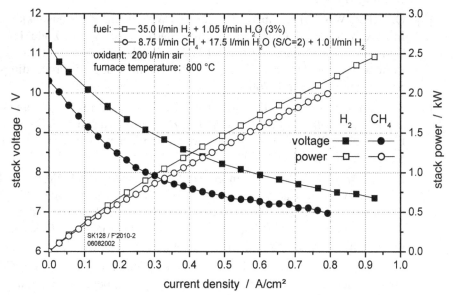

Figure 9. Stack voltage and power vs. current density for an F-design stack
(10 cells 20 x 20cm²) operated with hydrogen and methane

ACKNOWLEDGEMENTS

The authors gratefully thank all members of the Jülich anode substrate SOFC development team.

The work reported on here was partly funded by the German Federal Ministry of Economics and Technology (BMWi) within the project 'Zellen und Stackentwicklung für planare SOFC (ZeuS)', partly by the EC within the project 'Decentralised Power Generation Plants based on Planar SOFC Technology - Proof of Concept (ProCon)'.

REFERENCES

[1]   L. Blum, L.G.J. de Haart, I.C. Vinke, D. Stolten, H.-P. Buchkremer, F. Tietz, G. Blaß, D. Stöver, J. Remmel, A. Cramer and R. Sievering, Planar Anode Substrate Type SOFC kW-Class Stack Development, in *5th European Solid Oxide Fuel Cell Forum*, J. Huijsmans, Editor, Vol. 2, p. 784, European Fuel Cell Forum, Oberrohrdorf, 2002.

[2]   J.P. Abellán, V. Shemet, F. Tietz, L. Singheiser, W.J. Quadakkers and A. Gil, Ferritic Steel Interconnect for Reduced Temperature SOFC , in *Solid Oxide Fuel Cells VII*, H. Yokokawa and S.C. Singhal, Editors, PV 2001-16, p. 811, The Electrochemical Society Proceedings Series, Pennington, NJ, 2001.

[3]   O. Teller, W.A. Meulenberg, F. Tietz, E. Wessel and W.J. Quadakkers, Improved Material Combinations for Stacking of Solid Oxide Fuel Cells, in *Solid Oxide Fuel Cells VII*, H. Yokokawa and S.C. Singhal, Editors, PV 2001-16, p. 895, The Electrochemical Society Proceedings Series, Pennington, NJ, 2001.

[4]   K. Ahmed, J. Love, R. Ratnaraj, High Performance Cell Development at CFCL, in *Solid Oxide Fuel Cells VII*, H. Yokokawa and S.C. Singhal, Editors, PV 2001-16, p. 904, The Electrochemical Society Proceedings Series, Pennington, NJ, 2001.

[5]   E. P. Murray, S. A. Barnett, Improved Performance in $(La,Sr)MnO_3$ and $(La,Sr)(Co,Fe)O_3$ Cathodes by the Addition of a Gd-Doped Ceria Second Phase, in *Solid Oxide Fuel Cells VI*, H. Yokokawa and S.C. Singhal, Editors, PV 1999, p. 369, The Electrochemical Society Proceedings Series, Pennington, NJ, 1999.

[6]   K. Honegger, E. Batawi, Ch. Sprecher, R. Diethelm, Thin Film Solid Oxide Fuel Cell (SOFC) for Intermediate Temperature Operation (700°C), in *Solid Oxide Fuel Cells V*, H. Yokokawa and S.C. Singhal, Editors, PV 1995, p. 321, The Electrochemical Society Proceedings Series, Pennington, NJ, 1995.

CONFIGURATION OF TWIN WALLS IN LSGMO

D.I. SAVYTSKII[1], L. VASYLECHKO[1], U. BISMAYER[2], C. PAULMANN[2], M. BERKOWSKI[3]

[1]*Lviv Politechnic National University, 12 Bandera St., 79013, Lviv, Ukraine*
[2]*Min.-Petrogr. Institut, Universitat Hamburg, Grindelallee 48, D-20146 Hamburg, Germany*
[3]*Institute of Physics Polish Academy of Sciences, Al. Lotników 32/46, 02-668 Warsaw, Poland*

Abstract: The twin structure of $La_{0.95}Sr_{0.05}Ga_{0.9}Mg_{0.1}O_{2.925}$ perovskite-type crystals was studied using the white beam x-ray Laue diffraction technique in the temperature range 300-1000 K. The samples show twinning which occurs in the Laue patterns as peak splitting. Reflections from different domains were indexed separately. The twin structure of the orthorhombic as well as of the trigonal phase reveals that the system is composed of four different domain states. In both phases domain walls form a "chevron-like" structure with boundaries parallel to the crystallographic axes of the perovskite-like cell. This allows for a stress-free intergrowth of the four domain states in both ferroelastic phases. Temperature cycling through the orthorhombic – trigonal – orthorhombic phases showed that the "chevron-like" domain patter is fully reversible.

Key words: Perovskite/Twinnning/Laue Technique/Solid Oxide Fuel Cells

1. INTRODUCTION

Twin domains and their boundaries can form complex microstructural patterns in ferroelastic materials. It has been shown theoretically and experimentally that diffusion of impurity ions or vacancies can be enhanced within twin domain walls [1-4]. Furthermore, impurities [5] or oxygen vacancies [6-8] may be pinned in the wall or in its close vicinity. The fact that walls and bulk material show different dielectric transport and diffusion rates could explain the high ionic conductivity in doped heavy twinned lanthanum gallate, which is of interest as potential electrolyte in solid oxide

135

N. Sammes et al. (eds.), Full Cell Technologies: State and Perspectives, 135-147.
© 2005 *Springer. Printed in the Netherlands.*

fuel cells [9-10]. The aim of the present study is to determine the configuration of twin walls in the $La_{0.95}Sr_{0.05}Ga_{0.9}Mg_{0.1}O_{2.925}$ (LSGMO) solid solution and their temperature evolution using the Laue method.

2. THEORETICAL CONSIDERATION

$La_{0.95}Sr_{0.05}Ga_{0.9}Mg_{0.1}O_{2.925}$ is a perovskite-type material. At high temperature LSGMO undergoes three phase transitions: orthorhombic-monoclinic ($Imma - I2/a$) at 520 K - 570 K, monoclinic-trigonal ($I2/a - R\text{-}3c$) at 720 K and trigonal-trigonal ($R3c - R\text{-}3c$) at ca 870 K [11]. The prototype phase of LSGMO is cubic with space group Pm3m. From the temperature behavior of the perovskite-type cell parameters we conclude that LSGMO undergoes a phase transition to the cubic phase at a temperature close to the melting point of the compound.

Three different complementary techniques were used to analyze the LSGMO twin structure. Based on the condition of a stress-free domain wall we calculated the wall orientation using spontaneous strains tensors of contiguous domains [12]. Group theory allows to determine possible domain variants formed during a phase transition due to the symmetry change [13]. Symmetry breaking at the phase transition leads to energetically equivalent domain states in the ferroelastic phase. The domains are related to each other by symmetry elements which have been lost with respect to the prototype symmetry. The domain states include twin (orientation) states resulting from the loss of point-group symmetry elements as well as anti-phase domains which occur due to the loss of the translational symmetry. Only twin domain states can be macroscopically distinguished using X-Ray Laue diffraction technique. Twinning as a mechanical deformation had been observed long time ago by Mügge [14]. In this paper we consider the LSGMO ferroelastic domain structure and domain re-orientation in terms of mechanical twinning.

Parameters of the twin structure are shown in Table 1 for both, the orthorhombic [15] and trigonal phase.

Our theoretical analysis of the twin structure of LSGMO allows to correlate each twin law with a transformation matrix connected to the orthorhombic or trigonal basis vectors of neighbouring domains. Using twin models and the analytical geometry approach, the relationship of the basis vectors was obtained for all symmetry allowed domain pairs of the trigonal and orthorhombic phase. The basis vectors $\bar{a}^{t}, \bar{b}^{t}, \bar{c}^{t}$ of the domain D_t can be expressed by the basis vectors $\bar{a}^{0}, \bar{b}^{0}, \bar{c}^{0}$ of domain D_l according to:

Table 1. Parameters of the twin structure of LSGMO

Phase	Domain pair	Wall type	Domain wall orientation		Twin shift, μ_l	Transformation T_t matrix
Trigonal	D_1-D_2	W	$(001)_p$	(011)	$[110]_p$	$\begin{bmatrix} 1 & 0 & 0 \\ 0.990 & 0 & -1 \\ 0.990 & -1 & 0 \end{bmatrix}$
		W	$(110)_p$	(211)	$[001]_p$	$\begin{bmatrix} -0.990 & -0.010 & -0.010 \\ -0.995 & 0.995 & -0.005 \\ -0.995 & -0.005 & 0.995 \end{bmatrix}$
	D_1-D_3	W	$(100)_p$	(101)	$[011]_p$	$\begin{bmatrix} 0 & 0.990 & -1 \\ 0 & 1 & 0 \\ -1 & 0.990 & 0 \end{bmatrix}$
		W	$(011)_p$	(121)	$[100]_p$	$\begin{bmatrix} 0.995 & -0.995 & -0.005 \\ -0.010 & -0.990 & -0.010 \\ -0.005 & -0.995 & 0.995 \end{bmatrix}$
	D_1-D_4	W	$(010)_p$	(110)	$[101]_p$	$\begin{bmatrix} 0 & -1 & 0.990 \\ -1 & 0 & 0.990 \\ 0 & 0 & 1 \end{bmatrix}$
		W	$(101)_p$	(112)	$[010]_p$	$\begin{bmatrix} 0.995 & -0.005 & -0.995 \\ -0.005 & 0.995 & -0.995 \\ -0.010 & -0.010 & 0.990 \end{bmatrix}$
Orthorhombic	D_1-D_2	W	$(100)_p$	$(1\,0\,1)$	$[010]_p$	$\begin{bmatrix} -0.007 & 0 & -0.993 \\ 0 & -1 & 0 \\ -1.007 & 0 & 0.007 \end{bmatrix}$
		W	$(010)_p$	$(1\,0\,\bar{1})$	$[100]_p$	$\begin{bmatrix} -0.007 & 0 & 0.993 \\ 0 & -1 & 0 \\ 1.007 & 0 & 0.007 \end{bmatrix}$
	D_1-D_3	W	$(101)_p$	$(1\,2\,1)$		$\begin{bmatrix} 0.497 & -0.501 & -0.496 \\ -1.006 & -0.001 & -0.992 \\ -0.503 & -0.501 & 0.504 \end{bmatrix}$
		S	$(1k\bar{1})_p$	$(\overline{k2},1,k2\text{-}1)$	$[101]_p$	$\begin{bmatrix} -0.503 & 0.497 & 0.497 \\ 0.999 & -0.001 & 0.999 \\ 0.504 & 0.504 & -0.496 \end{bmatrix}$
	D_1-D_4	W	$(10\bar{1})_p$	$(1\,\bar{2}\,1)$		$\begin{bmatrix} 0.497 & 0.501 & -0.496 \\ 1.006 & -0.001 & 0.992 \\ -0.503 & 0.501 & 0.504 \end{bmatrix}$
		S	$(1k1)_p$	$(k2,1,\overline{k2-1})$	$[10\bar{1}]_p$	$\begin{bmatrix} -0.503 & -0.497 & 0.497 \\ -0.999 & -0.001 & -0.999 \\ 0.504 & -0.504 & -0.496 \end{bmatrix}$
	D_1-D_5	W	$(011)_p$	$(1\,2\,\bar{1})$		$\begin{bmatrix} 0.497 & 0.501 & 0.496 \\ 1.006 & -0.001 & -0.992 \\ 0.503 & -0.501 & 0.504 \end{bmatrix}$
		S	$(k1\bar{1})_p$	$(k2,\bar{1},k2\text{-}1)$	$[011]_p$	$\begin{bmatrix} -0.503 & -0.497 & -0.497 \\ -0.999 & -0.001 & 0.999 \\ -0.504 & 0.504 & -0.496 \end{bmatrix}$
	D_1-D_6	W	$(01\bar{1})_p$	$(\bar{1}\,2\,1)$		$\begin{bmatrix} 0.497 & -0.501 & 0.496 \\ -1.006 & -0.001 & 0.992 \\ 0.503 & 0.501 & 0.504 \end{bmatrix}$
		S	$(k11)_p$	$(k2,1,k2\text{-}1)$	$[01\bar{1}]_p$	$\begin{bmatrix} -0.503 & 0.497 & -0.497 \\ 0.999 & -0.001 & -0.999 \\ -0.504 & -0.504 & -0.496 \end{bmatrix}$

k=5.31 and k2=3.15 at room temperature.

$$
\begin{bmatrix} \overline{a}^t \\ \overline{b}^t \\ \overline{c}^t \end{bmatrix} = T_t \times \begin{bmatrix} \overline{a}^0 \\ \overline{b}^0 \\ \overline{c}^0 \end{bmatrix}
$$

(1)

where T_t is a are second rank tensor. Calculated transformation matrixes T_t for all permitted twin pairs of the trigonal and orthorhombic phase are listed in Table 1.

3. EXPERIMENTAL DETAILS

$La_{1-x}Sr_xGa_{1-2x}Mg_{2x}O_{3-\delta}$ solid solutions were grown using the Czochralsky method [8]. The Laue technique was used to investigate the twinning of LSGMO crystal samples with $0,2 \times 0,2 \times 0,2$ mm^3 dimensions. Experiments were carried out at the experimental beamline F1 at HASYLAB (DESY, Hamburg) using white X-ray synchrotron radiation. The equipment used includes a STOE 4-circle diffractometer and a BRUKER CCD detector with a $6,25 \times 6,25$ cm CCD matrix (1024×1024 pixel resolution). The sample-detector distance could be changed from 58 mm to 358 mm. Angular scanning of the sample was carried out around two axes using the conditions, $\omega = 0°$, ϕ axis = $0°-180°$ with $15°$ step, χ axis = $0°-90°$ with $15°$ step width. The temperature of the sample was adjusted using liquid nitrogen gas heated to the selected temperature.

OrientExpress V3.3 freeware was used to index the Laue diffraction patterns. We developed a special algorithm to measure the sample-detector distance (d) and to determine the coordinates of the projection of the sample position in the recording plane of the CCD detector. The algorithm implies the analysis of two Laue diffraction patterns detected at two different distances d_{FI} at the same angular position of the sample. A *Pearson VII* two-variable function was used to approximate the profiles of the Bragg reflections.

Fig. 1a shows a section of a Laue diffraction pattern of a LSGMO crystal detected at room-temperature with a crystal-detector distance of $d=58$ mm after the 4th thermal cycle. The Laue pattern shows multiplets splitting into 4 reflections, each one generated by Bragg reflections from its corresponding orientation state. The symbols *TO1*, *TO2*, *TO3* and *TO4* (Fig. 1a) indicate different ferroelastic orientation states and corresponding reflections.

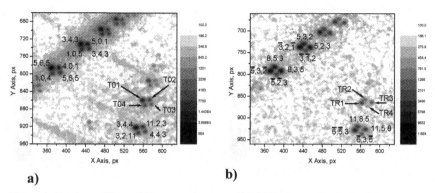

Figure 1. Sections of Laue diffraction patterns of LSGMO detected at 300 K (a) and 753 K (b) (d=58 mm□=87°, □=15°). Reflections are indexed according to the best solution determined for each of the four domain states.

Laue diffraction patterns show that heating the sample to the trigonal phase causes the multiplets to split into 4 reflections too. The symbols *TR1*, *TR2*, *TR3* and *TR4* (Fig. 1b) indicate the corresponding orientation state respectively. All Laue diffraction patterns were separately indexed using the reflections of each of 4 domain states. Positions of reflections (up to 30 in total) were used to refine the orientation matrix and sample-detector distance for the trigonal and orthorhombic phase.

4. CONFIGURATION OF TWIN WALLS

Our identification of the twin walls was based on the setting of one selected domain, referred to as "reference" domain. Using the transformation matrices given in Table 1 we determine the orientation matrices of all allowed domain states according to equation (1). We take for example the domain *TR1* as a "reference", i.e. D_1, and perform the calculations with respect to its orientation matrix. With the orientation matrices determined in this way we calculate the positions of the Bragg reflections of all domain states in the Laue diffraction pattern.

Fig. 2a shows a section of the Laue diffraction pattern obtained at a sample-detector distance of 358 mm. Additionally to the Bragg reflections from four *TR* domains, the calculated positions of reflections from allowed domain states in this phase are plotted as round spots. As seen in Fig. 2 the reflection positions caused by twinning due to the planes (011) and (121) coincide with the position of the *2 1 1* reflection of the domains *TR2* and *TR3*.

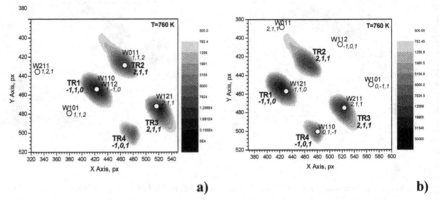

a)   b)

Figure 2. Enlarged section of the Laue diffraction pattern observed at 358 mm and 753 K
($\varphi=8^\square$, $\psi=15^\square$). Additionally to experimental Bragg reflections of four TR domains,
positions of reflections from other domain states in the trigonal phase are shown: (a)
positions calculated using the orientation matrix of the domain TR1; (b) positions
calculated using the orientation matrix of the domain TR3.

Fig. 2a shows that some calculated positions of reflections from symmetry allowed domains do not coincide with observed reflections of domain *TR4*. Therefore, we proceed to calculate the orientation matrix of domain *TR3* (previously determined with respects to *TR1*), and taking this domain as a "reference". Positions of reflections are given in Fig. 2b which shows that the domain *TR3* is connected with domain *TR1* via (121), and it is also connected with the domain *TR4* via the plane (110). However, there is no stress-free wall between the domains *TR3* and *TR2*. Based on the identification of domain walls between 4 observed orientation states we can now assume that the domain pattern of LSGMO crystal has a "chevron-like" configuration in the trigonal phase (Fig. 3).

A "chevron-like" twin structure is also observed in the orthorhombic phase. Figure 4 shows a section of the Laue diffraction pattern observed at room temperature. Besides Bragg reflections of four TO domains, the reflection positions calculated with respect to the orientation state TO3 from all allowed domain states in the orthorhombic phase are shown. As seen in Fig. 4 reflection positions due to twinning by $(10\bar{1})$ and $(\bar{1}21)$ planes regarding TO3 coincide with positions of the 1,2,1 and 2,0,0 reflections from domain states TO1 and TO2 correspondingly.

Note that the position of the reflection 0,0,2 of domain TO4 connected with TO3 by the mirror plane $(12\bar{1})$ rather poorly agrees with the position of the observed reflection of the TO4 state. We calculated the reflection positions due to the mirror plane $(10\bar{1})$ with respect to domain orientations determined by the reference TO3 through the reflections $\bar{1}21$ and $12\bar{1}$. Hence we calculated the orientation of domains connected with TO3 via two subsequent reflections with respect to $(\bar{1}21)$ (or $(12\bar{1})$) and $(10\bar{1})$ mirror planes represented as the sum W(-121)+W(10-1) shown in Fig. 4. Based on our analysis we suggest a twin structure depicted in Fig. 5. Similar to the trigonal phase, the perovskite direction $<100>_p$ is common for all four TO domains.

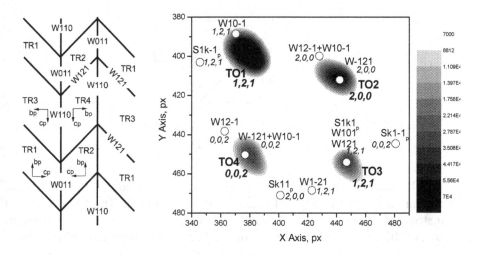

Figure 3. Model of the "chevron-like" twin structure of the trigonal phase formed by W-type domain walls (011), (110) and (121).

Figure 4. Section of a Laue diffraction pattern measured at RT and crystal-detector distance of 358 mm ($\varphi=80^o$, $\psi=15^o$). Besides the Bragg reflections of four TO domains, positions of reflections from all possible domain states in the orthorhombic phase are presented (as circles), calculated with respect to the orientation matrix of domain TO3.

Since 4 orientation states co-exist in the ferroelastic phases of LSGMO, the condition of a "stress-free" crystal can be written:

$$T_{41} \times T_{34} \times T_{23} \times T_{12} \equiv I, \qquad (2)$$

where T_{12} is the transformation matrix related to the domain pair TS_1 - TS_2, the T_{23} - transformation matrix is related to the domain pair TS_2 - TS_3, the T_{34} - transformation matrix is related to the domain pair TS_3 - TS_4, and the T_{41} - transformation matrix is related to TS_4 - TS_1. If the result of equation (2) is not the unitary matrix, although this twin configuration occurs, the domain configuration must be under stress because the basis vectors of the orientation states cannot be linked without additional strain.

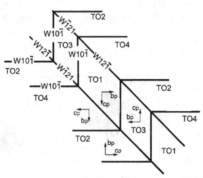

Figure 5. Model of the twin structure of the orthorhombic phase formed by W-type domain
walls namely ($\bar{1}21$) and ($10\bar{1}$) mirror planes.

All possible combinations were tested to verify condition (2) for W-type
domain walls in the trigonal and orthorhombic phase. The transformation
matrices T_i given in Tables 1 were used in our calculations. Three different
cases fulfilling condition (2) were derived for four trigonal orientation states:

$$W110 \times W121 \times W011 \times W121 = I \qquad (3.1)$$

$$W011 \times W112 \times W101 \times W112 = I \qquad (3.2)$$

$$W101 \times W211 \times W110 \times W211 = I, \qquad (3.3)$$

where the symbols of the domain walls between the relevant orientation
states are used instead of the symbols of the transformation matrices. Eq.
(3.1) was derived for the domain structure of the LSGMO crystal in the
trigonal phase (Fig. 3).

Two different cases fulfilling condition (2) were derived for four
orientation states of the orthorhombic phase in the co-existence test:

$$W121 \times W101 \times W1\bar{2}1 \times W101 = I \qquad (4.1)$$

$$W12\bar{1} \times W10\bar{1} \times W\bar{1}21 \times W10\bar{1} = I. \qquad (4.2)$$

Eq. (4.2) corresponds to the domain structure of our LSGMO crystal
which has been observed in the orthorhombic phase at room temperature
(Fig. 5).

Mechanical twinning theory [16] implies that a twin lamella intersects a
twin boundary without strain if the displacement direction μ_l is parallel to
the plane of the intersected wall. Our analysis of the twin configuration in
LSGMO shows that this condition holds for both, the orthorhombic and
trigonal phase. Thus, in the trigonal structure the crystal is divided into two
domains with walls (110) or (011) (or $(010)_p$ or $(001)_p$ using perovskite-type
setting (conventionally named as l type walls) because certain areas of the
crystal shift along $[101]_p$ or $[110]_p$ directions (Table 1). Hence, "chevron-

like" or "zig-zag-shaped" lamellae with walls (121) (or $(011)_p$ using the perovskite-type setting (conventionally named as the *2* type walls)) emerge due to the shift of certain areas along the direction $<100>_p$, parallel to the previously emerged boundaries. "Chevron-like" domain walls $(10\bar{1})$ are formed in the orthorhombic phase in a similar way through displacements along the same perovskite direction $<100>_p$.

5. REVERSIBILITY OF TWIN CONFIGURATION

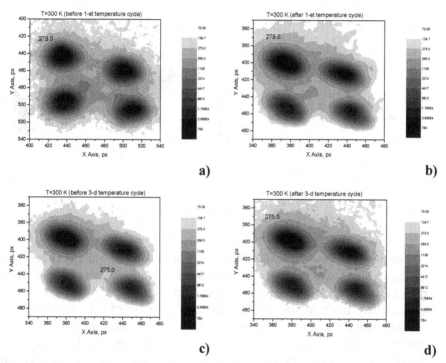

a)   b)

c)   d)

Figure 6. Sections of Laue diffraction patterns of a LSGMO crystal observed at 300 K before (a) and after (b) the 1ˢᵗ and before (c) and (d) after the 3ʳᵈ heating cycle (d=358 mm, φ=80°, ψ=15°).

Fig. 6 shows Laue diffraction patterns detected for a multiplet before and after heating the crystal above the trigonal transition point for the first and third time. After the first heating cycle (see Fig. 6b) the crystal maintains the initial orientation states (Fig. 6a), however the intensities of the reflections corresponding to the four orientation states indicate a change of individual state volumes because the ratio of the intensities changed considerably after the first thermal treatment. Fig. 6 also shows Laue diffraction patterns determined after the 3ʳᵈ thermal cycle, which are practically identical to the Laue diffraction pattern observed after the initial heating cycle. Reflection positions, shapes and intensities remain unchanged after the first thermal treatment.

Reversibility of orientation and configuration of twin walls was also observed in the high temperature rhombohedral phase as indicated by identical Laue diffraction patterns, detected above the ferroelastic phase transition point. Fig. 7 shows that the position, shape and intensity of reflections of four domains remain unchanged after the first heating and further thermal treatment.

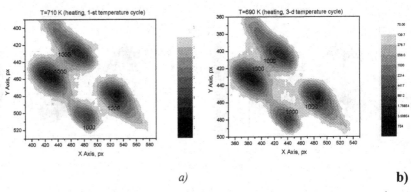

a)   **b)**

Figure 7. Laue diffraction patterns of LSGMO observed in the trigonal phase at the 1ˢᵗ (a)
and 3ʳᵈ (b) thermal cycles (d=358 mm, φ=80°, ψ=15°).

Our results show that orientation states and domain walls between them are reproduced without noticeable changes in the low-temperature ortho-rhombic and the high-temperature trigonal phase during the ferroelastic tran-sition. There is a reproducibility of the configuration of domain walls in the sample during thermal treatment (in other words, there is a domain wall location memory effect). Only the twin structure observed in the ortho-rhombic phase prior to the first phase transition is an exception. The corresponding twin structure is a "non-equilibrium" state because both, the crystal shape and its size changed after the sample heating. The "equilib-rium" twin structure is then formed in the trigonal phase on heating above the transition point. Subsequent cooling of the sample below the phase transition point results in a rearrangement of the walls according to the distribution of defects. The LSGMO crystals contain in fact a large number of defects because of the high concentration of bivalent cations and oxygen vacancies.

6. CONCLUSIONS

The present results show that heating from the orthorhombic to the trigonal phase and cooling from the trigonal phase to the orthorhombic causes a "chevron-like" twin structure in LSGMO crystals. All "chevron cells" are extended along the same perovskite axis $<001>_p$, in both, the orthorhombic and the trigonal phase. Besides four orientation states occur which perfectly match geometrically and no additional stress occurs at the intersections of domain walls throughout the full temperature range of the corresponding phases. The specific "chevron" twin structure allows for reproducibility of wall configurations in heavy defect LSGMO crystals, because the stress can completely relax forming phase-specific domain wall configurations, which take reorientations during thermal cycling.

In the trigonal and orthorhombic phases a "chevron-like" configuration is formed due to displacements along the perovskite direction $<001>_p$, parallel to the "chevron cells". This feature may be of practical use because the preparation of electrolyte and electrode ceramics for SOFC includes pressing as one of the synthesis stages. Pressure implies unidirectional mechanical stress. Ceramics can therefore be approximated by an ensemble of small crystallites and mechanical pressure imposed to an electrolyte pellet causes the formation or rearrangement of the twin structure in ceramic grains in the direction, which is parallel or nearly parallel to the imposed pressure. Hence, such pressure will cause texturing of the twins in the electrolyte layers along the direction of oxygen diffusion in the SOFC structure. Keeping in mind the influence of twin walls on the conductivity and the high density of twin walls in LSGMO [15], we suppose that texturing of the twin structure, e.g. the "chevron cells" increases the conductivity of the perovskite-type electrolyte LSGMO in the cathode-anode direction.

The twin structure in small LSGMO crystals tends to form "chevron-like" wall configurations that allows for a stress-free co-existence of four different orientation states. This pattern of domain walls is expected to be characteristic also for other perovskite-type compounds with a sequence of ferroelastic phase transitions related to those of LSGMO. Examples are mixed conductivity perovskites, which are used as electrodes and interconnectors in SOFC batteries.

ACKNOWLEDGMENTS

The work was supported by WTZ (UKR 01/012), the Ukrainian Ministry of Science (Projects "Cation" and M/85-2003), and the Polish Committee for Scientific Research (Grant N 7 T08A 00520).

REFERENCES

1. Aird A., Domeneghetti M.C., Mazzi F., Tazzoli V., Salje E.K.H. Sheet superconductivity in twin walls: experimental evidence of WO_{3-x}. J. Phys.: Condens. Matter. 1998; 10: L377-80.

2. Lee W.T., Salje E.K.H., Bismayer U. Structure and transport properties of ferroelastic domain walls in a simple model. Phase Transitions 2003; 76: 81-99.

3. Calleja M., Done M.T., Salje E.K.H. Anisotropic ionic transport in quartz: the effect of twin boundaries. J. Phys.: Condens. Matter. 2001; 13: 9445-54.

4. Calleja M., Done M.T., Salje E.K.H. Trapping of oxygen vacancies on twin walls of $CaTiO_3$: a computer simulation study. J. Phys.: Condens. Matter. 2003; 15: 2301-07.

5. Bartels M., Hagen V., Burianek M., Getzlaff M., Bismayer U., Wiesendanger R. Impurity-induced resistivity of ferroelastic domain walls in doped lead phosphate. J. Phys. Condensed Matter. 2003; 15: 957-62.

6. Klie R.F., Ito Y., Stemmer S., Browning N.D. Observation of oxygen vacancy ordering and segregation in perovskite oxides. J. Solid State Ionics 2000; 130: 289-302.

7. Harrison R.J., Redfern S.A.T. The influence of transformation twins on the seismic-frequency elastic and anelastic properties of perovskite: dynamical mechanical analysis of single crystal $LaAlO_3$. Physics of the Earth and Planetary Interiors 2002; 134: 253-72.

8. Harrison R.J., Redfern S.A.T. The effect of transformation twins on the seismic-frequency mechanical properties of polycrystalline $Ca_{1-x}Sr_xTiO_3$ perovskite. American Mineralogist 2003; 88: 574-82.

9. Feng Man, Goodenough J.B. A superior oxide-ion electrolyte. Eur. J. Solid State Inorg. Chem. 1994; 31: 663-72.

10. Ishihara T., Matsuda H., Takita Y. Doped $LaGaO_3$ perovskite type oxide as a new oxide ionic conductor. J. Am. Chem. Soc. 1994; 116: 3801-03.

11. L.Vasylechko, Vashook V., Savytskii D., Senyshyn A., Niewa R., Knapp M., Ullmann H., Berkowski M., Matkovskii A., Bismayer U. Crystal structure, thermal expansion and conductivity of anisotropic La_{1-}

$_x Sr_x Ga_{1-2x} Mg_{2x} O_{3-y}$ (x=0.05, 0.1) single crystals. J. Solid State Chem. 2003; 172: 396-411.

12. Sapriel J. Domain-wall orientations in ferroelastic. Phys.Rev. 1975; B12: 5128-39.

13. Janovec V. Group analysis of domains and domain pairs. Czech. J. Phys. 1972; B22: 974-94.

14. Mügge O. Twinning. Neues Jahrb. Mineral. Geol. 1889: 98-108.

15. Savytskii D.I., Trots D.M., Vasylechko L.O., Tamura N., Berkowski M. J. Twinning in $La_{0.95} Sr_{0.05} Ga_{0.9} Mg_{0.1} O_{2.92}$ crystal studied by white-beam (Laue) X-ray microdiffraction. Appl. Cryst. 2003; 36: 1197-203.

16. Klassen-Nekludova M.V. Mechanical twinning of crystals. Moscow, 1960 (in Russian).

STRUCTURAL STABILITY AND ION TRANSPORT IN LaSr$_2$Fe$_{1-y}$Cr$_y$O$_{8+\delta}$

V.L. KOZHEVNIKOV, I.A. LEONIDOV, M.V. PATRAKEEV, J.A. BAHTEEVA

Institute of Solid State Chemistry, Ural Branch of RAS, Ekaterinburg 620219

Abstract: Ceramic samples of LaSr$_2$Fe$_{3-y}$Cr$_y$O$_{8+\delta}$ are characterized with X-ray powder diffraction and total conductivity measurements in the temperature range 750-950°C and at oxygen partial pressures between 10^{-22} and 0.5 atm. It is shown that partial replacement of iron for chromium is an effective means to prevent structural transformation at the loss of oxygen. The isothermal ion conductivity is obtained from the analysis of the total conductivity, and it is shown to decrease with the increase in chromium doping to y = 0.6. However further increase in the chromium content to y = 1 results in an increase in the oxygen ion conductivity level. This behavior is explained with a model where the oxygen ions coordinated to chromium are not excluded entirely from the transport mechanism.

Key words: Oxygen Conductivity/Electron Conductivity/Lanthanum Strontium Chromium Ferrite/Perovskite

1. INTRODUCTION

The lanthanum-strontium ferrites with mixed, oxygen ion and electron, conductivity are considered as promising materials for development of oxygen separating membranes having a potential for partial oxidation of natural gas into syngas, important feedstock in chemical industry and a fuel for high-temperature SOFCs. One of the oxides in the ferrite family, LaSr$_2$Fe$_3$O$_{8+\delta}$, was described by Battle et al.[1]. The room temperature structure of the oxide with δ = 0 can be envisioned as derived from the

149

N. Sammes et al. (eds.), Full Cell Technologies: State and Perspectives, 149-155.
© *2005 Springer. Printed in the Netherlands.*

perovskite lattice where regular arrangement of oxygen vacancies results in tripling of the elementary unit with double layers of iron-oxygen octahedra (FeO_3) that alternate with the single layers of iron-oxygen tetrahedra (FeO_2). Interestingly, high level of oxygen ion conductivity (0.35 S/cm) was observed recently in this oxide at 950°C. It is important to notice that this value is larger than in the brownmillerite-like $Sr_2Fe_2O_5$, where oxygen vacancies tend to strong ordering, while it is smaller than in the perovskite-like $La_{0.5}Sr_{0.5}FeO_{2.75}$ (0.45 S/cm at 950°C [2]), where oxygen vacancies are heavily disordered. Considering that the amount of oxygen vacancies is larger in $LaSr_2Fe_3O_8$ ($La_{0.33}Sr_{0.67}FeO_{2.67}$) than in $La_{0.5}Sr_{0.5}FeO_{2.75}$ the conclusion can be made that heating to rather high temperatures cannot overwhelm completely partial ordering of the oxygen vacancies in $LaSr_2Fe_3O_8$. Disordering of the iron sub-lattice by chromium doping was utilized in this work in an attempt to enhance disordering on the oxygen sub-lattice and to increase ion conductivity and structural stability.

2. EXPERIMENTS AND DISCUSSION

The samples of $LaSr_2Fe_{3-y}Cr_yO_{8+\delta}$ were prepared by solid-state reactions in air at 750 to 1300°C. Phase purity control and determination of the lattice parameters were carried out with the help of X-ray diffraction using a STADI-P (STOE) diffractometer. The oxygen stoichiometry was studied with a Setaram TG-DTA-92 thermoanalyzer. The four probe d.c. conductivity measurements were carried out using rectangular $2 \times 2 \times 15$ mm sintered bars with the density about 93% of theoretical value. The electrical parameters were measured with a high-precision voltmeter Solartron 7081.

2.1. SAMPLE CHARACTERIZATION

The samples $LaSr_2Fe_{3-y}Cr_yO_{8+\delta}$, where y = 0, 0.3, 0.6 and 1, were equilibrated at 1200°C and slowly cooled in the air. The X-ray spectra of the chromium doped specimens were indexed with the tetragonal R-3c space group, Table 1. The oxygen content in the air-prepared samples was found to correspond to $\delta = 1$, i.e. to the formula $LaSr_2Fe_{3-y}Cr_yO_9$ regardless of the doping level. The lattice parameters in $LaSr_2Fe_{3-y}Cr_yO_9$ tend to increase with chromium content, which seems somewhat surprising. Indeed, chromium substitution for iron in the parent oxide $LaSr_2Fe^{3+}Fe_2^{4+}O_9$ is expected to proceed as replacement of Fe^{4+} for Cr^{4+} with formation of $LaSr_2Fe^{3+}Fe_{2-y}^{4+}Cr_y^{4+}O_9$. Therefore, one can foresee some contraction of the crystal lattice and a decrease in the lattice parameters because of the cation size difference. The observed increase of the lattice parameters in the oxidized samples can be attributed to 3d metal - 2p oxygen covalent overlap smaller for Cr^{4+}-O^{2-} than for Fe^{4+}-O^{2-} chemical bonds.

Table 1. The elementary unit parameters for $LaSr_2Fe_{3-y}Cr_yO_{8+\delta}$ equilibrated in different atmospheres

y	air		10% H_2/90% He	
0	5.4784(3)	13.3928(4)	*	*
0.3	5.4894(2)	13.4156(6)	5.5272(1)	13.5310(3)
0.6	5.4939(1)	13.4189(5)	5.5217(2)	13.5182(1)
1.0	5.5094(5)	13.4476(8)	5.5202(4)	13.5082(6)

* $LaSr_2Fe_{3-y}Cr_yO_8$ (s.g. Pnma) a=5.5095(1), b=11.8845(5), c=5.6028(1) [1].

The smaller covalence results in decrease of the bonding energy and in respective increase of the average length of metal-oxygen bonds, which overwhelms the cation size effect and gives rise to an increase in the lattice parameters. The treatment of the chromium doped specimens in the 10%H_2/90%He gas mixture at 570-590°C results in formation of oxygen depleted $LaSr_2Fe_{3-y}Cr_yO_8$ samples having X-ray spectra similar to the initial $LaSr_2Fe_{3-y}Cr_yO_9$. However, unlike in the oxidized $LaSr_2Fe_{3-y}Cr_yO_9$ form the increase in the amount of chromium in the reduced $LaSr_2Fe_{3-y}Cr_yO_8$ is accompanied with a decrease in the elementary unit parameters. Such a change can be explained as reflecting increasing ionicity of the chemical bonds on reduction and replacement of the larger Fe^{3+} for smaller Cr^{3+} cations in $LaSr_2Fe_{3-y}Cr_yO_8$.

2.2. ELECTRICAL CONDUCTIVITY

Logarithmic plots of the equilibrium total conductivity versus the partial pressure of oxygen for $LaSr_2Fe_{3-y}Cr_yO_{8+\delta}$ at 950°C are shown in Fig.1 as an example. The low-pressure segments of the isotherms near the minima, where deviations of the oxygen content from the nominal stoichiometric composition are very small, i.e. $\delta \ll 1$, are important for the analysis of the oxide ion contribution. Such minima are typical in oxides when oxygen pressure variations at constant temperature result in nearly equal concentrations of electron- and hole-like carriers. There may be another, pressure independent contribution in the conductivity and its presence is signaled by a smooth shape of the minima. Therefore the experimental results near the conductivity minima were approximated with the known relation

$$\sigma = \sigma_i + \sigma_n^\circ \cdot pO_2^{-1/4} + \sigma_p^\circ \cdot pO_2^{+1/4} \qquad (1)$$

where σ_i stands for the ion-oxygen contribution while parameters σ_n° and σ_p° denote electron σ_n and hole σ_p contributions to the total conductivity, respectively, at $pO_2 = 1$ atm.

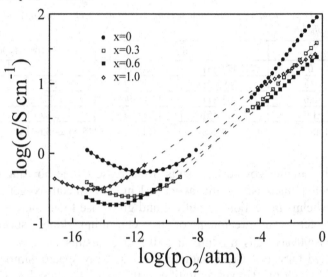

Figure 1. The logarithmic plots of the total conductivity versus oxygen pressure for the oxides $LaSr_2Fe_{3-y}Cr_yO_{8+\delta}$ Solid lines show fitting results according to (1). Dashed lines serve as guides to an eye to the high-pressure parts of the isotherms.

The doping dependent behavior of the ion conductivity at different temperatures is shown in Fig. 2 as obtained from the experimental data with the using of relation (1). Considering crystal structure of $LaSr_2Fe_{3-y}Cr_yO_{8+\delta}$ ($\delta \ll 1$) and taking in view preference of chromium to six-fold coordination, the oxygen migration pathway seems most probable that involves mainly oxygen ions and vacancies in the network of iron-oxygen octahedra and pyramids, while the oxygen ions coordinated by chromium and the oxygen structural vacancies in the first coordination sphere of the tetrahedrally coordinated iron do not take part in the transport. On the other hand heating may result in displacement of an oxygen ion from a FeO_3 octahedron to the structural vacancy near a FeO_2 and in formation of two $FeO_{2.5}$ pyramids in place of them. Denoting the equilibrium amount per formula unit of the thermally redistributed oxygen as α, the local crystallo-chemical formula of solid solution may be represented as $LaSr_2(FeO_3)_{2-2y-\alpha}(CrO_3)_y(FeO_{2.5})_{2y+2\alpha}(FeO_2)_{1-y-\alpha}$ where the influence of both temperature and dopant on the relative fractions of structural polyhedra can be seen. Assuming the effective mass law, the equilibrium amounts of the oxygen ions and vacancies in iron-oxygen polyhedra are interrelated as

$$\frac{(2y+2\alpha)^2}{(2-2y-\alpha)\cdot(1-y-\alpha)} = K(T) \tag{2}$$

where $K(T)$ is the equilibrium constant. The parameter α follows from Equation (2) as a function of the doping level and temperature. Then it can be used in order to determine the isothermal dependence of the ion conductivity from the doping according to

$$\sigma_i \sim (2-2y-\alpha)\cdot(2y+2\alpha) \tag{3}$$

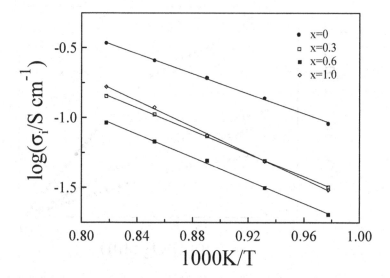

Figure 2. The Arrhenius plots for the ion conductivity in the oxides LaSr₂Fe$_{3-y}$Cr$_y$O$_{8+δ}$

The first multiplier shows the amount of iron-oxygen octahedra, which, as supposed, is proportional to the amount of the oxygen ions taking part in the transport, while the second one corresponds to the amount of the iron-oxygen pyramids, i.e. oxygen vacancies available for the jumps of the oxygen ions. However, the model calculations result in the monotonous decrease of the ion conductivity within the doping level utilized in this study, which does not agree quite well with the results in Fig. 2 where the conductivity increase is seen with the doping increase from $y = 0.6$ to $y = 1$. Better correspondence of the calculated and experimental results can be obtained assuming that the chromium coordinated oxygen ions are not excluded entirely from the transport, at least at the doping levels used, and some fraction of these oxygen ions can take part in the transport. Certainly, this "ion drag" may have viable effect upon ion transfer only at relatively small concentrations of the dopant and when chromium-oxygen octahedra are rather isolated from each other.

2.3. OXYGEN PERMEABILITY

The applicability of oxygen membrane materials depends essentially on their oxygen permeation fluxes. The dependence of the flux density jO_2 on the oxygen pressure difference across the membrane of thickness L can be calculated by the using of the measured total conductivity σ and calculated values σ_i for the ion conductivity.

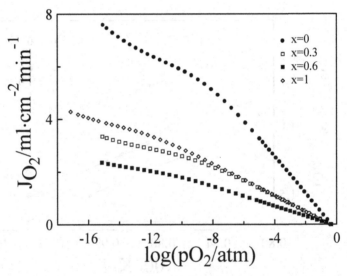

Figure 3. The estimated oxygen permeation fluxes through 0.1 cm thick membranes of the $LaSr_2Fe_{3-y}Cr_yO_{8+\delta}$ as functions of the permeate-side oxygen partial pressure at 950 °C.

These estimates give only an upper limit of oxygen permeation rate because surface exchange reactions may result in some suppression of the overall transport. For the conditions typical for syngas generation ($pO_2' = 0.21$ atm, $pO_2'' = 10^{-16}$ atm, 950°C), the results for the membranes with L = 0.1cm and different chromium contents are shown in Fig. 3. In the calculations at high pressures the ion conductivity values were assumed to be nearly equal to those at low pressures, Fig. 2. It is seen that the permeation rate in chromium doped samples is smaller than in the parent ferrite. Nonetheless, it may achieve a value of about 4 ml·cm^{-2}·min^{-1} in the sample with y = 1, which corresponds to the syngas production rate of about 20-25 ml·cm^{-2}·min^{-1} in the methane partial oxidation process. In combination with good thermodynamic and structural stability this permeation rate would indicate that the oxide $LaSr_2Fe_{3-y}Cr_yO_{8+\delta}$ is a promising candidate membrane material for the partial oxidation of natural gas.

This work was partially supported by the NATO SfP Program (award #978002), INTAS (grant #01-278) and by the research program of the Russian Academy of Sciences #26.

REFERENCES

1. P. D.Battle, T. C.Gibb, P. J. Lightfoot, *Solid State Chem.* 84 (1990) 237.

2. M.V.Patrakeev,   J.A.Bahteeva,   E.B.Mitberg,   I.A.Leonidov,
V.L.Kozhevnikov, K.R.Poeppelmeier, *J.Solid State Chem.* 172 (2003)
219.

ELECTRIC POWER GENERATION BASED ON COAL SLURRY ELECTROLYSIS WITH SUBSEQUENT USAGE OF PRODUCED HYDROGEN IN FUEL CELLS

B.A. TROSHEN'KIN, V.B. TROSHEN'KIN

Institute for Mechanical Engineering Problems of the National Academy of Sciences of Ukraine, Pogharskogo Dm. Str., Kharkiv, 61046, Ukraine

Abstract: Current mining practice allows extracting slightly over one-half of commercial coal reserves, whereas the other half remains underground due to technological and technical reasons. Only 10 to 15 % of the coal extracted to the surface is used in electric power generation. The remaining part is lost with coal mining waste, coal improvement, and in ash dumps in the form of non-combusted carbon. But the major portion of energy extracted from coal is dissipated to no effect to the environment with combustion and condensation products because the efficiency of TPPs does not exceed 33 to 35 %.

A known solution to this problem is creating and utilizing power installations with coal-fired fuel cells (FC) [1-3]. As a rule, coal is first gasified, and then the producer gas is fed to fuel cells. However, carbon oxide, which is formed by partial oxidation of coal and hydrocarbons, is an extremely unstable compound that recombines, under specific conditions, to solid carbon, i.e. fuel soot, which clogs the FC pores and damages the electrochemical equipment.

Soot formation can be eliminated by increasing the gasification pressure to 15-20 MPa, and the temperature to 200-250 °C. In so doing, shifting the reaction of interaction of carbon with water to formation of decomposition-resistant hydrogen and carbon dioxide is effected by coal slurry electrolysis. Electric current reduces the carbon activation energy, making it possible to run the gasification process at relatively low temperatures. Carbon dioxide is readily absorbed from the mixture with alkali solutions. Having separated carbon dioxide, pure hydrogen is fed to the fuel cell.

Key words: Electric Power Generation/Coal Slurry Electrolysis/Hydrogen/Fuel Cells

N. Sammes et al. (eds.), Full Cell Technologies: State and Perspectives, 157-161.
© 2005 *Springer. Printed in the Netherlands.*

Coal comprises the major part of Ukraine's energy resources.

Generation of electric power and heat involves hard and hazardous labor of miners underground.

Current mining practice allows extracting slightly over one-half of the commercial coal reserves, whereas the other half remains underground due to technological and technical reasons. Only 10 to 15 % of the coal extracted to the surface is used in electric power generation. The remaining part is lost with coal mining waste, coal improvement, and in ash dumps in the form of non-combusted carbon. But the major portion of energy extracted from coal is dissipated to the environment with combustion and condensation products to no effect because the efficiency of TPPs does not exceed 33 to 35 %.

A known solution to this problem is creating and utilizing power installations with coal-fired fuel cells (FC) [1-3]. As a rule, coal is first gasified, and then the producer gas is fed to fuel cells. However, carbon oxide, which is formed due to partial oxidation of coal and hydrocarbons, is an extremely unstable compound that recombines, under specific conditions, to solid carbon, i.e. fuel soot, which clogs the FC pores and damages the electrochemical equipment.

Soot formation can be eliminated by increasing the gasification pressure to 15-20 MPa, and the temperature to 200-250 °C. In so doing, shifting the reaction of interaction of carbon with water to formation of decomposition-resistant hydrogen and carbon dioxide is effected by coal slurry electrolysis. Electric current reduces the carbon activation energy, making it possible to run the gasification process at relatively low temperatures. Carbon dioxide is readily absorbed from the mixture with alkali solutions. Having separated carbon dioxide, pure hydrogen is fed to the fuel cell.

Taking into account that the share of coal in the total reserves of carbon and hydrocarbon fuel in Ukraine is 95 %, production of hydrogen employing the method discussed can be considered as the most promising one.

Presently, at IPMash NAS of Ukraine, an experimental installation has been developed and erected. It is intended for developing the regimes of coal-based production of high-pressure hydrogen.

At the same time, the Institute for Materials Science Problems of NAS of Ukraine (IMSP) has made an oxide zirconium-based fuel cell. By agreement between IPMash and IMSP, joint tests of the equipment developed will be conducted to receive source data for designing the experimental complex "Mine-TPP" (ECMT) with a capacity of 6 MW.

The equipment of the underground part of the ECMT is focused to safe working of coal beds. High temperatures and high pressure of underground water, which is used as an oxidant in deep mines, allows conducting coal slurry electrolysis directly in the coal faces. Naturally, in this case underground water is not pumped to the surface. Being located directly in the underground water layer, the coal in-seam miner with FCs can break, grind and gasify the coal seam, and supply high-pressure hydrogen (and carbon dioxide) to the surface to generate electric power. The coal miner uses part of the hydrogen obtained for its own functioning. To be completely self-contained, the coal miner that moves along the coal seam being worked requires oxygen in addition to hydrogen. For this, the coal miner should carry on board a reserve of oxygen. During coal slurry electrolysis, a definite amount of oxygen is produced in addition to hydrogen and carbon dioxide. This oxygen can be separated from the mixture by using membranes, and fed to the fuel cells.

Fuel cells are also the basic equipment of the surface part of the complex, but in this case air enriched with oxygen serves as the oxidant of hydrogen delivered from the mine.

The cost-performance indicators of the electric power generation technology being offered, as compared to existing conventional and alternative technologies, have the following advantageous features:

- the efficiency of a power plant operated to the method offered is 52 to 54 %. The efficiency of operating thermal power plants is 35 to 40 %;

- the unit cost of generated electric power is 30 % lower than that offered currently in the energy market;

- the coal-based method of generating electric power being developed is more reliable than other generation methods based on fuel cells;

- coal consumption for generating a unit of electric power (1 kW) is reduced 2-fold (from 330-350 to 165-175 g/kW) because, firstly, fuel cell efficiency exceeds the efficiency of turbogenerators approximately by 1.6 to 1.7 times, and, secondly, the methane contained in coal, rather than being discharged to the atmosphere, is fed to the fuel cell along with the hydrogen flow;

- non-commercial coal reserves, amounting to about 45 billion tons in Ukraine, are put into use. Because of the high content of the inorganic fraction, up to 40-80 %, this coal has found no application neither in metallurgy for coke production, nor in power engineering for generating electric power. There are theoretical assumptions that high ash-content coal can be subject readily to electrolysis and therefore used for electric power generation;

- coal presently left in mines due to technical and technological reasons (for instance, at present, extracting coal from seams less than 0.2 to 0.3 m thick is a challenge) can be reclaimed for production. In so doing, a robot coal miner with fuel cells can be designed to any size, including the micro size option);

- no electric power is required for surfacing the coal, pumping out water and ventilating the drifts; and the metal capacity of underground coal mining equipment is reduced. Since the underground water pressure is equal to the pressure of the surrounding rock, no special casings are required for securing the underground working. Conveyors, railways, cables, etc., are also not needed;

- metal consumption for manufacturing power equipment is somewhat lower. The fuel cells are surface-mounted apparatus, whereas turbogenerators are space-mounted equipment. And, at first sight, it seems that the former should be significantly more metal-intensive than the latter are. But the power equipment in the second case comprises not only turbogenerators, but also boilers, recuperative heat exchangers, condensers, pipelines and pumps. Systems for coal storage and preparation, slag and ash handling, and even railroads for transporting coal to TPPs should also be included herein.

The development, manufacturing and testing of a stationary 6 MW thermal power plant with fuel cells will take no more than 3 years; the project cost will be 23 mln USD. The R&D activities cost will be 3.5 mln USD.

A 6 MW thermal power plant with fuel cells fed with hydrogen from a coal slurry electrolyzer is, in essence, a module of future commercial and energy power plants. By combining modules, one can set up power plants of any capacity.

From the social and ecological viewpoint, implementing the results of research activities as per this project in the energy economy can yield the following:

- free the miners from hard and hazardous labor underground in very involved mining and geological conditions;

- dramatically reduce labor input for complex and labor-intensive mining-and-construction work, and also eliminate such processes as transportation, storage and preparation of mineral stock for combustion at power plants from the production cycle.

For the electric power generation technology being developed, emission to the atmosphere of carbon dioxide, which causes the green-house effect, is three times less than that in conventional technologies. This is because underground gasification of coal makes coal mining and coal improvement wasteless; and eliminates ash dumps with non-combusted carbon whose oxidation in atmospheric conditions produces additional amounts of carbon dioxide. In the method being offered, carbon dioxide emission is reduced by

30 to 40 %, and it can be completely eliminated by pumping liquefied or water-absorbed carbon dioxide into the deep layers of the Earth. In addition, fuel cells, per unit of generated electric power, expend about one-half of the carbon since their efficiency is 60 to 70 %.

Dust formation during conventional electric power generation is a source of occupational diseases. The method offered is free of this drawback since dust is not formed during the production process.

Hazardous substances (such as sulphur compounds, benz[a]pyrene, etc.) emitted to the atmosphere by operating TPPs remain underground in the process suggested. Hazardous substances (heavy metals, rare and trace elements, etc.) concentrated in the inorganic part of coal also remain in the mines.

In the conventional technology, waste piles and ash dumps are located next to mines and TPPs respectively. Because waste piles and ash dumps are liable to intensive wind erosion, self-ignition and dump leaching, they are extremely potent environmental contamination sources. In the advanced electric power generation technology, these objects simply do not emerge. Thus, this technology is not only environmentally safe, but also allows reclaiming vast farming land areas.

REFERENCES

[1] Vasiliev O. Fuel store //Elektropanorama. – Kiev, 2000. – 3. – pp. 18–20.

[2] U.S. Pat. 4226683. U.S. Class: 204/101; 204/129. IPC: C25B 001/02, C25B 001/00. Method and apparatus for hydrogen production in an absorber liquid by electrochemical of coal and water/Vaseen, Vesper A. (U.S.A.). –№ 065210; Filed 07.10.80; Publ. 09.09.79. –14 p.

[3] U.S. Pat. 4670113. U.S. Class: 204/80; 204/129; 204/101; 0423/415A. IPC: C25B 001/00. Electrochemical activation of chemical reactions / Lewis, Arlin C. (U.S.A.). –№ 788148; Filed 02.06.87; Publ. 16.10.85. – 12 p.

RESULTS OF ORGANIC FUEL CONVERSION AT FUEL CELL TEST INSTALLATION

O.M. DUDNIK, I.S. SOKOLOVSKA

Coal Energy Technology Institute of National Academy of Sciences and Ministry of Fuel and Power of Ukraine (CETI)

Abstract: CETI has carried out research and design activities to develop combined circulating fluidized bed and fuel cell technology. The first in Ukraine Fuel Cell Test Installation (FCTI) for research of polymeric, molten carbonate, phosphoric acid, and solid oxide fuel cells with electrical capacity up to 3 kW at atmospheric and elevated pressure was created in CETI. Molten carbonate fuel cell was developed in CETI and tested via natural gas reforming at FCTI. Non-catalytic and catalytic steam gasification of Ukrainian, American and Australian coals in dense and fluidized beds was studied for generation of synthesis-gas with the high hydrogen content (up to 62 vol. %). The software for fast estimation of the installation operation was created. CETI and State Rocket Design Office "Southern" have prepared the design documentation for Ukrainian coal pyrolysis & gasification PCFB pilot plant with thermal capacity up to 10 MW for future application in new hybrid co-generation SOFC power plant.

Key words: Fuel Cell/Reforming/Coal/Steam Gasification/Fluidized Bed/Hybrid Co-Generation Power Plant/Fuel Cell Test Installation

1. INTRODUCTION

Fuel cell power plants operate with high electrical efficiency (up to 60%) and lower emissions to environment (up to full emission absence). Fuel cell power plants, as rule, convert hydrogen, natural gas and methanol.

The fields of fuel cells application are co-generation power plants with electric capacity from 1 kW to 11 MW, fuel cell vehicles with electric capacity from 20 kW to 250 kW, small portable power generators with

N. Sammes et al. (eds.), Full Cell Technologies: State and Perspectives, 163-174.
© 2005 *Springer. Printed in the Netherlands.*

electric capacity from 3 kW to 5 kW, fuel cell power packs (cellular phones (1-3 W), computers (5–50 W), camcorders (2–5 W), cordless tools (20–200 W)), and special application.

CETI has experience in the development of experimental and pilot plants for conversion of various organic fuels, including Ukrainian high-ash coal with high metamorphism degree, for the purpose of power generation and in creation of software for estimation of power plant operation. CETI has carried out research and design activities to develop new combined cycle technology.

Since 1992, Fuel Cell Research Group of CETI has carried out scientific activities in the fields of fuel cells:

1. Analysis of modern fuel cell power technologies (1992-1994)

2. Creation of installation for research of fuel cells at atmospheric pressure (1994-1996)

3. Research of natural gas and petrol reforming (1996-2004)

4. Creation of fuel cells (1996-2004)

5. Research of gasification of Ukrainian high-ash coal, USA and Australian coals for generation of synthesis gas with high content of hydrogen (1997-2004)

6. Tests of fuel cells (1997-2004)

7. Modernization of Fuel Cell Test Installation for test and certification of fuel cells at elevated pressure (2001-2004).

The basic requirements to synthesis gas composition for fuel cell test at FCTI are shown in Table 1.

Table 1. The basic requirements to synthesis gas composition for fuel cell tests at FCTI

Fuel cell type	H_2	CO	H_2S + COS	C and C_nH_m
Low-temperature (up to 100°C): - with polymeric membrane (PEM)	Fuel	10 ppm	10 ppm	-
Medium-temperature (up to 220°C): - with phosphoric acid electrolyte	Fuel	< 1.0 vol. %	200 ppm	100 ppm
High-temperature: - with molten carbonate electrolyte (up to 650...750°C);	Fuel	Fuel	10 ppm	100 ppm
- with solid oxide electrolyte (up to 800...1000°C).	Fuel	Fuel	10 ppm	100 ppm

2. FUEL CELL TEST INSTALLATION

Fuel Cell Test Installation of CETI [1] consists of fuel processor and fuel cell test module. The fuel processor includes system of hot sulfur cleaning, reactor, water tank, feed water supply, steam generator, synthesis gas cooler, and drop separator (see Fig. 1a). In case of natural gas or petrol conversion for generation of synthesis gas with high content of hydrogen, the fuel processor consists of hot desulphurization system, reformer, steam generator, and synthesis gas cooler (see Fig. 1b). In case of coal or coke conversion for generation of synthesis gas with high hydrogen content, the fuel processor consists of steam fluidized bed gasifier, cyclone, steam generator, cooler, dryer, and hot desulphurization system (see Fig. 1c). Molten carbonate fuel cell test module includes furnace with air and synthesis gas heaters, electrical furnace with fuel cell cassette, and air compressor.

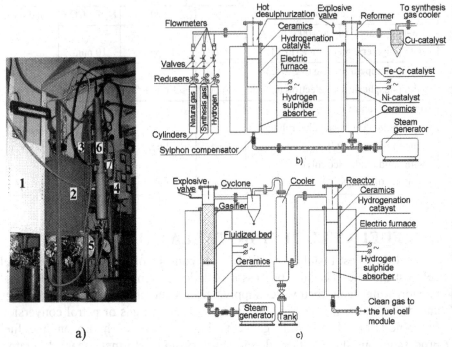

Figure 1. Fuel Cell Test Installation: a) fuel processor: 1- system of hot desulphurization, 2 – reactor, 3 – water tank, 4- feed water supply, 5 – steam generator, 6 – synthesis gas cooler, 7 – drop separator; b) circuit of reformer with hot desulphurization; c) circuit of coal steam gasifier with hot desulphurization

3. RESULTS OF NATURAL GAS REFORMING

FCTI simulates the operation of a fuel cell power plant. Natural gas reforming with use of only nickel catalyst was carried out at FCTI for generation of synthesis gas for high-temperature fuel cell operation (see Fig. 2). The steam/methane mass ratio reached 1.5 during the experiments. There was no formation of soot in these modes of operation. The system of cooling and drying provided complete removal of moisture from synthesis gas, and the hydrogen content in dry synthesis gas was from 71 to 78 % at the methane conversion degree of 96 to 98 %. The thermal capacity of FCTI was 4.5 kW.

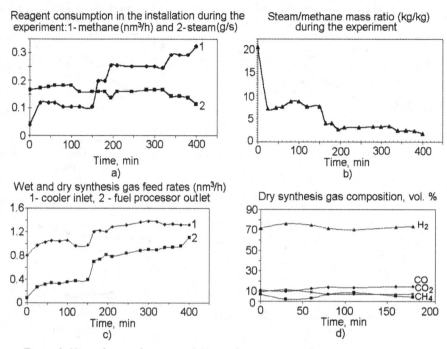

Figure 2. Natural gas reforming with Ni-catalyst use: input characteristics (a, b), output characteristics (c, d)

Medium-temperature (Fe-Cr) and low-temperature (Cu) catalysts were used for a decrease in CO content by shift reaction (see Fig. 3). The use of these catalysts allowed reducing CO concentration in synthesis gas for the test of polymeric fuel cell stack. Electric capacity of this stack can be up to 400 W. In this case, complete conversion of natural gas and CO is reached.

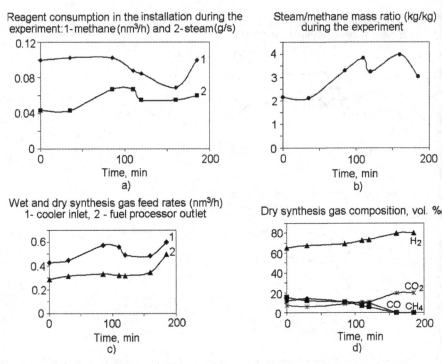

Figure 3. *Natural gas reforming with use of Ni, Fe-Cr, and Cu catalysts: input characteristics (a, b); output characteristics (c, d)*

4. RESULTS OF COAL STEAM GASIFICATION IN FLUIDIZED BED

Steam gasification of coals in dense and fluidized beds showed possibilities of the use of coals with different metamorphism degree for generation of synthesis-gas with high hydrogen content. Coals with various metamorphism degrees from lignite (brown coal) to anthracite culm, both with and without catalysts, were converted at FCTI. In the case of steam gasification of coals in fluidized bed, the temperature of fluidized bed was changed in the range from 600 up to 1050 °C. The main analyzable characteristics during the experiments were the composition of synthesis gas, the rate of carbon conversion, the degree of carbon conversion, and the feed rate of synthesis gas (see Fig. 4 and 5).

For the increase in rate and carbon conversion degree, in case of anthracite culm use, two types of catalysts were used: fluidized bed Ni-catalyst and K_2CO_3. Application of catalysts allowed increasing carbon reaction rate by 15 to 20 % and achieving complete conversion of carbon.

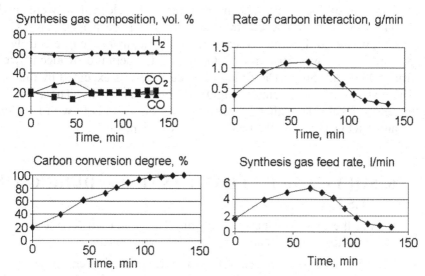

Figure 4. Steam gasification of coke of bituminous coal in fluidized bed

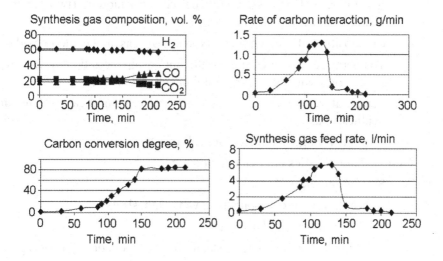

5. MOLTEN CARBONATE FUEL CELL

CO_2 was used for the initial heating of the fuel cell with molten carbonate electrolyte. The rate of heating was $5°C/min$ for the uniform fuel cell heating. After heating of the fuel cell to the necessary temperatures, synthesis-gas and air were added into the streams of carbon dioxide and then CO_2 consumption was reduced.

Samples of fuel cells with molten carbonate electrolyte were made in CETI. Molten carbonate electrolyte on the basis of Na_2CO_3, K_2CO_3, and Li_2CO_3 was used. For the decrease in melting temperature of electrolyte, the part of Li_2CO_3 was 50 mass %. Operating temperature of electrolyte was from 600 up to $650°C$. The fixing of electrolyte was executed by the matrix. The matrix made of ceramics (MgO) with a 40% porosity. The anodes were made of nickel. The cathodes were made of NiO and NiO with Li addition. Ni and NiO were put on the matrix by plasma deposition.

6. SOFTWARE FOR DATA ANALYSIS DURING THE TEST

During the experiments, software of CETI was used for the analysis of the obtained data. Input characteristics for calculation are:

- Consumption of organic fuel, water, and hydrogen or synthesis gas for desulphurization (in case of natural gas reforming), synthesis gas for use in fuel cell;
- Electricity usage for supply of hot desulphurization system, steam generator, fuel processor, and fuel cell test module (two electrical furnaces and compressor);
- Composition of organic fuel, dry synthesis gas, and cathode oxidant;
- Temperatures and pressures in the hot desulphurization system, fuel processor, steam generator, and fuel cell test module;
- Temperatures and pressures of water, steam, natural gas, wet and dry synthesis gas;
- Current, voltage, load, and fuel cell resistances.

Output characteristics after calculation are:

- Steam/fuel ratio;
- Concentration of components in wet synthesis gas;
- Feed rate of wet synthesis gas;
- Enthalpies of water, heating steam, fuel, wet and dry synthesis gas;
- Fuel conversion degree;
- CO conversion degree;
- Formulas of fuel interactions;
- LHV of synthesis gas;
- Excess oxidant coefficient for fuel conversion in fuel cell;
- Thermal capacities of hot desulphurization system, fuel processor, fuel cell test installation;

- Thermal capacities of installation in recalculation on methane and synthesis gas LHV;
- Chemical, thermal, and electrical efficiencies.

7. DEVELOPMENT OF PCFB-1.0 PILOT PLANT FOR CONVERSION OF HIGH-ASH COAL

The design documentation on the pilot plant for combustion and gasification of Ukrainian high-ash coals in circulating fluidized bed (PCFB) under pressure up to 2.5 MPa with thermal capacity up to 10 MW was developed by the "Southern" State Design Office according to technical task and calculations of CETI. Circuits of this pilot plant in two modes of operation (PCFB combustion and pyrolysis-gasification) are shown in Figure 6.

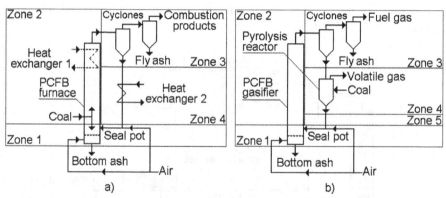

Figure 6. Circuits of PCFB Pilot Plant: a) coal combustion;

b) coal pyrolysys and gasification

In the case of bituminous coals pyrolysis and gasification in PCFB, the plant consists of pyrolysis reactor, gasifier, two cyclones, combustion chamber, air compressor, and throttling device (or gas turbine).

Software, developed in CETI for calculation of PCFB-1.0 plant, includes the following program complexes: CFB Creator, PCFB Combustion, and PCFB Gasification.

The PCFB Gasification software package was used for calculation of following characteristics: specific characteristics of processes of coal pyrolysis and gasification under pressure in CFB (consumption of air, gasification products, LHV of pyrolysis and fuel gases, heat of gasification, used heat in heat exchanger, circulation factor etc. in recalculation on 1 kg of initial coal), redistribution of the initial heat brought by coal and air in the gasifier basic elements (gasifier, cyclones, pyrolysis reactor, and heat exchanger) for generation of fuel and pyrolysis gases, coal and air consumption and feed rates of pyrolysis and fuel gases under pressure up to 2.5 MPa, thermal capacities of gasifier, pyrolysis reactor, and heat exchanger for pilot plant of CETI.

8. CONCEPT OF HYBRID CO-GENERATION POWER PLANT

After the analysis of PCFB-1.0 plant design documentation, the circuit of new hybrid co-generation power plant with use of PCFB gasifier, solid oxide fuel cells, and gas turbine power plant with built-in air recuperator was proposed (see Fig. 7). Thermal capacity of power plant will be 1.14 MW at gasifier operation under pressure of 0.35 MPa and Ukrainian bituminous coal consumption of 222.6 kg/h. Electric capacity of solid oxide fuel cell module will be 375 kW and of electric capacity of high-speed gas turbine plant will be 125 kW.

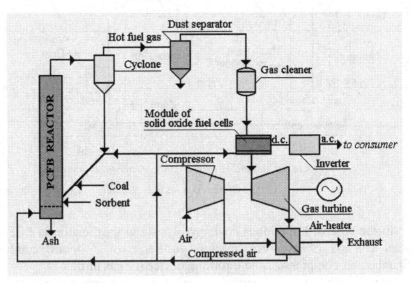

Figure 7. Principal circuit of hybrid power plant with gasification of Ukrainian bituminous coals, gas-turbine plant with built-in recuperator and solid oxide fuel cells

After the analysis of the experimental and calculation studies and the analysis of operation of foreign solid oxide fuel cell hybrid power plants, a conclusion about developments of solid oxide fuel cell hybrid power plants was reached (see Fig. 8). It was determined that modern hybrid power plants operated at the air use factor of 20 to 30 % with the electrical efficiency of fuel cell stacks from 45 to 62 % (without taking losses in inverter into account). The future increase in efficiency of solid oxide fuel cell stacks can be provided due to the improvements of fuel cell constructions with an increase in factor of air use. In this case, the electrical efficiency of solid oxide fuel cell stack operation can be more than 62%. The theoretical value of the electrical efficiency of SOFC stacks at the use of natural gas conversion products and excess-air factor of 1.0 is 84%.

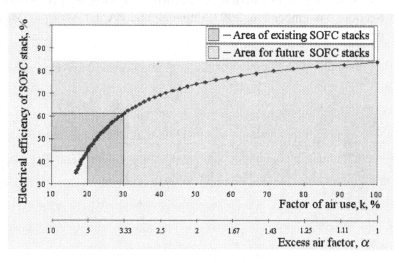

Figure 8. Results of calculations of hybrid power plants with the tube
SOFC: dependence of SOFC stack electrical efficiency on the factor of air use

9. CONCLUSIONS

1. The installation for test of fuel cells of 4 types (with polymeric, molten carbonate, solid oxide, and phosphoric acid electrolytes) is created in Coal Energy Technology Institute of National Academy of Sciences and Ministry of Fuel and Power of Ukraine.

2. The conversion of natural gas and coal was carried out at the installation with the purpose of generation of synthesis gas with high hydrogen content.

3. At the Fuel Cell Test Installation, it is possible to test stacks of fuel cells with polymeric electrolyte with electric capacity up to 1 kW, with molten carbonate and solid oxide electrolytes with electric capacity up to 3 kW.

4. The technology of conversion of Ukrainian high-ash coals in pressurized circulating fluidized bed was developed in Coal Energy

Technology Institute. This new technology can be used for the design of new hybrid power plants.

REFERENCES

[1] Dudnik A. Fuel Cell Test Installation/ Conference proceedings - Partnership for Prosperity & Security, Hydrogen Tech Fuel Cell Section, November 5-6, 2003, Philadelphia, PA, USA-17 p. - Materials on CD.

CIP-BASED FABRICATION OF SOFC CERAMIC COMPONENTS FROM OXIDE NANOCRYSTALLINE POWDER MATERIALS

G.YA. AKIMOV, V.M. TIMCHENKO, E.V. CHAYKA

Donetsk Phys.Tech.Institute, National Academy of Science of Ukraine, R.Luxemburg St. 72, 83114 Donetsk, Ukraine, akimov@host.dipt.donetsk.ua

Abstract: Cold isostatic pressing approach has been applied to fabricate three-layered ceramic sandwich membranes for SOFC from nanocrystalline oxide powders. Advantages and problems inherent to the method are revealed and discussed, the ways of the problem solutions are envisaged. Prototypes of the membranes are manufactured.

Key words: Solid Oxide Fuel Cells/Zirconia/Cold Isostatic Pressing.

1. INTRODUCTION

Energy consumption and, in particular, electric power consumption continues to grow around the world. Because of it, novel more efficient technologies of energy production are needed. Fuel cells, chemical energy sources able to convert directly the energy of fuels into the electric power, are among the most promising ones in this respect. Although fuel cells were discovered about 150 years ago, a related market has been formed only during last two decades on the base of intensive research and development activities in all technological fields including simulation, design, and development of new materials.

N. Sammes et al. (eds.), Full Cell Technologies: State and Perspectives, 175-180.
© 2005 *Springer. Printed in the Netherlands.*

Solid oxide fuel cells (SOFC) constitute one class of the chemical sources. SOFC ensure much more effective electric power production and environmental safety as compared with conventional power generation technologies [1,2,3].

The main component of a solid oxide fuel cell is a three-layered sandwich consisting of anode, electrolyte, and cathode, each being made from a different oxide ceramic material. Such ceramic structures can be fabricated by various methods including slip or tape casting, injection molding, ceramic coverings, etc. [1]. Whatever the method applied is, it should provide the best able microstructure and specified performance of materials besides the desired shape of a SOFC membrane. However, layers of the membrane have different properties that requires combination of two or more different methods of ceramic engineering in the component fabrication.

Cold isostatic pressing (CIP) can be one of such methods [1]. In the case of SOFC components, it has advantages of:

- easy microstructure management using CIP pressure optimization;
- porosity control in anode and cathode materials;
- perfect contact between layers of the sandwich;
- no need in organic binders (with an exception of pore-forming agents) in initial powders and, as a result, a shorter sintering time;
- high production rate of the process;
- fabrication of both flat and cylindrical shapes for planar and tubular fuel cells, respectively.

Moreover, it is known that the CIP procedure involves lowering of the sintering temperature by 100 to 200°C for many ceramic materials including those applied in fuel cells [1].

The goal of the present work was CIP application for manufacturing three-layered ceramic structures from nanocrystalline powders and determination of characteristics of the involved ceramic materials immediately affecting on performance of resulting membranes.

2. MATERIALS AND PROCEDURES

The expected layout of the work consisted in the fabrication of three-layered ceramic membranes with 30 to 35 mm in diameter, see Fig.1.

All three layers, namely porous anode, electrolyte, and cathode, were manufactured from agglomerated ceramic powder ZrO_2 + 8 mol.% Y_2O_3 with a crystallite size from 10 to 20 nm. The anode and cathode materials were doped with 50 wt.% nickel oxide and 50 wt.% lanthanum manganate, respectively. Taking into account that the sintered electrolyte material should be gas-tight

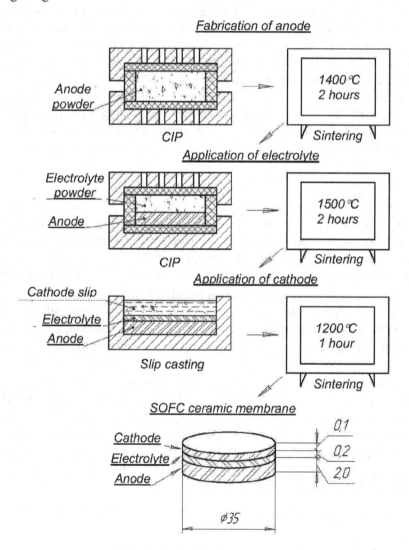

Figure 1. Flow chart of SOFC three-layered membrane manufacturing process based on CIP.

while anode and cathode materials have to be gas permeable, a powder material of anode was mixed with 10 or 30 wt.% of starch as a pore-forming agent, whereas the cathode layer was made porous using low sintering temperature.

Selection of the most perfect and reliable electrolyte-anode adjoining method based on CIP techniques was performed as follows. Anode and electrolyte layers in various conditions (press-powder, green compact, partially sintered compact, or sintered ceramics) were jointly isostatically pressed at pressure values from 0.05 to 0.8 GPa. Since the electrolyte layer should be much thinner than the anode one, such joint isostatical pressing, as a rule, had a nature of adpressing the electrolyte powder onto a pre-formed anode compact. The two-layered plate produced by such a way was sintered at 1500°C. Thereafter a cathode layer was build up in a similar manner. Alternatively the cathode layer was applied by molding a slip prepared from a mixture of the powder and alcohol. A green cathode layer on the plate was sintered at 1200°C.

3. RESULTS

It was found that the layer adjoining performance was always unsatisfactory at adpressing either electrolyte onto sintered anode or cathode onto sintered electrolyte. It was caused probably by high hardness of the sintered ceramics preventing good adhesion of adpressed powder particles.

Satisfactory results were obtained by adpressing an electrolyte layer onto a pre-compacted anode layer or at slip painting the powder cathode onto a partially sintered electrolyte. A flowchart of the developed process for manufacturing ceramic membranes using the CIP procedure is presented in Fig. 1.

Experimental studies have shown that a number of problems can occur due to different shrinkage values in anode and electrolyte compacts during joint CIP or sintering, including bending, buckling, delamination, splitting along anode and electrolyte layers interface, and/or radial cracking of plates.

Bending or buckling are caused generally by mismatching shrinkages of the adpressed materials. It was found experimentally that these phenomena can be fully or partially eliminated by means of:

- strengthening the green anode with preliminary partial sintering at temperatures below 1500°C;
- thinning the adpressed layer of electrolyte;
- matching shrinkages of both jointly pressed powder materials;
- implementing the composite rubber/metal pressforms strictly confining the compacts and precluding any bending or buckling.

On the contrary, splitting or delamination are obviously connected with weak cohesion between anode and electrolyte layers. Numerous experiments have shown that such delamination at pressing can be precluded in the following ways:

- increasing a porosity of anode (in the present work the content of the pore-forming agent was increased from 10 to 30 wt.%);

- weakening granules of the adpressing electrolyte powder or, in other words, reducing coherence of individual particles, by means of modifying processes of re-granulation and drying of initial powder materials.

In its turn, delamination at sintering originates from the low strength of the anode material. It follows from splitting the anode layer instead of separating anode and electrolyte as a cause of delamination.

Experimental process optimization by selecting the best CIP pressure, anode sintering temperature, and pore-forming agent content has enabled to fabricate prototypes of two-layered (anode/electrolyte) and three-layered (anode/electrolyte/cathode) membranes, see Fig. 2.

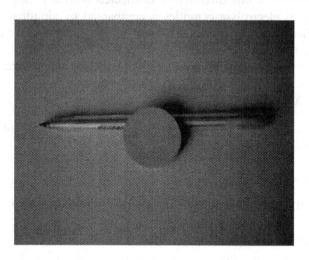

Figure 2. Three-layered SOFC membrane.

Final examinations and measurements have shown that anode and cathode have open porosity at the level of about 35% while the electrolyte is virtually poreless. Scanning electron microscopy of the sandwich fracture surface has revealed only small closed pores below 0.01 μm in the electrolyte layer.

4. CONCLUSIONS

The present work has confirmed good perspectives for the CIP application to fabricate multi-layered membranes for fuel cells by building-up the electrolyte layer onto green anode.

Successful application of the developed process for membranes production requires:

1. Matching shrinkages of the powder materials in use.

2. Providing free-flowing powders suitable to form very thin layers.

3. Providing powders with 'soft' granules.

4. Designing rigid pressforms preventing bending or buckling of compacts at pressing.

5. Providing sintering of virtually poreless electrolyte ceramics at relatively low (about 0.15 GPa) CIP pressures.

Implementation of the CIP route at the SOFC membrane production ensures a good cohesion between anode and electrolyte layers and high strength and thermal shock resistance of the electrolyte.

Further work in this direction is connected with lowering the sintering temperature of electrolyte ceramics accompanied by simultaneous increase of the isothermal exposure time. Lower sintering temperature can result in smaller grain sizes at the same high density of ceramics that could improve such characteristics of membranes as current density and thermal stability.

ACKNOWLEDGMENT

Many thanks to Prof. *V.G.Vereschak* and Dr. *A.Yu Koval* for their invaluable contributions.

REFERENCES

[1] M. Feng, and J.B. Goodenout. J. Solid State Inorg. Chem., 1994, v.31, p.663.

[2] T. Ishihara. Solid State Ionics, 1996, v.86-88, p.197.

[3] J.A. Kilner. Solid State Ionics, 2000, v.129, N(1-4), p.13.

[4] W.G. Coors, and D.W. Ready. J. Am. Ceram. Soc., 2002, v. 85, p.2637.

THE STATE OF FUEL CELLS AND ITS DEVELOPMENT IN UKRAINE

V.YU. BAKLAN, M. V. UMINSKY, I. P. KOLESNIKOVA

I. I. Mechnikov National University of Odesa, Research Laboratory of Fuel Cells
Dovzhenko Sr. 7.A, UA – 65058, Odesa, Ukraine

Abstract: The paper is the result of development, making and tests of various fuel cells modifications. Several types of catalysts for oxygen (air) and hydrogen electrodes have been worked out and were investigated in Laboratory of Fuel Cells. They are activated by carbon, pare and modified, complex oxides of transition metals and skeletal nickel-Raney catalyst. Hydrogen-oxygen battery of fuel cells (power 1 kW) was made with catalysts without noble and deficient materials. Zn, Mg, Al and their alloys were used in metal-air sources of electric current as anodic materials. Aluminum-air sources of electric current have been designed, made and tested. Al-air battery with power 1kW was intended for electrocar.

Key words: Catalysts For Oxygen And Hydrogen Electrodes/Electrochemical Generator/ Aluminum-Air Cells

The Laboratory of Fuel Cells of Odesa I. I. Mechnikov National University is one of the scientific organizations of Ukraine, which was developing the electrochemical power sources: hydrogen-oxygen fuel cells, half-fuel cells of system metal-air and accumulators. The laboratory was founded by famous scientist in area of fuel cells, Prof. Davtjan O.K in 1962.

For Several years the Laboratory of Fuel Cells investigated catalysts without noble and deficient materials for electroreduction process of oxygen (oxygen (air) electrode) and electrooxidation of hydrogen (fuel electrode). High electrochemical activity, corrosion stability, stable work and non-high cost were as main requirements for electrodes of fuel cells [1, 2].

181

N. Sammes et al. (eds.), Full Cell Technologies: State and Perspectives, 181-186.
© 2005 *Springer. Printed in the Netherlands.*

Platinum and silver are active catalysts in reaction of electroreduction of oxygen. However wide use of such catalysts is limited by their high cost and deficiency. Several types of catalysts for oxygen (air) electrodes have been elaborated and investigated in Laboratory of Fuel Cells. They are activated by carbon (coal)-pure and modified, promoted by soot, complex oxides of transition metals, especially by nickel-cobalt spinel [3, 4].

Gas-diffusion electrodes with these catalysts are prepared by a few methods: metalceramic porous hydroclosing layer used to work with enhanced pressure, as well as to work on air hydroclosing layer is hydrophobic Teflon or hydrophobic soot. Thus three-phase boundary (reaction's zone) created by pressure difference between electrolyte and gas hollow or by gradient of layer's wetting. Nickel net 400 micron is current conductor for all types of electrodes.

The life time of oxygen (air) electrodes, prepared of these catalysts was up to 10000 hours in alkaline electrolytes at room temperature (polarization not exceed 0,2 V). The life time of electrodes decreased up to 1000 hours at temperature 60 °C.

Advantages of oxygen (air) electrodes made of nickel-cobalt spinel are high electrochemical activity, relatively low cost and satisfactory thermodynamic stability. Life time reached 1000 hours at room temperature for air electrodes from nickel-cobalt spinel in alkaline electrolyte at a current density equal 100 mA/cm^2. Developed air (oxygen) electrodes ensure current density equal 100-150 mA/cm^2 at polarization 0,2 V at room temperature.

Multicomponent alloys of nickel and aluminum activated by Ti, Mo are most widespread and wide by used materials for hydrogen electrodes of low temperature alkaline fuel cells. To make hydrogen electrodes skeletal nickel prepared by alkali-soluble of alloy with composition 50 % Ni + 47 % Al + 3 % Ti is used. Raney catalyst is processed by 20 % suspense of Fluoroplast F-4 D with following drying in vacuum at 50 °C that permits pyrophoric catalyst to protect against self combustion and serves hydrophobic binder to form electrodes [5].

Current's density of half hydrophobic hydrogen electrodes with hydroclosing layer from power Fluoroplast F-4D was 90-100 mA/cm^2 and for electrodes with closing layer from porous nickel was 70-80 mA/cm^2 at polarization 0,1 V and test temperature 70 °C.

Hydrogen-oxygen battery of fuel cells (power 1 kW) with alkali electrolyte have been developed and made using catalysts without noble and deficient materials. The electrodes with working surface 350 cm^2 were made on section's current conductors with dimension's sections 40-45 cm^2. Common dimensions of electrode with a frame are 200x320 mm, it has 8 sections. Active mass was pressed in ready framework.

250 electrodes were made to assemble the electrochemical generator (ECG). Current density is 500 A/m^2 it voltage of 0,75 V at temperature 50-70 °C. Hydrogen and oxygen were fed at pressure less than 20-25 kPa. ECG was assembled from 4 block battery of fuel cells (BFC) with 250 Wt each. The electrolyte's circulation was compulsory with circulative pump. The product of reaction is water – 0,476 liters in hour, in shift -3,5 liters. Special system removal of water (SRW) was used in order to prevent dilution of electrolyte and support its concentration as constant [1, 2].

Parameters of the ECG are presented in table 1. The developed BFC is tested in laboratory and in customer's plant to smooth maximal loading of system electrical ensuring of atomic electrical power station (All Union Scientific Research Institute of Atomic Machining, Moscow).

Series of tests of elements, blocks from several elements and batteries 100, 200 and 500 Wt were made before assembly and test of the ECG-1000. Volt-ampere characteristics were measured at different regimes.

Electrochemical generator consists of systems as follows:

1) battery of fuel cells (four blocks);

2) pneumonic system of feeding by fuel and oxidizer, and removal of waste reagents;

3) system of circulation of electrolyte;

4) system of water removal;

5) system of temperature control;

6) scheme of energy stabilization required by customers;

7) system ECG parameters control.

Under projection of Knot's battery much attention has been paid to the questions of hermetical sealing to work in aggressive electrolyte (30 % mass KOH) with gases. Special crosspieces allowed considerably diminishing contact resistance at combination of separate fuel cells into battery. Each block united 36 cells consistently.

Metal-air chemical power sources call also peculiar interest among numerous electrochemical systems that elaborate autonomous power sources. It is connected with high coal energy, simplicity of service, reliability of metal-air power sources. Wide using of such systems are connected with development of active air electrode based on cheap and effective catalyst.

Zinc, magnesium, aluminum and their alloys are used in metal-air power sources as anodic material. The use of aluminum is the most perspective because its high energy-capacity (3.3 A hour/g) that allows to get high specific energy 400 Wt*hour/kg at relatively small cost [6]. Aluminum-air cells had been developed and investigated with anode of aluminum A-995 and A-99.

The design of aluminum-air cell allows assembling such cells in block from 10 pieces. Aluminum-air batteries have been developed, made and tested. The cells assembled in blocks have side slots across which air (oxygen) is fed in compulsory regime. Working area of electrode is 350 cm^2. Feeding electrolyte in each cell is carried out across system of pipe-wire from circulate pump stirring compensation of electrolyte.

Electrochemical tests of aluminum-air battery consisting of 24 cells are made in solution of KOH (20 %). The voltage of battery was 18-20 V, working voltage was 10 V at load current 100-105 A. Aluminum-air battery with power 1 kW was elaborated for electrocar by "Rotor" (Cherkasy, Ukraine).

Microalkaline zinc-air power sources had been developed and made. Its shape is similar to button; its diameter and height is 11 mm and 3.6 mm respectively. Their voltage is 1,5 V, capacity is 0,15 A*hour, specific energy is 150 Wt*hour/kg. They might be used for different electronic devices.

Reserve magnesium-air power sources with saline electrolyte had been developed to feed life-belt on water and in mines. Specific energy such sources is 230-250 Wt*hour/kg.

The sensor of glucose in man's blood was developed, made and tested for another application of fuel cells. The cathode was carbon activated organic complexs (ftalocyanine of Co and Fe), the anode was Pt, Pd. The electrolyte was solution of Krebbs-Ringer (pH=7,4). Such cell can work in static and flowing regime. The value of current at known load was determined by amount of fuel (glucose) at unlimited amount of oxidizer (air).

REFERENCES

[1] Author's certificate USSR N 1515977 from 15.06.1989, IC H 01 M 8/04. The claim 4374105 from 02.01.1988. Fuel cell. Baklan V. Yu., Tetelbaum S. D., Djablo V. V.

[2] Author's certificate USSR N 1528262 from 08.08.1989, IC H 01 M 8/04. The claim 4375938 from 22.12.1987. Electrochemical generator. Baklan V. Yu., Tetelbaum S. D., Djablo V. V.

[3] Author's certificate USSR, IC V01 11/50. The method of receipt catalyst of air electrode. Uminsky M. V., Trunov A. M., Koceruba A. M., Presnov V. A. Discoveries and inventions, 1975, B. N 4.

[4] Uminsky M. V., Makordey F. V., Khitrich V. F. The influence of conditions synthesis of electrochemical property acetylene soot

activated by dioxide manganese // Ukrainian chemical journal. – 2001. – V. 67, N 10, P. 94-97.

[5] Author's certificate USSR. IC H 01. The claim 4157635 from 08.12.1986. The method of making hydrophobic hydrogen electrode. Uminsky M. V., Ponomarenko T. I., Stupichenko L. N., Rogachko M. M.

[6] Uminsky M. V., Makordey F. V., Ivanova T. A., Tkachenko N. M. The cathodes for Al-air sources of current with neutral electrolite // Ukrainian chemical journal. – 1997. – V. 63, N 6, P. 118-122.

Table 1. Parameters of electrochemical generator (ECG-1000)

Time's work, hour	Pressure of Fuel and oxidizer				Compensation of volume	Temperature, °C				Current loading, A	EMF of BFC, V	Voltage of BFC, V
	on entrance in ECG, mPa		on entrance in BFC, mPa			Entrance into block 1-2	Going out in block 1-2	Entrance into block 3-4	Going out in block 3-4			
	H_2	O_2	H_2	O_2								
1-20	0,24	0,24	14,7	14,7	70	70	73	70	73	45 – 42	36	25,2
21-20	0,24	0,24	14,7	14,7	70	70	73	70	73	45 – 42	34,4	25,2
41-60	0,24	0,24	14,7	14,7	70	70	73	70	73	45 – 42	34,4	25,2
61-80	0,24	0,24	14,7	14,7	70	70	73	70	73	45 – 42	34,6	25,2
81-100	0,24	0,24	14,7	14,7	70	70	73	70	73	45 – 42	35,0	25,2
101-150	0,24	0,24	14,7	14,7	70	70	70	70	70	38	32	23
150-217	0,24	0,24	14,7	14,7	70	70	70	70	70	38	32	23

EMF – electro driving force, BFC – battery of fuel cells

SYNTHESIS OF CERMET Sr(Ti,Fe)O$_{3-\delta}$-PtRu BY COMBUSTION

E. CHINARRO[A], J.C. PEREZ[B], B. MORENO[A], M. CARRASCO[A], J.R. JURADO[A]

[A]Instituto de Cerámica y Vidrio, CSIC, Campus de Cantoblanco, 28049 Madrid, Spain
[B]AJUSA Pol. Ind. Campollano, C/ C, n°102007, Albacete, Spain

Abstract: High specific surface area metallic, ceramic and cermet powders may be active electrocatalyst materials, particularly for hydrocarbon oxidation in the anode compartment of a PEMFC, DMFC or SOFC. Several synthesis techniques are available for electrocatalyst preparation. Sol-gel and coprecipitation are considered dependable methods but are time-consuming and complex. Combustion synthesis is a rapid and reliable route that allows powders of metals, ceramics and cermets free of impurities, with nanoparticle scale and high specific surface area to be obtained. In this work, cermet material in the system SrTiFeO$_{3-\delta}$/ Pt–Ru was prepared for electrochemical applications by combustion synthesis.   The perovskite (SrTiFeO$_{3-\delta}$support for the Pt/Ru particles may decrease the Pt poisoning by CO. The Pt-Ru alloy particle size was around 15 nm. The material has been tested as anode electrocatalyst in a protonic exchange membrane fuel cell; polarization curves have been obtained with power output of 25 mW/cm^2.

Key words: Combustion Synthesis/Strontium Titanate/Cermet/Electrode Fuel Cell

N. Sammes et al. (eds.), Full Cell Technologies: State and Perspectives, 187-192.
© 2005 *Springer. Printed in the Netherlands.*

1. INTRODUCTION

A large number of compounds with interesting conducting electronic properties, such as ceramic metallic conduction, a metal-semiconductor transition and ionic-electronic mixed conduction, with the perovskite oxide structure can be prepared by combining A and B in the general formula ABO_3, A= Sr, La, Sm and B= Ni, Fe, Ti [1]. Beside, numerous modifications of these materials, in the form of either solid solutions or doping additions, have been researched with the aim of obtaining improved dielectric and piezoelectric properties, compared with those of the simple perovskite. Complex perovskites with A-site or B-site substitutions have been studied, though relatively less work has been done on A-site modifications.

These materials with suitable composition changes may also yield good mixed conductors for prospective high temperature applications [2]. Perovskites in the system $SrTi_{1-x}Fe_xO_{3-\delta}$ are mixed conductors, generally p-type at high p_{O2}, n-type at low, and oxygen vacancy ionic conductors in an intermediate range of p_{O2}. They can be good candidates for electrochemical applications (e.g. semipermeable membranes, electrode or interconnector materials in solid oxide fuel cells, resistive gas sensors). This kind of material usually are synthesised by the solid state reaction method, mixing oxides with the desired stoichiometry and then calcined at elevated temperatures. The combustion synthesis is a rapid and confidence method that let to obtain metals, ceramics and cermets with low particle size and high specific surface area. The combustion synthesis technique consists in bringing a saturated aqueous solution of the desired metal salts and a suitable organic fuel to boil, until the mixture ignites and a self-sustaining and raher fast combustion reaction takes off, resulting in a dry, usually crystalline, fine powder [3]. To produce a mixed oxide, a mixture containing the desired metal ions in the form of, for instance, water soluble nitrate salts and a fuel, such as urea can be used. While redox reactions such as this are exothermic in nature and often lead to explosion if not controlled, the combustion of metal nitrates-urea mixtures usually occurs as a self-propagating and non-explosive exothermic reaction. The large amounts of gases formed can result in the appearance of a flame, which can reach temperatures in excess of 1000°C.

The metal and ceramic electrocatalysts can be described as active materials, particularly on hydrocarbon oxidation reaction that is carried out on the anode of a fuel cell [4]. The aim of this work:

- Synthesis by combustion of $Sr(Ti,Fe)O_{3-\delta}/PtRu$ cermets.

- Catalysis powder preparation for electrode fabrication in PEMFC and DMFC.

- Membrane preparation for gas separator membranes (GSM).

- Anodes and/or cathodes for ITSOFC.

This paper tries to link different properties of a $Sr(Ti,Fe)O_3$ perovkite with PtRu metal catalyst for high and low temperature applications. The ceramic/metal $(SrTi_{0.85}Fe_{0.15}O_{3-d})_{0.5}(PtRu)_{0.5}$ cermet is prepared by combustion synthesis, as well as the material achieved are studied and characterised as electrocatalyst.

2. EXPERIMENTAL

For the combustion synthesis of the $(SrTi_{0.85}Fe_{0.15}O_{3-d})_{0.5}(PtRu)_{0.5}$ cermet (perovskite/alloy), $Sr(NO_3)_2$ (Merck), liquid $Ti[OCH(CH_3)_2]_4$ (Aldrich), $Fe(NO_3)_3.9H_2O$ (Merck), $[CH_3COCH=C(O-)CH_3]_2Pt$ (Aldrich), $H_2PtCl_6.xH_2O$ (Aldrich), $[CH_3COCH=C(O-)CH_3]_3Ru$ (Aldrich) were used as cation precursors and $CO(NH_2)_2$ (urea) (Aldrich) as fuel. Iron and strontium and urea were first melted in a wide-mouthed vitreous silica basin by heating at 150°C on a hot-plate inside a cupboard under ventilation. Platinum acetilacetonate or hexacloroplatinate acid and ruthenium acetilacetonate and liquid titanium isopropoxide were then added. The temperature was raised to 250°C. The reaction lasted less then 5 min and produced a dry, dark black, fragile foam, which readily crumbled into powder. The colour of the foam provides a good indication of the homogeneity and yield of the reaction. The powders obtained from $H_2PtCl_6.xH_2O$, as Pt precursor, were washed with deionised water to remove chloride salts (water soluble) formed during the combustion.

As-prepared combustion reaction powders were characterised by XRD, BET and TEM. These powders were press and sintered (1400 °C for 12 h) in air to obtain a pellet to be characterised by SEM. As-prepared powders (using Pt Acac as precursors) were used as anode in a membrane-electrode assembly (MEA) in a protonic exchange membrane fuel cell, and its polarisation curve was obtained at 60, 70 and 80 °C cell temperature, with 20 and 10 cc/min gas flow in anode and cathode respectively.

3. RESULTS AND DISCUSSION

These materials are normally prepared by mechanical mixing of oxides, but the combustion method is also used to produce impurity-free materials with high specific surface area. A perovskite ($SrTiFeO_{3-δ}$)-alloy (PtRu) cermet is obtained directly by combustion synthesis (fig.1). It presents a low crystallinity, low Pt crystal size (~15nm) and a specific surface area of 48.5 m^2/g.

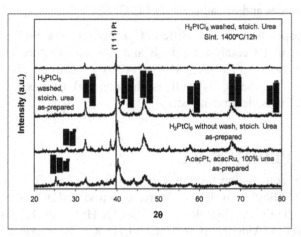

Figure 1. XRD of the $(SrTi_{0.85}Fe_{0.15}O_{3-\delta})_{0.5}(PtRu)_{0.5}$ sample: as-prepared powders and sintered pellet.

At low temperature (~300 °C) Ru is in solid solution with Pt forming an alloy, this can be seen because Pt peak is shifted to higher 2θ angles respect Pt metal peak. And Ru is volatilized and cermet only has Pt (the Pt peak shift has disappeared) at high temperatures (1400 °C).

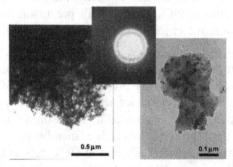

Figure 2. TEM of the $(SrTi_{0.85}Fe_{0.15}O_{3-\delta})_{0.5}(PtRu)_{0.5}$ as-prepared powders, and electron diffraction.

Fig. 2 shows TEM micrographs of the samples, and a nanometric morphology of the as-prepared powder were noticed. The electron diffraction (rings and dispersed dots) indicates a low particle size. In both of the micrographs can be observed the agglomerates of fine powder formed during the combustion synthesis. The highest magnified micrograph (0.1μm) lets perceive the structure of the powder: Pt with spherical shape (the darkest points) supported on a ceramic $(SrTi_{0.85}Fe_{0.15}O_{3-\delta})$ matrix (clearer part). That fact is also observed in SEM micrographs (Fig. 3).

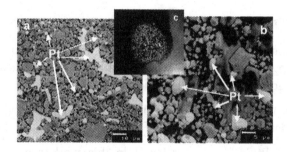

Figure 3. (SrTi₀.₈₅Fe₀.₁₅O₃₋δ)₀.₅(PtRu)₀.₅ sintered pellet (1400°C): (a, b) SEM micrographs and (c) optical microscopy.

In these SEM micrographs (Fig. 3) a cermet SrTiFeO₃₋δ/Pt is clearly obtained and both phases are completely compatible. Using EDS analysis it has been checked PtRu alloy is no longer present. As it is expected, the Pt "particles" are being percolated because of high temperature (1400 °C) effect. The pellet obtained is very porous, as you can see in the optical micrograph (fig.3c), that is of suitable skeleton microstructure for use as electrode in ITSOFC. In order to know the electrocatalyst activity of as-prepared powders some polarisation curves were achieved. A PEM MEA, using as-prepared (SrTi₀.₈₅Fe₀.₁₅O₃₋δ)₀.₅(PtRu)₀.₅/C as anode, was prepared and measured at three different temperatures (Fig. 4).

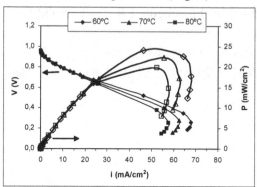

Figure 4. Polarisations curves at 60, 70 and 80°C, using as-prepared (SrTi₀.₈₅Fe₀.₁₅O₃₋δ)₀.₅(PtRu)₀.₅ as anode in a PEM electrode-membrane assembly.

The polarisation curves (fig.4) show an important and promising result since an appreciable power (25mW/cm²) is observed. The high amount of ceramic phase (50% of total) can be responsible for the low power observed. However the presence of this phase can reduce the poisoning of Pt by CO.

4. CONCLUSIONS

1. The combustion as-prepared powders are a two compatible phases on alloy/ceramic cermet, with a small particle size, low crystallinity and a good PtRu alloy dispersion.

2. As-prepared powder of ($SrTiFeO_{3-\delta}$)/PtRu may be a promising candidate for catalyst in PEMFC and DMFC.

3. Cermet $Pt/SrTiFeO_{3-\delta}$ prepared in this work can be a promising candidate for anodes/cathodes in both GSM and ITSOFC.

ACKNOWLEDGEMENT

Authors are grateful for the financial support from postdoc contracts I3P, CSIC, and CAM 07N/0102/2002 and APOLLON ENK5-CT-2001-00572 projects.

REFERENCES

[1] R.J.H. Voorhoeve, D.W. Jonson Jr., J.P. Remeika, P.K. Gallagher, Science 195 (1977) 827.

[2] J.R. Jurado, F.M. Figueiredo, J.R. Frade, Solid State Ionics 122 (1999) 197.

[3] D.A. Fumo, J.R. Jurado, A.M. Segadaes, J.R. Frade 32 (1997) 1459.

[4] J.R. Jurado, E. Chinarro, M.T. Colomer, Solid State Ionics 135 (2000) 365.

HYDROGEN STORAGE IN METAL HYDRIDE UNDER ACTION OF SUNLIGHT

YU.M. SOLONIN, D.B. DAN'KO, O.Z. GALIY, I.A. KOSSKO, G.YA. KOLBASOV*, I.A. RUSETSKII*.

Frantcevych Institute for Problems of Materials Science, National Academy of Sciences of Ukraine, 3 Krzhyzhanovsky st., UA-03142 Kiev, Ukraine,

*Institute of General and Inorganic Chemistry, National Academy of Sciences of Ukraine, 32/34 pr. Palladina, 03680 Kiev 142, Ukraine.

Abstract: The new construction of the photoelectrochemical cell with a higher efficiency of hydrogen production and storage is discussed. It is first proposed to use metal-hydride electrode as a cathode and GaAs electrode modified by Pt as a photoanode in the cell for water splitting under action of sunlight. The operation of the cell has been investigated, and promising results have been obtained.

Key words: Photoelectrochemical Cell/Hydrogen Production/Water Splitting/Metal-Hydride/Sunlight Conversion.

1. INTRODUCTION

The common photoelectrochemical cell for the conversion of solar energy to the chemical energy of hydrogen consists of a semiconductor photoanode and a metal cathode immersed in the solution of electrolyte. Oxygen is released at the photoanode and hydrogen – at the cathode under illumination of the photoanode by sunlight. In our work it is first proposed to

N. Sammes et al. (eds.), Full Cell Technologies: State and Perspectives, 193-198.
© 2005 Springer. Printed in the Netherlands.

use $LaNi_5$ based alloys which form compounds with hydrogen as the cathode. In this case hydrogen is not released at the cathode, but accumulates in it according reaction:

$$Me + H_2O + e^- = OH^- + MeH. \qquad (1)$$

The adventure of our construction lies in the fact that the hydrogen forming energy, ΔZ is subtracted from the total energy of water splitting. Then the hydrogen storage process goes on under a lower voltage than the hydrogen releasing. Therefore it is possible to use narrow band semiconductors ($Eg < 1.5$ eV) as the photoanodes. This permits to increase the efficiency of sunlight-to-hydrogen conversion as compared with the traditional TiO_2 and WO_3 photoanodes. One more advantage of our construction is the fact that hydrogen is obtained as metal-hydride, i.e. in a suitable technological form.

2. EXPERIMENTAL

The single-crystal GaAs ($Eg = 1.42$ eV) with high photosensitivity and small losses for volume recombination was used as the photoanode. The anode and cathode areas were separated by cation-exchange membrane [1] (Fig. 3). The photoanode was in polysulphide electrolyte where it was not practically oxidized. The cathode was in 30 % solution of KOH. The cathode was charged from GaAs electrode under action of sunlight. The $LaNi_{2.5}Co_{2.4}Al_{0.1}$ alloy was used as the hydride forming alloy which shows itself well as the cathode of nickel-metal-hydride accumulators. It has high electrochemical activity at room temperature and possesses discharge capacity of about 250 mAh/g and good cyclic stability. It practically does not require spatial activation for the achievement of standard capacity. The active mass of cathode (0.2 g) contains 5 % of binding paste. The cathodes were prepared as the pressed tables 8 mm in diameter coated by nickel net.

For the effective accumulation of hydrogen, it is necessary to use GaAs photoanode with photopotential Ef closed to the charge potential of metal-hydride cathode. The GaAs n-type photoanodes with electron concentration of $1-2 \times 10^{16}$ cm^{-3} were investigated. The surface of photoanodes was modified by Pt nanoparticles in order to increase the sunlight conversion efficiency. The surface of photoanode was covered with Pt by using the electrodeposition method. The presence of Pt on the surface was checked by Auger-spectroscopy in the electron spectrometer JAMP-10S. The distribution of Pt particles on the surface was investigated in TEM EM-200 under accelerated voltage of 100 kV. The spectral characteristics of photoelectrochemical current were measured with the help of the experimental setup analogous described in [2] by using monochromator MDP-2 and Xenon high pressure lamp. The radiation of lamp was modulated with frequency of 20 Hz.

Electrochemical cell with quartz window and saturated calomel electrode as a reference electrode was used (Fig. 3). Photoelectrochemical measurements were conducted with PI-50-1 potentiostat under illumination power density of 75 mW/cm^2. At first the efficiency of energy accumulation (in the form of absorbed hydrogen) was estimated from the cathode discharge curves and from the hydrogen volume released under cathode heating. The volume of hydrogen released was measured in the tailor-made setup. The discharge capacity measurements were performed in electrochemical cell with nickel counter electrode.

3. RESULTS AND DISCUSSION

The presence of Pt on GaAs surface was confirmed by Auger spectra (Fig. 1).

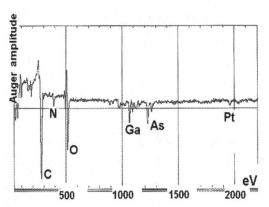

Figure 1. The Auger spectrum of GaAs photoanode after Pt modification.

From TEM investigations it was obtained that the diameter of Pt particles falls in the range between several nanometers and tens of nanometers, and the maximum of particle size distribution was at about 10 nm (Fig. 2).

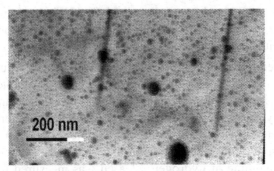

Figure 2. The TEM image of GaAS photoanode.

The typical discharge curves for $LaNi_{2.5}Co_{2.4}Al_{0.1}$ electrode are shown in Fig. 3.

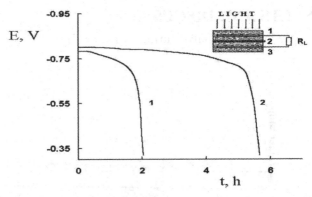

Figure 3. The discharge curves of $LaNi_{2.5}Co_{2.4}Al_{0.1}$ electrode when GaAs photoanode was used for charging. Charging time was equal to 140 h under illumination. 1 – initial GaAs electrode, charging current - 100 μA; 2 – modified GaAs electrode, charging current - 300 μA. The insert shows the photoelectrochemical cell with GaAs electrode (1), cation-exchange membrane MF-4SK (2), and metal-hydride electrode for hydrogen storage (3); R_L – resistive load.

The cathode was charged from initial GaAs electrode (curve 1) and from GaAs electrode modified by Pt (curve 2). It is clear that the Pt modification significantly improves discharge characteristics. The photopotential of GaAs electrode, E_f has to be -1.15 ÷ -1.25 V (relative to the saturated calomel electrode) for effective hydrogen storage. At the same time for the unmodified electrode E_f = -0.85 ÷ -0.95 V because the charge regime was not optimum. The value of E_f increases by 0.25 ÷ 0.3 V after Pt deposition that corresponds to the optimum charge regime.

It was obtained that the efficiency of sunlight-to-current conversion amounted to 11-15 %. The efficiency of hydrogen accumulation estimated from discharge curves and from hydrogen volume released under heating reached 50 %. The efficiency of sunlight-to-hydrogen conversion comprised consequently 6 - 8 %.

The spectra of quantum yield of photoelectrochemical current η_i with the potential of GaAs electrode as a parameter were investigated to establish the reason of E_f increasing after Pt modification (Fig. 4).

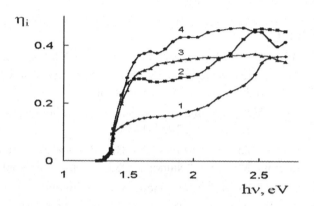

Figure 4. Relationship between the quantum yield of photoanode η_i and the quantum energy of light for initial GaAs electrode (1, 2) and for GaAs electrode modified by Pt (3, 4).

Potentials, E: 1, 3 ÷ -0.6 V; 2, 4 ÷ - 0.5 V. 1 n. KCl solution.

As shown in Fig. 4, η_i increase with increasing anode potential. In this case Pt modification causes the essential rise of η_i in the visible spectral range (curves 3, 4). The spectra were analyzed under condition that charge falls significantly in space charge region near surface according relation [3]:

$$\eta_i = \frac{k_s^{\,a}}{s_p + k_s^{\,a}}\left[1 - \frac{\exp(-KL_{sc})}{KL_p + 1}\right],$$ (2)

where $k_s^{\,a}$ and S_p are the velocities of anode electrochemical reaction and surface recombination, respectively, K is the coefficient of light absorption of semiconductor dependent on wavelength of light [4,5], L_p is the diffusion length of minority carriers (holes), L_{sc} is the width of space charge region. The rise of K results in an increase of η_i in shortwave part of the spectra. The increase of η_i observed under anode potentials is connected with a decrease of surface recombination velocity S_p for n-type semiconductors [3]. In this case, as is evident from Fig. 4 (curves 3, 4), the significant decrease of S_p (by a factor of 3 - 4) was observed after modification of GaAs by Pt nanoparticles.

4. SUMMARY

The use of metal-hydride as the cathode and GaAs modified by Pt as the photoanode in the photoelectrochemical cell for water splitting under action of sunlight permits to conduct the reaction more effectively and to obtain hydrogen in the suitable technological form.

REFERENCES

[1] Kuzmiski E.B., Kolbasov G.Ya., Tevtul Ya.Yu., Golub N.B. Nontraditional electrochemical systems for energy conversion. Kyiv, "Akademperiodyka", 2002, 182 p, in Ukrainian.

[2] Kublanovsky V.S., Kolbasov G.Ya., Litovchenko K.I. Polish J. Chem., 1996, 270, № 11, p. 1453-1458.

[3] Gurevich Yu.Ya., Pleskov Yu.B. Photoelectrochemistry of semiconductors. Moskow, "Nauka", 1983, 312 p., in Russian.

[4] Optical properties of semiconductors (semiconductor compounds A^3B^5) red. Willardson P., Bir A., Moskow, "Mir", 1970, 370 p., in Russian.

[5] Kolbasov G.Ya., Gorodyski A.B. The processes of photoexited charge transfer in semiconductor-electrolyte system. Kiev, "Naukova dumka", 1993, 192 p., in Russian.

HYDRIDES OF TRANSITION METALS AS HYDROGEN ACCUMULATORS FOR POWER ENGINEERING

YU.M. GORYACHEV, V.I. DEKHTERUK, M.I. SIMAN, L.I. FIYALKA.
Frantsevych Institute for Problems of Materials Science, National Academy of Sciences of Ukraine, 3 Krzhyzhanivskoho str., Kyiv-142, 03680, Ukraine

Abstract :   On the basis of the information available and the authors" previous investigations , the main regularities of change in the tendency of various transition metals to hydride formation have been established. It has been shown in detail that the ultimate metal hydrogenation changes periodically along the rows and monotonously down the columns of the periodic table. Quantitative relationships between the density and Young"s modulus, on the one hand, and the hydride composition and pressure, on the other hand, have been deduced. They are in agreement with the fact that ultimate metal hydrogenation increases with increasing pressure.

The investigation carried out has shown that transition metal hydrides as well as boron hydride and fullerenes are promising as basic materials for high-performance hydrogen accumulators.

Keywords: Hydride/Hydrogen Accumulator/High Pressure/Hydride Synthesis/ Hydrogenation

Hydrides of transition metals (TM) and their alloys are expected to be promising hydrogen accumulators for power engineering. To clarify which of them are most efficient and under which conditions, we have collected the data available including our previous results [1,2] and analyzed them. In Table 1, some characteristics of group-1V,-Y and -Y1 TM hydrides with an

N. Sammes et al. (eds.), Full Cell Technologies: State and Perspectives, 199-204.
© 2005 *Springer. Printed in the Netherlands.*

ultimate value of x=H/Me obtained in the case of metal hydrogenation under atmosphere pressure are listed.

These data show a periodical change in the ultimate characteristics of the hydrides studied along the rows and their monotonic increase within the groups of the periodic table with increasing the main quantum number of metal.

The results obtained permit one to draw conclusion that using Hf and W and their alloys for saturation with hydrogen may result in obtaining high-performance hydrogen accumulators.

Table 1. Characteristics of group-IY-YII transition metal hydrides with ultimate hydrogenation (T dec is the decomposition temperature)

MeH$_x$	TiH$_x$	VH$_x$	CrH$_x$	MnH$_x$
x = H / Me	2	0.3	2÷3	0.03
Tdec,^{o} C	640	20	12	20
MeH$_x$	ZrH$_x$	NbH$_x$	MoH$_x$	TcH$_x$
x = H / Me	2	0.4	1.0	0.55
T dec, ^{o}C	870	300	−55	80
MeH$_x$	HfH$_x$	TaH$_x$	WH$_x$	ReH$_x$
x = H / Me	2.1	0.3÷0.8	≈3	0.7÷1.0
T dec, ^{o}C	1000	315	130÷ 150	≈100

Different hydrides have different hydrogen accumulation capacities. We evaluated the capacities taking into account that accumulated hydrogen released into standard balloons at a pressure of 150 atm and room temperature. We used the data from Ref. [3] and deduced the following relations:

$$V=2.24 \cdot 10^{4} \cdot u \cdot d/(2 \cdot P \cdot (A \cdot v+u));$$

$$M=2.24 \cdot 10^{4} \cdot u/(2 \cdot P \cdot (Av+u));$$

$$G= u/(A \cdot v + u).$$

where V,M and G denote volume, weight and mass capacities, respectively; v and u are indexes it the hydride formula A_vH_u; A is the metal atomic weight and d is the hydride density.

The obtained data on the relative capacities for different solid hydrides are presented in Table 2.

Table 2. Relative hydrogen capacities for some solid hydrides

Hydride	TiH_2	$Hf H_2$	CrH_2	WH_3	PdH_{07}	BH	$C_{60}H_{60}$
Volume capacity V, $cm^3(H_2)/cm^3$(hydride)	13.47	11.01	19.64	23.14	5.566	14.56	11.483
Weight capacity W, $cm^3(H_2)/g$ (hydride)	2.994	0.828	2.767	1.199	0.488	6.331	5.742
Mass capacity M, $g(H_2)/g$(hydride)	.04	.01	.037	.016	.007	.085	.077

These data evidence that tungsten and boron hydrides are the most promising hydrogen accumulators of the hydrides studied; the former being efficient in volume accumulation, the latter in weight one. The capacity of $C_{60}H_{60}$ is close to that of boron tungsten. Both the hydrides may meet the demands of BMW car production to the mass capacity of hydrogen accumulators, namely it is to be not less than 10 %.

As a rule, data on the hydride density are not available. This characteristic was determined in accordance with the method described in Ref. [4]:

$$d_1/d_0 = [1 - a (P_0/P_1)^{b} \cdot x],$$

where d_1 is the hydride density, d_0 is the metal density; $a = 4.25 \cdot 10^{-2}$; b = 0,22; d_0 is the metal density at pressure P_1; $P_0 = 10^{-4}$ GPa (standard pressure). Some pressure and composition dependences of the relative density of hydrides (d_1/d_0) are given in Table 3.

Table 3. Pressure and composition dependences of the ratio d_1/d_0 for metal hydrides MeH$_x$

P,GPa	10^{-4}	10^{-3}	10^{-2}	10^{-1}	10^2
x					
0	1.000	1.000	1.000	1.000	1.000
1	0.952	0.971	0.982	0.989	0.998
2	0.913	0.950	0.970	0.970	0.997
3	0.873	0.925	0.953	0.971	0.994
4	0.838	0.903	0.940	0.962	0.991

As seen, an increase in pressure leads to increasing the hydride density and, on the contrary, an increase in the hydrogen content leads to decreasing the hydride density.

In addition to the hydride density, there are two more characteristics which determine fitness of a hydride as a hydrogen accumulator, namely its mechanical strength and form stability under repeated charge-discharge cycles, which in turn are connected with Young"s modulus. We have analyzed this connection, taking into account the fact that alongside with a density change, the hydride Young"s modulus changes, too. The authors have deduced the following expression for pressure and hydrogen content dependence of Young's modulus:

$$\beta_p^{(x)} = ((\beta_0^0 + \beta_0^1 \cdot x) / (1 + x)) \cdot \exp[\, 2\Upsilon(1+x)\cdot p /$$
$$(\beta_0^0 + \beta_0^1 \cdot x)],$$

where Υ is Griunizen's constant, β_0^0 is Young's modulus under the standard conditions ; β_0^1 is the reduced Young"s modulus for solid hydrogen equal to 38 GPa. Some data on the pressure and composition dependence of Young's modulus are summarized in Table 4.

Table4. Dependence of Young's modulus of titanium hydrides (TiH $_x$) on their composition and pressure

P,GPa	10^{-4}	10^{-2}	1.0	10	15
x					
0	145	145	149	191	219
1	92	92	96	146	176
2	74	74	78	126	166
3	65	65	69	120	164
4	59	59	64	116	163

These data reveal the fact that the influences of pressure and hydrogen content, which increases with increasing pressure, on Young"s modulus compensate each other, which results in practical constancy of Young"s modulus of hydrides with the ultimate composition. This feature of hydrides may prove to be useful in regarding the characteristic stability and strength of hydrogen accumulators under operation.

CONCLUSIONS

1. Periodicity of ultimate metal hydrogenation in the rows of the periodic table and its monotonous change within the table columns (groups) have been confirmed and detailed.
2. Quantitative relationships between the density and Young"s modulus, on the one hand, and hydride composition and pressure, on the other hand, have been deduced.
3. The relationships above have been shown to be in agreement with the fact that ultimate metal hydrogenation increases with increasing pressure.
4. The investigation carried out has shown that transition metal hydrides as well as boron hydride and fullerenes are promising basic materials for high-performance hydrogen accumulators.

REFERENCES:

[1] Goryachev Yu.M., Siman M.I., Fiyalka L.I.   et al.   Hydrogen in titanium hydride and GOLKAO method. Book "Electronic structure and properties of refractory compounds and alloys and their application in materials science". IPMS NASU, Kyiv, 2000, 85-89.

[2] Tonkov E. Yu .Phase transformation of   compounds under high pressure;M "Metallurgiya", 1988, v.1, 63-65.

[3] Lobach A. S., Shulga Yu. M. et al. Fullerene hydrides. Promising technologies, Moscow - Chernogolovka, 1955, v. 2, is. 14,20 - 22.

[4] Goryachev Yu. M., Dekhteruk V. I. et al. Electron mechanism of baric depedences of hydrogen - transition metal interaction . Proceedings of the Conference on hydrides and fullerenes, Kyiv - Yalta, IPMS NASU, 1999, 11 - 12.

MATHEMATICAL MODELING AND EXPERIMENTAL STUDY OF PROTON EXCHANGE MEMBRANE FUEL CELLS

S.A. GRIGORIEV, A.A. KALINNIKOV, V.I. POREMBSKY,
V.N. FATEEV

Hydrogen Energy and Plasma Technology Institute of Russian Research Center "Kurchatov Institute"; 1, Kurchatov Sq., Moscow, 123182, Russia

Abstract: At present time proton exchange membrane fuel cells are being intensively developed for vehicular, portable, stationary and other applications. However some fuel cell components such as gas diffusion layer and bipolar plate need an optimization. Both experimental methods and the mathematical models developed for this purpose by the authors are presented for review in this article.

Key words: Fuel Cells/Proton Exchange Membrane/Optimization

1. INTRODUCTION

Proton exchange membrane (PEM) fuel cells are very promising as direct energy conversion devices since they are highly efficient, safe for environment and they are being developed worldwide for vehicular, portable and stationary applications. Optimization of the stack's components such as bipolar plates, gas diffusion layers design is one of the main tasks for PEM fuel cell engineering. Both experimental methods and mathematic modeling have been used for this purpose. In particular, new test methods and test equipment have been developed, for example the electrodiffusion flow measurements; the special electrochemical cells, with a cathode consisted of several electrically independent "test" electrodes, each electrode is supplied

N. Sammes et al. (eds.), Full Cell Technologies: State and Perspectives, 205-210.
© 2005 *Springer. Printed in the Netherlands.*

with a gas concentration and temperature probes; the complex for water management study. In addition, the efficiency of electric contact between bipolar plates and gas diffusion layer was measured using the special test station comprising a hydraulic press with a temperature control, current supply and control systems. These test systems allow estimating the modes of gas flows, reagent and water concentration, current, potentials, and temperature distribution in the stack. The three-dimensional physical-chemical model of PEM fuel cell has been developed.

2. EXPERIMENTAL

2.1. ELECTRODIFFUSION METHOD

Electrodiffusion measurements have been provided for study of flow conditions [1]. The method of electrochemical flow diagnostics is based on the measurements of the limiting diffusion current on the microelectrodes, which are flash-mounted with the walls of bipolar plate channels. In accordance with p-theorem (equality of Reynolds number) the "real flow" (of hydrogen, oxygen or air) was replaced by a suitable electrolyte flow ($K_3Fe(CN)_6$, $K_4Fe(CN)_6$ and K_2SO_4 solutions). Both the mean current values on platinum microelectrode and their fluctuation were recorded. The comparison of the current value for different points in bipolar plate provide the information on its accessibility reagents; the magnitude of current fluctuations allows estimating the character of flow regimes (laminar, turbulent, and transient).

The electrodiffusion measurements were implemented with several types of bipolar plates: bipolar plate with parallel structure of channels; bipolar plate with meander-like channels; bipolar plate with spots-like channels (for instance, such design can be technologically implemented by cross milling of graphite or metal plate); bipolar plate manufactured with use of precision-expanded grid (the same as previous but protuberant paths (rectangular parallelepipeds) are replaced by precision-expanded grid made by machining from thin metal sheet). It was shown that flow regime is laminar everywhere on the hydrogen and oxygen sides of all types of bipolar plates when the pure gases are used at their stoichiometric consumption. However when the air is used as an oxidant the different flow regimes may occur (it is related to increasing of flow rate to reduce mass transfer limitations; the rate of air flow was calculated assumed that oxygen utilization is 15-50% at current density up to 3 A/cm^2). The flow regime is laminar for bipolar plate with parallel line channels. But for bipolar plate with meander-like and spots-like channels a transition from laminar to turbulent regime is possible with the increasing of flow rate. Furthermore, for the "spots" type of bipolar plate it is typical that so-called "dead zones" exist.

It was identified that the presence of carbon cloth (gas diffusion layer) stabilized the flow. In those positions, where without carbon cloth the vortexes were took place, at presence of carbon cloth at the same flow rates the laminar flow takes place.

The analysis of electrodiffusion measurements results, in terms of the mass transfer intensity, especially at high current densities, demonstrates that the design of bipolar plate with spots-like channels is preferable, regardless the possibility of existence "dead zones" in corners of bipolar plate. However, as the measurements of cell resistance caused first of all by contact area between bipolar plates demonstrate, the channel flow field structure generally leads to better performance of the fuel cell in comparison to spots- and meander-like structure.

2.2. DISCRETE ELECTRODE METHOD

The special electrochemical cells [2] have been developed for measurements of current density, reagent concentration and temperature distribution. The cathodes of these fuel cells consist of several electrically independent "test" electrodes. Each electrode is supplied with gas concentration and temperature probes. Such design enabled one to measure the mentioned above distribution both along the cell and in a plane, through which a diluted reactant, specifically, air, was passed.

In experiments, we used an ion-exchange membrane made of MF-4SK (Nafion-type) perfluorinated polymer with functional sulfo-groups for preparation of membrane-electrode assemblies (MEAs). The membrane thickness was 130 μm, the exchange capacity was 0.86 mg-equiv/g, the catalyst was 20 wt % Pt on a hydrophobic carbon carrier, the catalyst loading at the anode and cathode was 0.35 mg/cm^2.

2.3. THE EFFECT OF WIDTH OF BIPOLAR PLATE CURRENT TRANSFER RIBS AND CHANNELS ON FUEL CELL PERFORMANCE

The experimental dependence between the current density and width of air channels of bipolar plate has been obtained [3, 4]. Based on these data optimal width of channels 0.4-0.7 mm and current transfer prominent elements 0.2-0.7 mm was prescribed (exact values are determined by technological and material aspects of bipolar plate production, and also by gas diffusion layer parameters (thickness, porosity, mechanical characteristics, electric resistance).

One can suppose that the increase of channel width results in a sag of gas diffusion layer above a channel and, as a consequent, in impairment of electric contact efficiency between gas diffusion and catalytic layers. The increase of width of ribs results in aggravated limitations on reagent transport through gas diffusion layer to electrocatalytic layer, in air electrode zone furthermore complicated by counter water flow.

2.4. MEASUREMENTS OF MECHANICAL AND ELECTRICAL PARAMETERS OF BIPOLAR PLATES AND GAS DIFFUSION LAYERS

In addition, the efficiency of electric contact between bipolar plates and gas diffusion layers was measured using special test station comprising a hydraulic press with a temperature control, current supply and control systems. The purpose of provided measurements is the comparative analysis of parameters of different bipolar plates and gas diffusion layers, and also obtaining of the necessary data for use in calculations on mathematical model. Resistance tests of gas diffusion layers and bipolar plates were performed both in a longitudinal direction (four-contact method) and in a transverse direction.

2.5. INVESTIGATIONS ON WATER MANAGEMENT

Operation mode of fuel cell is strongly determined by water balance. Water production by electrochemical process and also water transport due to proton migration and diffusion were measured with use of special complex. For MEA based on MF-4SK proton exchange membrane with hydrophobic catalytic layer an effective water drag coefficient $n=0.28$ for air and $n=0.53$ for pure oxygen, water diffusion coefficient trough membrane $D_{max}=1.55\times10^{-10}$ m^2/s.

3. MATHEMATICAL MODEL

3.1. TWO-DIMENSIONAL DIFFUSION MODEL

The diffusion model [5, 6] has been developed for mass transfer study within the system "bipolar plate – gas diffusion layer (with micro-porous sublayer) – electrocatalytic layer". It was shown that the current density distribution is a complex function and depends mainly on the electrochemical parameters of the MEA (electrocatalytic layer activity) and on the transport properties of the gas diffusion layer, which may be significantly influenced by design of bipolar plate.

The analysis of spatial oxygen concentration and current distribution allows for the conclusion that the diffusion limitations across the channel for the current density up to 1 A/cm^2 are negligible (gradient of oxygen concentration < 1.25 vol %). The main obstacle for oxygen transport in cross direction is related to gas diffusion layer coated by micro-porous sublayer (gradient up to 6-7 vol % at 1 A/cm^2). For example, if the current density is 0.7 A/cm^2 and the integral porosity of gas diffusion layer with sublayer is 10%, the transverse concentration gradient there can reach 7 vol %, whereas the transverse concentration gradient in the channel is only 0.08 vol %.

This "transparent" model is very useful for the estimation of influence of any separately taken factor on mass transfer. Also a parameter allowing the simulation of the activity of electrocatalytic layer and thus providing calculations for catalysts of different types is present in the model.

3.2. THREE-DIMENSIONAL PHYSICAL-CHEMICAL MODEL

In addition to this, the three-dimensional mathematical model of heat and mass transfer [3, 4] has been developed. Stephan-Maxwell equation was used for mass transport calculations in gas channels and gas diffusion layers. Proton transport in membrane and electrocatalytic layer was described by Nernst-Planck equation. The diffusion and electroosmosis of water were taken into account for membrane potential distribution.

On the basis of the provided theoretical analysis the operation of fuel cell can be conditionally divided into three modes by physicochemical nature. The first mode is characterized by optimal fuel cell operation with good humidifying of membrane and good water removal from cathode zone. The nominal current density > 0.4 A/cm^2 (at 0.7 V) is provided dependant on a humidifying degree. The second mode: the dry anode, liquid water presents on the cathode, water transport by protons current is equilibrated by a reversible diffusion flow, current density < 0.4 A/cm^2. The third mode: the cathode is flooded by water, the current greatly decreases due to the mass transfer limitations. Depending on a ratio between anode and cathode gas flow parameters and operating current density the transition between these modes takes place. The variation of gas diffusion and electrocatalytic layers parameters, parameters and speed values of cathode and anode gas flows will lead to the optimization of fuel cell parameters. In particular, it was shown that at effective porosity of cathode gas diffusion layer 65% or more, ohmic losses in last one could play a significant role with increasing of width of bipolar plate channels. Thus, not only reactant diffusion but also ohmic losses do limit the process. However, for anode gas diffusion layer 40-30% porosity is recommended since diffusion limitations for anode are negligible and a certain role is played by electric conductivity.

Calculation of temperature distribution in the system was done. The consideration was given to several options of the cooling flows management, including the case when "air-reactant" is used as a heat-transport medium within both operating cells and additional cooling cells.

4. CONCLUSIONS

Thus, the complex of numerical and experimental investigations allowed the authors to recommend the values of operating parameters and some design requirements providing increase of PEM fuel cell efficiency. Developed methods of PEM fuel cell modeling can be useful for engineering of other electrochemical systems.

REFERENCES

[1] Grigoriev S.A., Martemianov S., Fateev V.N. Electrodiffusion diagnostics of flow regimes in PEM fuel cells. Magnetohydrodynamics, 2003, v.39, N4, p.475-483.

[2] Grigoriev S.A., Fateev V.N. Current Distribution, Mass- and Heat-Transfer in PEM FC. Proceedings of the 13th World Hydrogen Energy Conference (12-15 June 2000, Beijing, China), v.2, p.938-943.

[3] Grigoriev S.A., Kalinnikov A.A., Fateev V.N., Porembsky V.I., Blach R. Numerical and experimental study of current collector and bipolar plate design for PEM fuel cells. Book of abstracts of 1st European Hydrogen Energy Conference (2-5 September 2003, Grenoble, France), p.92.

[4] Grigoriev S.A., Kalinnikov A.A., Fateev V.N., Porembsky V.I., Blach R. Optimization of mass and heat transport in PEM fuel cells. Proceedings of 4th European Congress on Chemical Engineering (21-25 September 2003, Granada, Spain), v.3, p.3.1-015.

[5] Grigor'ev S.A., Alanakian Yu.R., Fateev V.N., Rusanov V.D. Modeling of the mass transfer in proton-exchange-membrane fuel cells. Doklady Physical Chemistry, 2002, v.382, N4-6, p.31-35.

[6] Grigoriev S.A., Alanakyan Yu.R., Fateev V.N., Rusanov V.D., Blach R. Bipolar plates and current collectors for PEMFC. Modeling of the mass transfer processes. Book of abstracts of the "2002 Fuel Cell Seminar" (18-21 November 2002, Palm Springs, USA), p.21-24.

EXPERIMENTAL AND THEORETICAL STUDY OF METAL-HYDRIDE REACTORS

Y. KAPLAN[A], M. DEMIRALP[A], T. N. VEZIROGLU[B]

[a]Mechanical Engineering Department, Nigde University, 51100 Nigde, Turkey
Tel: +903882252249, Fax: +903882250112, e-mail:ykaplan@nigde.edu.tr
[b]Clean Energy Research Institute, University of Miami, Coral Gables, FL 33124, USA

Abstract: Hydrogen absorption in $LaNi_5$-H_2 reactor is experimentally and theoretically investigated. Two different reactors were designed and hydrogen gas was charged at constant temperature at constant pressure. Temperature changes in the tanks during the hydrogen charge were measured at several locations and readings of the data were continued until the temperature has stabilized. In the theoretical program, a two dimensional mathematical model, which considers complex heat and mass transfer during this process is developed and numerically solved. The numerical results are compared with the measured data to validate the mathematical model. A reasonable agreement between the numerical results and experimental data is obtained.

Key words: Hydrogen Storage/Hydriding Time/Mathematical Model/Heat And Mass Transfer

1. INTRODUCTION

The chemically storage in a metal alloys in the form of metal hydrides is concerned, there has been great interest on many metals such as Mg, Na, La, Li etc [1-3]. The critical issues for storage materials are amount of hydrogen absorbed/desorbed, thermal stability of the hydride, hydrating/dehydrating kinetics, thermodynamic and thermo-physical properties, crystal structures, surface processes like segregation, carbonization. Therefore, efficient conditions to form metal alloys has been the main target for chemists, metallurgist, and engineers.

N. Sammes et al. (eds.), Full Cell Technologies: State and Perspectives, 211-222.
© *2005 Springer. Printed in the Netherlands.*

Theoretical studies also shed light on hydriding process and much useful information is obtained. Sun et al. [1], Ben Nasrallah and co workers [3,4], Mat and co-workers [5-8] developed several mathematical models for detailed analysis of metal-hydride bed operations. Jemni and Nasrallah [4] developed two mathematical models, one based on solid and gaseous phase and other consider solid and gaseous phases as a mixture. Based on Jemni and Nasrallah's continuum mixture model Mat and co-workers [5-8] carried out a detailed analysis and parameters effecting the hydriding process, and extended analysis to three dimensional cases. A prediction of heat and mass transfer in a closed metal hydrogen reactor is based on mathematical and numeric model besides measured evolutions of pressure in the reservoir which reported by Askri at al. [4]. This study contains development a theoretical model that makes it possible to predict the transient heat and mass transfer in a closed cylindrical reactor. They also discuss some governing operating conditions such as gaseous part volume, height to the radius ratio of the reactor and the initial hydrogen to metal atomic ratio, temperature.

Although there are several theoretical and experimental studies in the literature on the various aspects of the hydrogen storage processes. There are very limited studies validating theoretical models and investigating various bed geometries on hydriding process. Therefore, the objective of this study is to study theoretically and experimentally hydrogen absorption in metal-hydride reactors and analyze various bed geometries and configurations in order to reduce the storage time.

2. MATHEMATICAL MODEL

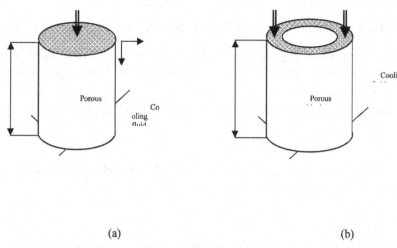

(a)   (b)

Figure 1. Schematic sketch of metal-hydride reactors a) reactor I, b) reactor II

Metal hydride formation is considered in two-dimensional metal hydride reactors shown in Fig. 1a and 1b. The differential equations governing the hydriding process are mass balance of hydrogen and metal, momentum and energy equations [5].

Mass Balance for Hydrogen

$$\varepsilon \frac{\partial \rho_g}{\partial t} + \frac{1}{r} \frac{\partial}{\partial r}\left(r \rho_g u_r\right) + \frac{\partial}{\partial z}\left(\rho_g u_z\right) = -m \tag{1}$$

Mass Balance for Metal

$$(1-\varepsilon)\frac{\partial \rho_s}{\partial t} = m \tag{2}$$

Energy Equation

For the fluid

$$\varepsilon \rho_g C_{pg} \frac{\partial T_g}{\partial t} = \frac{1}{r}\frac{\partial}{\partial r}\left(\varepsilon r \lambda_{ge} \frac{\partial T_g}{\partial r}\right) + \frac{\partial}{\partial z}(\varepsilon \lambda_{ge}\frac{\partial T_g}{\partial z}) - \left(\rho_g C_{pg} u_r\right)\frac{\partial T_g}{\partial r}$$

$$-\left(\rho_g C_{pg} u_z\right)\frac{\partial T_g}{\partial z} - H_{gs}(T_g - T_s) - \max(-m,0)C_{pg}(T_g - T_s)$$

$$\tag{6}$$

For the solid

$$(1-\varepsilon)\rho_s C_{ps} \frac{\partial T_s}{\partial t} = \frac{1}{r}\frac{\partial}{\partial r}\left((1-\varepsilon)r\lambda_{se}\frac{\partial T_s}{\partial r}\right) + \frac{\partial}{\partial z}((1-\varepsilon)\lambda_{se})\frac{\partial T_s}{\partial z}$$

$$+ H_{gs}(T_g - T_s) + m(\Delta H° + C_{ps}T_s - C_{pg}T_g) \tag{7}$$

where ε is the porosity, u_r and u_z are the velocity components in the r and z-direction respectively and ρ, C_p, λ and m are effective density, specific heat, effective thermal conductivity and rate of hydrogen absorption respectively.

The heat exchange coefficient between solid and gas is written as [3];

$$H_{gs} = \frac{\lambda_g}{d_p}(2 + 1.1 Pr^{1/3} Re^{0.6}) \tag{8}$$

where Pr and Re are Prandtl number and Reynolds number respectively and d_p is the particle diameter.

Auxiliary Equations

The amount of hydrogen absorbed is directly linked to the reaction rate, which is expressed as;

$$m = -C_a \exp\left(-\frac{E_a}{RT}\right)\ln\left(\frac{P_g}{P_{eq}}\right)(\rho_{ss} - \rho_s) \tag{9}$$

where C_a is material depended constant, ρ_{ss} is density of the solid phase at saturation and P_{eq} is equilibrium pressure calculated using the van't Hoff relationship;

$$\ln P_{eq} = A - \frac{B}{T} \tag{10}$$

where A and B are materials constants deduced from the experimental data.

Initial and Boundary Conditions

Hydrogen charge into metal bed and subsequent hydriding processes studied in the reactors shown in Fig. 1a and 1b. Initially hydride bed is assumed to have constant temperature and pressure. These conditions can be expressed mathematically as;

$$\text{at } t=0 \qquad P=P_0, \qquad T=T_0, \tag{11}$$

The boundary walls are assumed to be impermeable and no slip conditions are valid at the boundary walls. The reaction heat is removed from the boundary walls with a cooling fluid whose temperature is T_f. Hydrogen is charged at z = 0 with a constant pressure, P_0 and constant temperature, T_0. The boundary conditions can be expressed as;

$$\text{at } r=0 \qquad \frac{\partial T}{\partial r}(0,z,t)=0 \tag{12}$$

at z = 0 $-\lambda\dfrac{\partial T}{\partial z}(r,0,t)=h_1(T_0-T)$ (13)

at r = r$_0$ $-\lambda\dfrac{\partial T}{\partial r}(r_0,z,t)=h_2(T-T_f)$ (14)

at z = H $\dfrac{\partial T}{\partial z}(r,H,t)=h_2(T-T_f)$ (15)

where h_1 and h_2 are the heat transfer coefficients between hydride bed and hydrogen gas and boundary walls and cooling fluid respectively. Similar condition is applied to geometry II.

3. NUMERICAL METHOD

The partial differential equations are solved numerically with a fully implicit numerical scheme embodied in PHOENICS code [9]. This code solves following general differential equations,

$$\frac{\partial(\rho\phi)}{\partial t}+\nabla(\rho u\phi)=\nabla(\Gamma\nabla\phi)+S_h$$ (16)

where ϕ is a generic variable that representing the variable solved (i.e,u,v,T), Γ is the exchange coefficient. S_h represents the source terms. An important advantage of PHOENICS code is that it allows the user to incorporate additional source term that not available in the main program.

A 25×60 grid system is employed after a grid refinement test. A typical run until 3000 sec. takes about 6 hours in a Pentium III. P.C.

4. EXPERIMENTAL METHOD

Experimental set up for hydrogen absorption is schematically presented in Figure 2. Set up mainly consist of a reactor, a vacuum pump to evacuate the reactor before the experiment, a ball mill to produce fine particles, manometers, thermocouples, a data acquisition system, a tank containing 99.999% pure hydrogen gas and an argon tank to provide working under inert atmosphere environment. Two different reactors were designed for experiments. The first reactor is a cylindrical tank with 40 mm radius and 120 mm height (Fig. 1a). The second reactor consists of two co-eccentric cylinders and the space between the cylinders is used as a bed (Fig. 1b). The inner radius of the bed is 20 mm while outer radius and height were calculated as 40 mm and 160 mm respectively in order to have some volume with the first reactor. The main objective of the study is to see effect of heat transfer rate on hydriding time

The following procedure is followed during the experiments:

- LaNi$_5$ material is grounded for 3 hours. Grinding procedure is conducted by a ball mill which is located in a glow box of 60 mm diameter and 80 mm height. Argon gas is used during the process to prevent possible oxidation. Grinding speed was varied between 500 rpm and 890 rpm. During the grinding process, some graphite at a ratio of 1% of LaNi$_5$ is added as an anti-sticking agent. LaNi$_5$ material is used in the experiments.

- Grounded material is transferred into the reactor in a glove box. This processes carried out at argon atmosphere unless otherwise noted.

- The reactor is heated to almost 200°C for 2 hours under low pressure ($\approx 10^{-4}\, mmHg$). After the heating process the reactor is cooled down to room temperature and than hydrogen is charged to the reactor under 10 bar pressure. This heating and cooling process is repeated for 5 times for activation of LaNi$_5$ alloy.

- The experimental part was started after activation process had completed. Reactors were then charged with hydrogen at a range of pressure (6-10 bars). Hydriding process is monitored with temperature measurements obtained at several locations on the reactor. The temperature readings are recorded on a computer for further processing and interpretation of hydriding behavior.

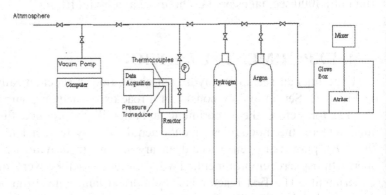

Figure 2. The experimental set-up

5. RESULTS AND DISCUSSIONS

The measured temperature evolution in z-direction under several charging pressures within the reactor I and II are given in Fig. 3 and Fig. 4, respectively. Experiments are performed under three charge pressures (6, 8 and 10 bars). Temperature changes on the bed are measured at 15, 30, 45, 60, 75 mm from the top of the bed. General characteristic of hydriding process is a sudden increase of bed temperature after charging is initiated. This increase is the result of exothermic hydriding reaction. It is observed that there is a perceivable temperature gradient in the bed within first 500 seconds for both reactors. This difference vanishes with time in Reactor I especially under higher charging pressure (see Fig. 3a,b and c). However, a significant temperature gradient observed in Reactor II (Fig. 4). This result may be attributed to the enhanced heat transfer in reactor I. Since hydriding reaction rate is strongly dependent on temperature.

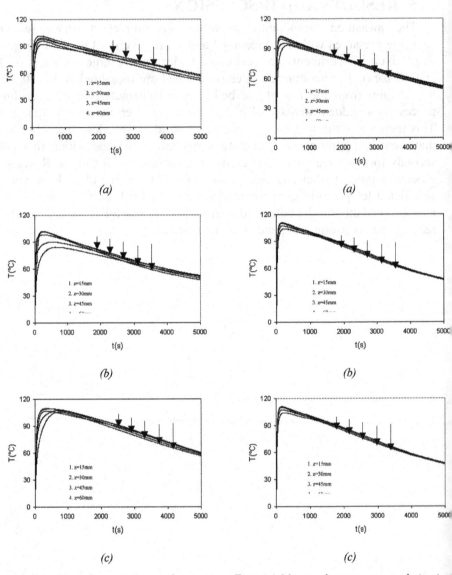

Figure 3. Measured temperature evolution in   Figure 4. Measured temperature evolution in
the Reactor I.   Reactor II.
a)6 bar, b)8 bar, c)10 bar   a)6 bar, b)8 bar, c)10 bar

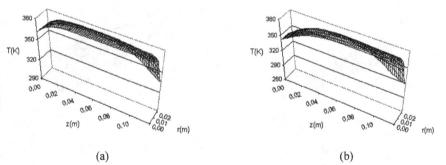

(a)   (b)

Figure 5. Temperature profile in the Reactor I at selected times. a) t=100 s, b) t=500 s.

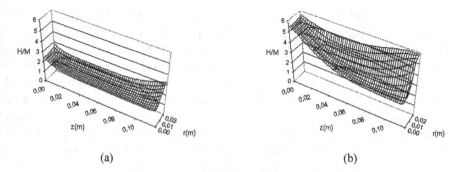

(a)   (b)

Figure 6. Evolution of hydride formation in the Reactor I. a) t=100 s, b) t=500 s.

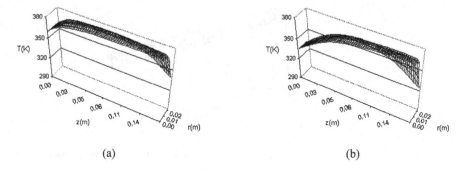

(a)   (b)

Figure 7. Temperature profile in the Reactor II at selected times. a) t=100 s, b) t=500 s.

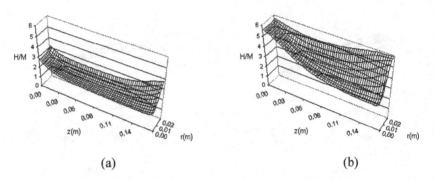

(a)   (b)

Figure 8. Evolution of hydride formation in the Reactor II. a) t=100 s, b) t=500 s.

Experimental data and model predictions are compared in Fig. 9 and 10. This figures also reflect the effect of inlet pressure on the hydriding process. The selected inlet hydrogen gas pressures are 6, 8 and 10 bar. There is a slight difference in experimental and theoretical values, which is somehow smaller in Reactor II as compared with Reactor I. It can be seen that evolved temperature values for Reactor II are greater than that of Reactor II. It can be concluded that temperature evolution is very sensitive to hydrogen gas inlet pressure. There is an increase in temperature evolution as increase in inlet hydrogen pressure. This effect is more pronounced between 6 bar and 8 bar in Reactor I. In Reactor II higher pressure (10 bar) is needed to see this effect.

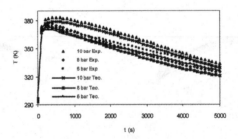

Figure 9. Influence of the hydrogen gas inlet pressure on the temperature evolution in the Reactor I.

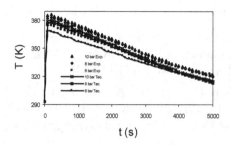

Figure 10. Influence of the hydrogen gas inlet pressure on the temperature evolution in the Reactor II.

6. CONCLUSION

Hydrogen absorption in metal hydride reactors experimentally and theoretically investigated. An experimental set up is designed to study main characteristics of hydriding process and effect of bed geometry and heat transfer on the hydriding process.hydriding process is characterized by exothermic reaction between LaNi$_5$ and H$_2$ and rapid temperature increase due the heat release. Hydriding time mainly depend on the successful heat removal from the bed. A bed geometry which provides more heat transfer area significantly reduces hydriding time.a mathematical model is developed to study hydriding process in detail and system optimization. Mathematical model considers coupled heat and mass transfer in the hydride bed. The governing equations are numerically solved and calculated results are compared with experimental data. It is found that mathematical model adequately capture the main physics of the hydriding process and can be employed for a better hydride bed design to reduce hydriding time.

ACKNOWLEDGEMENT

The financial support from Turkish State Planning Organization (DPT) under contract number 2002K120490 is gratefully acknowledged.

REFERENCES

[1] Sun DW, Deng SJ. A Theoretical Model Predicting the Effective Thermal Conductivity in Powdered Metal Hydride Beds. Int. J Hydrogen Energy 1990; 15(5):331-336.

[2] Chi H, Chen C, Chen L, An Y, Wang Q. Hydriding/Dehydriding Properties of $LaMg_{16}Ni$ alloy Prepared by Mechanical Ball milling in Benzene and Under Argon. Int. J Hydrogen Energy 2004; 29:737-741.

[3] Jemni A, Ben Nasrallah S., Lamloumi J. Experimental and Theoretical Study of a Metal-Hydrogen Reactor. Int. J Hydrogen Energy 1999; 24(1): 631-644.

[4] Askri F, Jemni A, Ben Nasrallah S. Prediction of Transient heat and Mass Transfer in a Closed Metal-Hydrogen Reactor. Int. J Hydrogen Energy 2004; 29: 195-208.

[5] Mat M, Kaplan Y. Numerical Study of Hydrogen Absorption in an La Ni_5 Hydride Reactor. Int. J Hydrogen Energy 2001; 26(9):957-963.

[6] Kaplan Y, Veziroglu T. N. Mathematical Modelling of Hydrogen Storage in a $LaNi_5$ Metal-Hydride Bed. Int. J. Energy Research, 2003, 27(11), 1027-1038.

[7] Dogan K., Kaplan Y., Veziroglu T.N., "Numerical Investigation of Heat and Mass Transfer in a Metal Hydride Bed" Applied Mathematics and Computation, 150(1), 169-180, 2004

[8] Mat M. D, Kaplan Y, Aldas K. Investigation of Three-Dimensional Heat and Mass Transfer in a Metal Hydride Reactor. Int. J. Energy Research, 2002; 26: 973-986.

[9] Rosten, H., and Spalding, D.B. PHOENICS Beginner's Guide and User's Manual, CHAM Limited (UK) Technical Report, TR/100, 1986.

CELLULOSE-PRECURSOR SYNTHESIS OF ELECTROCATALYTICALLY ACTIVE COMPONENTS OF SOFCs AND MIXED-CONDUCTING MEMBRANE REACTORS

E.V. TSIPIS[1], I.A. BASHMAKOV[2], V.V. KHARTON[1,2], E.N. NAUMOVICH[1,2], J.R. FRADE[1]

[1] *Department of Ceramics and Glass Engineering, CICECO, University of Aveiro, 3810-193 Aveiro, Portugal*
[2] *Institute of Physicochemical Problems, Belarus State University, 14 Leningradskaya Str., 220050 Minsk, Belarus*

Abstract: Developments of intermediate-temperature solid oxide fuel cells and electrocatalytic reactors require novel electrode and catalyst materials with high performance at 800-1100 K, low-cost processing technologies, and methods for nano-scale surface modification of ion-conducting ceramics. This work presents one promising technique for the synthesis of nanocrystalline powders and components of the electrochemical cells, based on the use of structure-modified cellulose containing metal cations.

Key words: Nanocrystalline Powder/SOFC Electrode/Catalyst/Cermet Anode

Decreasing operation temperature of solid oxide fuel cells (SOFCs) and electrocatalytic reactors down to 800-1100 K requires developments of novel materials for electrodes and catalytic layers, applied onto the surface of solid electrolyte or mixed conducting membranes, with a high performance at reduced temperatures. Highly-dispersed active oxide powders can be prepared and deposited using various techniques, such as spray pyrolysis, sol-gel method, co-precipitation, electron beam deposition etc. However, most of these methods are relatively expensive or based on the use of complex equipment. This makes it necessary to search for alternative synthesis and porous-layer processing routes, enabling to decrease the costs of electrochemical cells. Recently, one synthesis technique based on the use

223

N. Sammes et al. (eds.), Full Cell Technologies: State and Perspectives, 223-229.
© 2005 *Springer. Printed in the Netherlands.*

of structure-modified cellulose containing metal salts, was used for preparation of dense LaCoO$_3$ and SrCoO$_3$-based membranes [1, 2]. This method provides a series of obvious advantages, in particular low costs, simplicity, and possibility to synthesize powders with submicron particle size. As typical for solid oxide electrolytes, decreasing grain size of mixed-

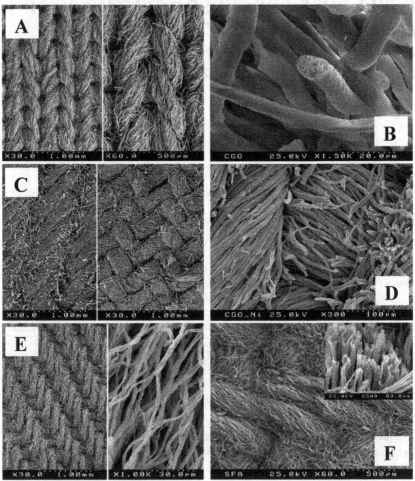

Figure 1. SEM images of CGO (A and B), Cu-CGO (C), Ni-CGO (D), metallic Ni (E) and SFA (F) fibers prepared by the cellulose-precursor technique.

conducting ceramics increases the role of grain-boundary resistance to ionic transport due to increasing boundary area. For porous catalytically-active layers, however, the performance is determined by surface morphology and exchange currents rather than ionic conductivity of the materials.

The present work was focused on the synthesis of nanocrystalline components for electrochemical cells via the cellulose-precursor technique. This method was used to prepare nanostructured intermediate-temperature (IT) SOFC anodes made of a series of cermets comprising gadolinia-doped ceria (CGO), yttria-stabilized zirconia (YSZ), $Gd_{1.86}Ca_{0.14}Ti_2O_{7-\delta}$ (GCTO) pyrochlore, metallic nickel and copper. Perovskite-type $SrFe_{0.7}Al_{0.3}O_{3-\delta}$ (SFA) powder, also obtained via the cellulose precursor, was applied onto membranes of the same composition to enhance specific surface area and electrocatalytic activity in the reactors for methane conversion [3].

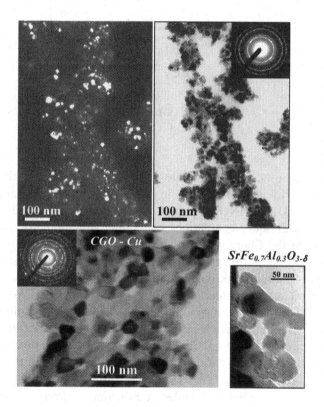

Figure 2. Dark- and bright-field TEM images and electron diffraction patterns of CGO, Cu-CGO cermet and SFA (F) powders prepared by cellulose-precursor technique.

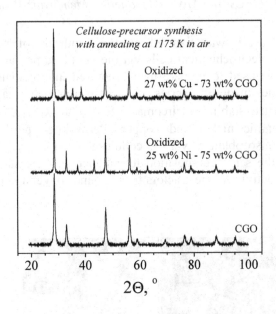

Figure 3. XRD patterns of CGO, Ni-CGO and Cu-CGO powders.

For the synthesis, starting cellulose fibers were reacted with a 68-70 % solution of nitric acid, resulting in formation of the so-called Knecht compound, $(C_6H_{10}O_5 \cdot HNO_3)_n$, which was subsequently hydrolyzed and impregnated with aqueous solutions containing stoichiometric metal nitrate mixtures. Then the fibers were dried and ignited, forming cermet or metal oxide fiber retaining the precursor texture (Fig. 1). Formation of single phases or two-phase composites in the combustion front was confirmed by electron diffraction, transmission electron microscopy (TEM) and X-ray diffraction (XRD) studies, Figs. 2 and 3. These analytic techniques showed that the size of oxide particles constituting highly-porous fibers is in the range 8-60 nm, and can be optimized choosing appropriate precursors and/or concentrations of nitrate solutions. Sintering of the powders onto the surface of $(La_{0.9}Sr_{0.1})_{0.98}Ga_{0.8}Mg_{0.2}O_{3-\delta}$ (LSGM) solid electrolyte or mixed-conductive SFA ceramics enables fabrication porous layers with a homogeneous microstructure and well-developed surface (Fig. 4). The measurements of anode overpotentials (η) as function of the current density (i) were performed by the 3-electrode technique in symmetrical cells with Pt counter and reference electrodes, as described elsewhere [4], in flowing wet 10 %H_2-90 %N_2 gas mixture with p(O_2) being controlled by an oxygen sensor. After the electrochemical tests, selected anodes were surface-modified by impregnation with a cerium nitrate solution in ethanol and annealing at 1073-1273 K; then the η-i dependencies were re-measured. The microstructure changes after the surface-modification are shown in Fig. 4A.

Figure 4. SEM micrographs of as-prepared and surface-modified Ni-GCTO-CGO anodes (A, right and left) and SFA layer deposited onto SFA membrane.

Testing of the Ni-Y8SZ-CGO and Ni-GCTO-CGO (50-30-20 wt%) anodes containing nanocrystalline CGO in single SOFC-type cells showed relatively high electrochemical performance at 873-1073 K (Fig. 5A). One-step cellulose-precursor synthesis of the anode compositions (Fig. 1C and D) was suggested to enhance the homogeneity of component distribution in the cermet and to improve mechanical stability of the electrode layers. The Ni-CGO (25-75 wt%) and Cu-CGO (27-73 wt%) anodes fabricated using this technique exhibit, however, a worse electrochemical activity compared to YSZ- and GCTO-containing electrodes. The overpotentials of Ni-CGO anode are lower than those of the Cu-containing analogue, due to poorer catalytic activity and high sinterability of copper [5, 6]. Surface modification of these layers considerably decreases their overpotentials to the similar level, e.g. η≈80 mV at 1073 K and 150 mA/cm². At the same time, the activation with ceria has a smaller effect on the performance of Ni-GCTO-CGO anode with respect to Ni-CGO, where the polarization at 1073 K decreases 3-5 times (Fig. 5). The observed behavior indicates an instability of Ni-CGO electrodes, particularly due to the volume changes on varying redox conditions, worsening the contacts at the electrode/electrolyte interface and between the grains constituting electrode layer. The results suggest a necessity to combine one redox-stable material with a catalytically-active ceria-based component in cermet anodes for IT SOFCs, with further surface modification. The effects of the latter include an enhancement of catalytic activity, improvement of intergranular contacts and, possibly, increasing electronic conduction at the electrolyte surface.

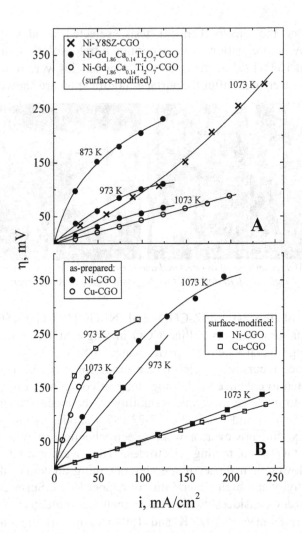

Figure 5. Overpotential vs. current density for various cermet anodes.

In summary, the cellulose-precursor technique is relatively simple and low-cost synthesis route, enabling to fabricate nanostructured single-phase and composite layers for the electrochemical applications.

ACKNOWLEDGEMENTS

This work was partially supported by the FCT, Portugal (projects POCTI/CTM/3938/2001 and BD/6827/2001) and the NATO Science for Peace program (project 978002).

REFERENCES

[1] Kharton V.V., Figueiredo F.M., Kovalevsky A.V., Viskup A.P., Naumovich E.N., Yaremchenko A.A., Bashmakov I.A., Marques F.M.B. Processing, microstructure and properties of $LaCoO_{3-\delta}$ ceramics, J. Eur. Ceram. Soc., 2001, v. 21, p. 2301-2309.

[2] Kharton V.V., Tikhonovich V.N., Shuangbao Li, Naumovich E.N., Kovalevsky A.V., Viskup A.P., Bashmakov I.A., Yaremchenko A.A. Ceramic microstructure and oxygen permeability of $SrCo(Fe,M)O_{3-\delta}$ (M= Cu or Cr) perovskite membranes, J. Electrochem. Soc., 1998, v. 145, p.1363-1374.

[3] Shaula A.L., Kharton V.V., Vyshatko N.P., Tsipis E.V., Patrakeev M.V., Marques F.M.B., Frade J.R., Oxygen ionic transport in $SrFe_{1-y}Al_yO_{3-\delta}$ and $Sr_{1-x}Ca_xFe_{0.5}Al_{0.5}O_{3-\delta}$ ceramics, J. Europ. Ceram. Soc., in press (2004).

[4] Mizusaki J., Tagawa H., Isobe K., Tajika M., Koshiro I., Maruyama H., Hirano K. Kinetics of the Electrode Reaction at the H_2-H_2O Porous Pt/ Stabilized Zirconia Interface. J. Electrochem. Soc., 1994, v. 141, p. 1674-1683.

[5] Gorte R.J., Park S., Vohs J.M., Wang C. Anodes for direct oxidation of dry hydrocarbons in a solid-oxide fuel cell, Adv. Mater., 2000, v. 12, p. 1465-1469.

[6] Kim H., Lu C., Worrell W.L., Vohs J.M., Gorte J.R. Cu-Ni cermet anodes for direct oxidation of methane in solid-oxide fuel cells, J. Electrochem. Soc., 2002, v. 149, p. A247-A250.

GLASS-CERAMIC SEALANTS FOR SOFC-BASED SYSTEMS

V.V. KHARTON[1,2], E.V. TSIPIS[1], A.P. CARVALHO[1],
A.V. KOVALEVSKY[2], E.N. NAUMOVICH[1,2], F.M.B. MARQUES[1],
J.R. FRADE[1], A.L. SHAULA[1]

[1]*Department of Ceramics and Glass Engineering, CICECO, University of Aveiro, 3810-193
Aveiro, Portugal*

[2]*Institute of Physicochemical Problems, Belarus State University, 14 Leningradskaya Str.,
220050 Minsk, Belarus*

Abstract: The studies of electrochemical, physicochemical and mechanical properties of
glass-ceramic sealants, based on SiO_2-Al_2O_3-CaO-BaO and SiO_2-MgO-BaO
systems, showed sufficient stability in contact with zirconia, lanthanum gallate
and ceria solid electrolytes at temperatures below 1100-1150 K. Thermal
expansion of glass-ceramics can be optimized, to a considerable extent, by
adding submicron powders of yttria-stabilized zirconia (YSZ). Excessive YSZ
additions result, however, in increasing ionic contribution to the total
conductivity, which is predominantly electronic. For moderately doped
sealants, the conductivity is essentially composition-independent and varies in
the range $(2-7) \times 10^{-7}$ S/cm at 1123 K.

Key words: Sealant/Glass-Ceramics/Thermal Expansion/Solid Oxide Electrolyte/Transport
Properties

Development of planar solid oxide fuel cells (SOFCs) and other
electrochemical devices, such as oxygen generators and sensors, makes it
necessary to elaborate sealant materials for hermetization in high-
temperature zone. The sealants should satisfy numerous requirements,
including chemical stability, good adhesion and thermal expansion.

231

N. Sammes et al. (eds.), Full Cell Technologies: State and Perspectives, 231-238.
© *2005 Springer. Printed in the Netherlands.*

Table 1. Composition and physicochemical properties of glass-ceramics

Designation and additions	Main components (weight ratio)	Additives	$T_{sealing}$ K	Average TECs	
				T, K	$\alpha \times 10^6$, K^{-1}
S1	SiO_2, Al_2O_3, CaO, BaO (100 : 9 : 38 : 53)	8YSZ (1.2 wt%)	1150-1170	350-990	6.46 ± 0.02
S2	SiO_2, Al_2O_3, CaO, BaO (100 : 9 : 38 : 53)	8YSZ (16.4 wt%)	1220-1240	350-1040	8.02 ± 0.02
S3	SiO_2, Al_2O_3, CaO, BaO (100 : 16 : 30 : 88)	TiO_2	1430-1460	350-1000	8.65 ± 0.02
S3 + 30 wt% 8YSZ			1440-1470	350-950	9.53 ± 0.03
S4	SiO_2, Al_2O_3, CaO, BaO (100 : 17 : 29 : 75)	MgO, TiO_2, P_2O_5	1260-1270	400-950	8.81 ± 0.03
S5	SiO_2, MgO, BaO (100 : 50 : 68)	CaO, 3YSZ	1470-1490	400-1000	11.47 ± 0.04
S5 + 30 wt% 8YSZ			1520-1550	350-1000	10.59 ± 0.03
S6	SiO_2, Al_2O_3, CaO, BaO (100 : 17 : 29 : 75)	MgO, TiO_2, P_2O_5, 8YSZ	1260-1280	380-950	8.42 ± 0.02
S7	SiO_2, MgO, BaO (100 : 17 : 64)	CaO, 3YSZ	1375-1385	490-970	8.45 ± 0.03

Table 2. Mechanical and ion transport properties of selected glass-ceramic materials

Sealant	Modulus of rupture, MPa	Vicker's hardness, GPa	$J(O_2)$, $mol \times s^{-1} \times cm^{-1}$ (1123 K)	E / E_{YSZ}, % (1123 K)
S1	75 ± 13	0.53 ± 0.04	$< 1 \times 10^{-11}$	< 0.1
S2	103 ± 19	0.69 ± 0.06	–	< 0.1
S3	85 ± 21	0.52 ± 0.07	$< 8 \times 10^{-12}$	< 0.1
S3 + 30 wt% 8YSZ	97 ± 16	–	3×10^{-11}	3
S5	121 ± 24	>1	$< 8 \times 10^{-12}$	< 0.1
S5 + 30 wt% 8YSZ	136 ± 30	>1	2×10^{-11}	2
S5 + 50 wt% 8YSZ	–	–	7×10^{-11}	18

Notes:

• The modulus of rupture and the Vicker's hardness were measured at room temperature using a LR 30K machine (Lloyd Instruments, UK) and M-type microhardness-tester (Shimadzu, Japan), respectively.

• The specific oxygen permeability $J(O_2)$ was measured under air/H_2-H_2O-N_2 gradients, corresponding to the e.m.f. of a YSZ sensor (E_{ysz}) of 1.10-1.15 V, using setup shown in Fig.2A. The glass-ceramic membrane thickness was 1.0 ± 0.1 mm; the data are averaged for 80-100 h of testing. The e.m.f. (E) was measured using the same conditions and setup, for the glass-ceramic samples with porous Pt electrodes

compatible with other cell components, minimum electrical conductivity and electrochemical permeability, and adequate mechanical and thermal properties. To date, a number of glasses and glass-ceramic sealants were tested ([1-3] and references cited); in most cases, however, the stability of seals was insufficient due to thermal expansion mismatch and/or interaction with SOFC components. The latter is characteristic, in particular, for most borate- and phosphate-based glasses.

Table 3. Vicker's hardness of glass-ceramic layers (~100 µm) deposited onto the surface of various solid-electrolyte ceramics under conditions, identical to sealing conditions

Sealant	Electrolyte	Vicker's hardness, GPa
S1	8YSZ	0.55 ± 0.12
	LSGM	0.50 ± 0.09
S3	8YSZ	0.46 ± 0.17
	CGO	0.49 ± 0.14
S3 + 30 wt% 8YSZ	8YSZ	0.56 ± 0.10
S5	8YSZ	0.73 ± 0.19
S5 + 30 wt% 8YSZ	8YSZ	0.68 ± 0.15

The present work was centered on the evaluation of glass-ceramics based on the multicomponent system SiO_2 (30-65 mol%) – Al_2O_3 (0-6%) – MgO (0-40%) – CaO (0-25%) – BaO (10-18%). Selected compositions and their designations are listed in Table 1. Moderate amounts of 8 mol.% Y_2O_3-stabilized ZrO_2 (8YSZ) and TiO_2 with submicron particle size, and also P_2O_5 were introduced into the compositions before melting in order to achieve external nucleation, to modify crystal growth rate, and to optimize physicochemical properties. In some cases, 8YSZ powder (10-50 wt%) was added into glass-ceramics in the course of final ball-milling. These additions make it possible to adjust thermal expansion coefficients (TECs, Table 1) close to those of well-known solid electrolytes, such as YSZ, $Ce_{0.8}Gd_{0.2}O_{2-\delta}$ (CGO) and $(La_{0.9}Sr_{0.1})_{0.98}Ga_{0.8}Mg_{0.2}O_{3-\delta}$ (LSGM).

Typical microstructures of glass-ceramics with and without YSZ additions, and their adhesion to solid electrolytes are illustrated by SEM micrographs shown in Fig. 1. For most cells, a relatively good adhesion and negligible interaction with YSZ and CGO are observed, whilst LSGM may form reaction layers when the sealing temperature is higher than 1450 K. Nonetheless, the sealants studied in this work provide a better stability in contact with LSGM if compared to Pyrex. One should also note that moderate additions of zirconia seem to increase mechanical strength (Tables 2 and 3), although this effect is within the limits of experimental uncertainty.

V.V. Kharton, E.V. Tsipis, A.P. Carvalho, A.V. Kovalevsky,
E.N. Naumovich, F.M.B. Marques, J.R. Frade, A.L. Shaula

Fig.3 presents the temperature dependencies of relative elongation and total resistivity of the glass-ceramic materials. The total conductivity was found essentially independent of composition; the impedance spectroscopy and e.m.f. measurements of oxygen concentration cells suggest that the role of electronic transport is predominant (Table 2 and Fig.4). Since excessive additions of YSZ increase the ionic conduction and oxygen permeability, zirconia amounts in sealants should be less than 30-35 wt%.

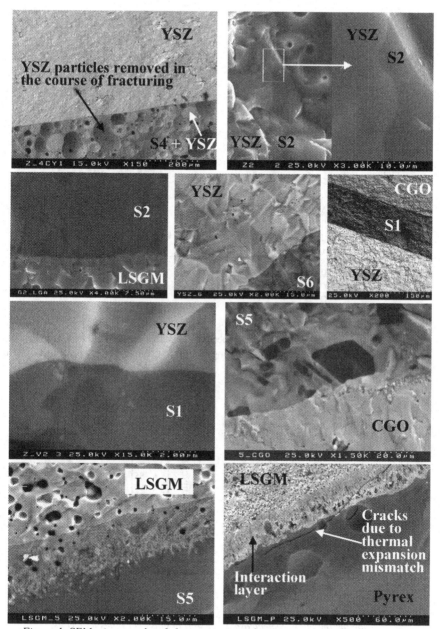

Figure 1. SEM micrographs of glass-ceramic sealants in contact with various solid electrolytes.

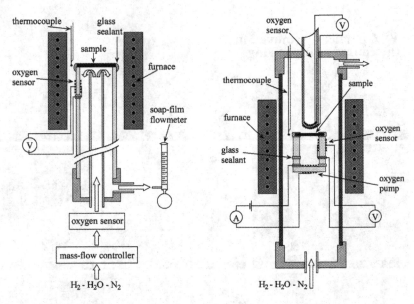

Figure 2. Schematic drawing of experimental setups used for the measurements of oxygen permeation and e.m.f. (left), and long-term testing under SOFC operation conditions (right). All electrochemical cells are made of YSZ ceramics.

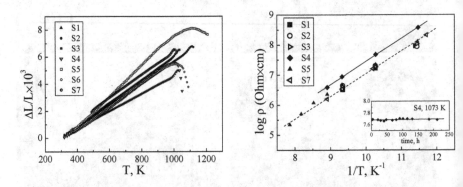

Figure 3. Dilatometric curves (left) and Arrhenius plots of the total resistivity measured by the impedance spectroscopy (right) of glass-ceramics materials.

The long-term stability tests under SOFC operation conditions were performed using YSZ electrochemical cells comprising one oxygen pump and a sensor (Fig.2, right). Gas-tight membranes, made of glass-ceramics or YSZ, were sealed onto the cell in air. Then these cells were placed into flowing H_2-H_2O-N_2 mixtures; the current through the pump was adjusted to provide time-independent sensor e.m.f., thus compensating leakages. Except for noise, the variations of sensor e.m.f. of the cells are related to the changes of oxygen partial pressure inside these cells. In all cases, no degradation at temperatures below 1150 K during 300-1000 h was found, as illustrated by Fig.5. At higher temperatures, however, the long-term stability of glass-ceramic sealants under O_2-N_2/H_2-H_2O-N_2 is often insufficient; the leakage drastically increases after 300-700 h. The XRD studies showed that the degradation phenomena are associated with different crystallization mechanisms in reducing and oxidizing conditions, leading to re-crystallization of the sealants under SOFC operation conditions. Such a behavior requires further modifications of composition and, possibly, processing conditions of the glass-ceramics.

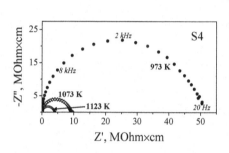

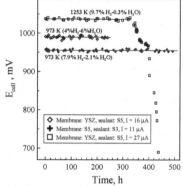

| Figure 4. Typical impedance spectra of the glass-ceramics. | Figure 5. Selected examples of e.m.f. vs. time dependencies of the YSZ cells, shown in Fig.2B, with fixed gas composition in the chamber and current through the oxygen pump. |

In summary, the properties of materials based on SiO_2-Al_2O_3-CaO-BaO and SiO_2-MgO-BaO systems, seem to enable their use as sealants in intermediate-temperature SOFCs. Possible drawbacks, including interaction with LSGM solid electrolyte and degradation in reducing atmospheres, can be suppressed decreasing the temperatures of SOFC operation and sealing.

ACKNOWLEDGEMENTS

This work was partially supported by the FCT, Portugal (projects SFRH/BD/6827/2001, SFRH/BD/6595/2001 and SFRH/BPD/15003/2004), the NATO Science for Peace program (project 978002), and the Belarus State University.

REFERENCES

[1] Bieberle A., Gauckler L.J., Glass seals. In: Oxygen Ion and Mixed Conductors and Their Technological Applications, Eds. H.L. Tuller, J. Schoonman and I. Riess, Kluwer, NATO ASI Series, Dordrecht-Boston-London, 2000, p. 389-397.

[2] Sakaki Y., Hattori M., Esaki Y., Ohara S., Fukui T., Kodera K., Kubo Y., Glass-ceramics sealants in $CaO-Al_2O_3-SiO_2$ system. In: SOFC-V, Ed. U. Stimming, the Electrochemical Society, 1997, PV-97-40, p. 652-660.

[3] Yang Z., Stewenson J.W., Meinhardt K.D., Solid State Ionics, 2003, v. 160, p. 213-225.

EFFECT OF "UPHILL" DIFFUSION IN SOLID ELECTROLYTES ZrO_2+8mol.%Sc_2O_3 AND ZrO_2 +8mol.% Y_2O_3

V.I. BARBASHOV, YU. A. KOMYSA

Donetsk Phys.Tech.Institute, National Academy of Science of Ukraine,R.Luxemburg St. 72, 83114 Donetsk, Ukraine, yurkom@inbox.ru

Abstract: Kinetic properties of the intrinsic defects in solid electrolytes on the basic of zirconia ZrO_2 have been determined experimentally. DC measurements of the composition ZrO_2+8 mol. % Sc_2O_3 and ZrO_2+8 mol. % Y_2O_3 show an electrical response of the system at the applying of external load. Indicated effect is discussed as a result of "uphill" diffusion of oxygen vacancies in the field of bending stress.

Key words: Solid Electrolytes/Zirconia/"Uphill" Diffusion.

1. INTRODUCTION

Zirconia (ZrO_2) stabilized in its high temperature, cubic form by addition of 8 to 12 mol.% yttria or scandia is currently the materials of choice in devices utilizing a solid-state oxide ion conducting electrolyte, *eg* oxygen sensors, oxygen pumps and solid oxide fuel cells (SOFCs), because of its good oxide ion conductivity and redox resistance [1-4]. Improvements in conductivity, however, are necessary to enhance theirs performance and efficiency.

In the given study authors try to consider this problem from such poorly studied side as influencing of mechanical stress on conducting properties of material being a solid electrolyte.

239

N. Sammes et al. (eds.), Full Cell Technologies: State and Perspectives, 239-244.
© *2005 Springer. Printed in the Netherlands.*

2. "UPHILL" DIFFUSION IN SOLIDS

The phenomenon of "uphill" diffusion or motion of point defects in an inhomogeneous field of stresses was predicted by Gorsky [6] as early as 1934. The essence of this effect is that in an inhomogeneous field of stresses (for example, created under bending of the crystal representing substitution solid solution) the impurity ions are offered by the force proportional to the difference of volumes of impurity atom and lattice atom. In this case, the atoms having the greater ionic radius transfer into crystals stretched region, and the atoms with smaller ionic radius diffuse in the region of compression.

The given effect is reversible, i.e. after the external loading was removed the equalization of concentration of point defects over the specimen's volume takes place. Later on it was shown by Kosevich [7] that the intrinsic defects of a crystal (vacancies and interstitials) can also diffuse in an inhomogeneous field of external stress, and the forces that act on them are equal accordingly to

$$f_v = \Omega_0 \, \nabla P, \; f_i = - \, \Omega_0 \, \nabla P,$$ (1)

where Ω_0 - volume of a point defect; ∇P - pressure gradient.

This effect is often revealed both in experiments on the analysis of diffusion process of different point defects in solids, and at study of dislocation's creep phenomenon. In the latter case, by methods of chemical etching or TEM electron microscopy it was possible to establish not only qualitative development of the effect, but also to make some quantitative estimation.

There are some obstacles to studying the kinetic peculiarities of this process: on the one hand, the problems of defects detection, and the necessity of preservation of high strength without the dislocation mechanism of plastic deformation at increased temperatures, on the other hand.

In the case of a four-point bending in the field of a sample arranged between internal bearings, the pressure gradient is $\nabla P = 2 \, \sigma / h$ (σ is normal stress on a surface of a sample, h is the depth of a sample). As a result of the external loading on compressed surface of the sample the vacancy concentration will be increased, and the value of the increasing is:

$$\delta c_v = \frac{\Omega_0 \sigma}{kT} \, c_v,$$ (2)

were c_v is equilibrium concentration, T is temperature, k is Boltzmann's constant. Accordingly, on the side of stretching there will be a decreasing of vacancy concentration on the same value. For interstitials the sign of effect will be inversed in comparison with vacancies. If to suppose that the indicated point defects are electrically charged, and there is a mechanism of charge compensation (for example conduction electrons in metals) in the system, on the inverse surfaces of the sample the electric charge should be induced

$$\delta q = z\,e\,\frac{\Omega_0 \sigma}{kT}\,c_v, \tag{3}$$

were z is the valence of the point defect.

It is necessary to emphasize that the effect of "uphill" diffusion should be distinguished from the directional diffusion of point defects that can also take place under conditions of a homogeneous stressed state (diffusive creep). Nevertheless, the indicated mechanisms of plastic deformation have a lot in common, because they both are diffusive mechanisms of plastic deformation of crystalline solids.

3. SAMPLE PREPARATION AND EXPERIMENTAL DETAILS

In our experiments the polycrystals of zirconia doped by scandium and yttrium oxide were studied (ZrO_2 – 8 mol. % Y_2O_3 and ZrO_2 – 8 mol. %Sc_2O_3). Samples of 2x15x50 mm³ in size were produced by the sintering of a powder (sizes of crystallites were less than 10 nm) and had the density of $5.97\cdot10^3$ kg·m⁻³ at a characteristic grain size of approximately 0.2-0.4 microns. After mechanical polishing on the greatest opposite edges of the specimen a silver contact was sintered. The so-obtained electrolytic cell was placed into the device for a four-point bending located in compact furnace allowing the heating of samples to temperature of about 1000^0 C (Fig. 1). The monitoring of temperature was implemented with the help of the thermocouple arranged directly near the sample. The mechanical effort to the sample in the high-temperature zone was transmitted with the help of a special bar on which the relevant load was hanged. The electric potential on the sample was registered with the help of a recorder working in the time sweep mode.

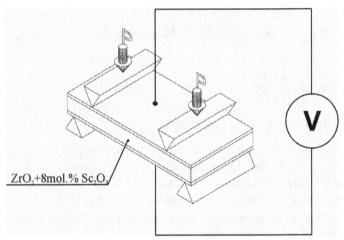

Figure 1. Scheme of mechanical-electrical tests (at different temperatures)

4. RESULTS AND DISCUSSION

In the Fig.2 the typical polarization curves demonstrating the change of potential on electrodes of an electrolytic cell during its heating with speed of about 10 K/min at different values of bending stress are shown. The applying of bending stress ("+") to the sample on different segments of a polarization curve uniquely testifies to increasing of charge change speed on the cell electrodes. The removing of load ("-") is accompanied by returning of the kinetic characteristics of an initial (unloaded) state.

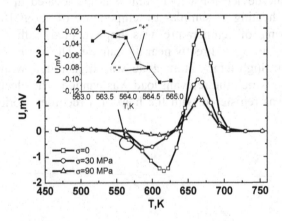

Figure 2. Polarization curves for different values of the bending stress.

Besides of qualitative experiments testifying to an electrical response of a solid electrolyte on action of inhomogeneous mechanical stress, the quantitative measurements were executed at fixed temperature. In Fig.3 the potential change curves on electrodes of electrolytic cell in time at the temperature of 400 ^{0}C are presented. The arrows indicate the moments of applying and removing of the load by value σ = 50 MPa. Changing of potential ΔU on electrodes of the sample was about 0.025-0.030 mV. The sign of the effect testifies to the motion of negatively ionized atoms (oxygen ions) in the direction of stretched surface of the sample and positively charged vacancies in the direction of the compressed side. In the context of existence of different diffusive mechanisms of plastic deformation it is necessary to note that the observed effect is reversible and does not correspond to plastic deformation of the sample. The marked circumstances give the basis to suppose that the observed effect is the result of "uphill" diffusion of vacancies in the direction of compressed surface of the sample that accompanied by a reflux of oxygen ions transferring the negative charge.

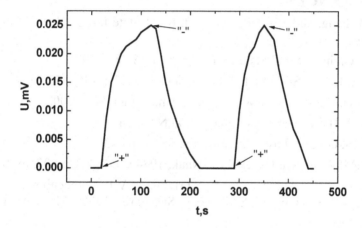

Figure 3. Time dependence of the electric potential change at the applying ("+") and removing ("-") of mechanical bending stress

One of the goals of given research was detection of differences in value of mechanical-electrical effect for polycrystals ZrO_2 - 8 mol.% Y_2O_3 and ZrO_2 - 8 mol.% Sc_2O_3. The basis for this supposition is the stabilization mechanism of zirconia's high temperature cubic phase: the stabilization is realized due to the replacement of zirconium atoms by impurity atoms having the greater ionic radius (r_{Zr4+} = 0.82 Å, r_{Y3+} = 0.97 Å, r_{Sc3+} = 0.83 Å [8]), that is the analogue to creation of high pressure or temperature. As a result, the energy of crystal cell in the field of impurity ion will be increased owing to the presence of local stresses. Stabilization of a cubic phase needs smaller quantity of an impurity which ions have the greater volume (Y^{3+}).

Other relevant property of such materials should be more sharp sensitivity to external influence of cubic zirconia stabilized by an impurity with smaller ionic radius (Sc^{3+}). Last statement guesses grater value of the effect of "uphill" diffusion in polycrystals ZrO_2 - 8 mol.% Sc_2O_3 as contrasted to ZrO_2 - 8 mol.% Y_2O_3. However, measured value of mechanical-electrical effect for both materials within the framework of error of experiment has appeared practically identical that is the result of smallness of the found effect.

ACKNOWLEDGMENT

The authors thank Dr Vereshak for presented powders, Dr Akimov and Dr Timchenko for useful discussions.

REFERENCES

[1] M. Feng, and J.B. Goodenout. J. Solid State Inorg. Chem., 1994, v.31, p.663.

[2] T. Ishihara. Solid State Ionics, 1996, v.86-88, p.197.

[3] J.A. Kilner. Solid State Ionics, 2000, v.129, N(1-4), p.13.

[4] W.G. Coors, and D.W. Ready. J. Am. Ceram. Soc., 2002, v. 85, p.2637.

[5] S.M. Haile. Materials Today, 2003,N3, p.24.

[6] W.S. Gorsky. Phys. Zs. Sow., 1935, v.8, p.443.

[7] A.M. Kosevich. Uspekhi fiz. Nauk, 1974,v.114, N3, p.507, in Russian.

[8] B.K. Vainshtein, V.M. Fridkin, V.L. Indenbom. Modern Crystallography. V.2. Crystals Structure. 1979, Moscow, p.79, in Russian.

NANOPARTICLE SYNTHESIS OF LSM SOFC CATHODE MATERIALS

I.A. DANILENKO, T.E. KONCTANTINOVA, V.N. KRIVORUCHKO, G.E. SHATALOVA, G.K. VOLKOVA, A.S. DOROSHKEVICH, V.A. GLAZUNOVA

The Donetsk Physical and Technical Institute named after A.Galki of the, National Academy of Science of Ukraine,. 72, R.Luxemburg Str., Donetsk, 83114, Ukraine

Abstract: There is reported the synthesis of the $La_{0.7}Sr_{0.3}MnO_3$ nanopowder by co-precipitation method. The nanopowder obtained consists of particles of two types, 40 and 100-200 nm in size, respectively, and of a different morphological form. The phase transformation of the synthesis products have been studied by the X-ray diffraction (XRD) and transmitting electron microscopy (TEM) in the range from 300 to 900^0C. The genesis of a nano-particle formation was being traced and the reasons of their bimodal size distribution have been explained. A comprehensive investigation of both the Mn NMR spectra and a nuclear relaxation has been carried out to obtain microscopic information about the magnetic structure of $La_{0.7}Sr_{0.3}MnO_3$ fine particles.

Key words: Manganites/SOFC Cathode Materials/Synthesis/Co-Precipitation/Particle Size

1. INTRODUCTION

There have been intensively studied the nanometer scale oxide ceramic materials for some years with respect to their specific properties arising from the high ratio of an interface area to the volume. Many interesting effects have been found as a consequence of a particle size reduction, and many

N. Sammes et al. (eds.), Full Cell Technologies: State and Perspectives, 245-251.
© *2005 Springer. Printed in the Netherlands.*

fields of their applications have come to existence now. One of them is the manufacture of solid oxide fuel cells (SOFC) cathode and anode materials as a cell design requires a large-area interface between an electrolyte and electrodes. Maximizing a cathode material surface, we increase the number of pores that affect the catalytic properties of SOFC cathode materials. The nanosized mixed electronic/ionic conductor perovskite system $La_{0.7}Sr_{0.3}MnO_3$ that is now the most-used SOFC cathode material with respect to their specific electrical, magnetic, mechanical and chemical properties.

Our main aim is to understand the influence of a nano-structure manipulation via chemical methods and pulse wave effects on the grain size. Moreover, a comprehensive study of [55]Mn NMR spectra, nuclear relaxation, measurements of the resistance and magneto-resistance have been performed to obtain a microscopic information on the magnetic structure of LSM fine particles. The investigation of the grain size effect is of a crucial importance if magnetic properties of manganites are expected to be used for a fuel cell reaction control.

2. EXPERIMENTAL

In recent years, several methods have been described in the literature concerning the so-called wet-chemical routes, including the co-precipitation method [1]. This method has been used for oxide nanopowders preparation [2,3] to replace the existing ceramic procedure not only by homogenous precursors, but also by increasing a specific surface area. We have prepared the $La_{0.7}Sr_{0.3}MnO_3$ nanopowder by the co-precipitation method supplemented with a microwave heating and ultrasound treatment [4]. In the course of preparation, stoichiometric amounts of La_2O_3, $SrCl_2 2H_2O$, $Mn(NO_3)_3 6H_2O$ and $(NH_4)_2CO_3$ were used as starting materials. All chemicals used in the synthesis procedure were of a high grade. La_2O_3 was dissolved in nitric acid, and ammonium carbonate was used as a precipitant. The final solution was being mixed for 1 hour at room temperature. The precipitate was then washed more than once to remove the by-products of the reaction. After washing and filtration, the hydro-gel was dried in the thermo-box. The powder obtained was dried in a pulsed magnetic field at 120°C in a special setup, with the magnitude of magnetic field $H=10^4–10^5$ A/m and frequency $f=0.5–10$ Hz. The precursor was being further milled by an ultrasound treatment [5] and decomposed by heating for 1 hour at 300°C. A nanometer scaled mono-phase $La_{0.7}Sr_{0.3}MnO_3$ powder was obtained by annealing at 850°C.

The phase structure of the $La_{0.7}Sr_{0.3}MnO_3$ powder and the size of powder particles have been determined by XRD with Cu $K\alpha$ radiation. The morphological features of crystallization have been investigated by TEM. In parallel with the study of the $La_{0.7}Sr_{0.3}MnO_3$ synthesis each of carbonate of all metals included were obtained and investigated with same procedures.

3. RESULTS AND DISCUSSION

A number of phase transformations was observed in the precipitate of a powder La-Sr manganite with the$La_{0.7}Sr_{0.3}MnO_3$ formation. A set of X-ray patterns of the samples obtained at various temperatures of a successive calcination is presented in Fig.1.

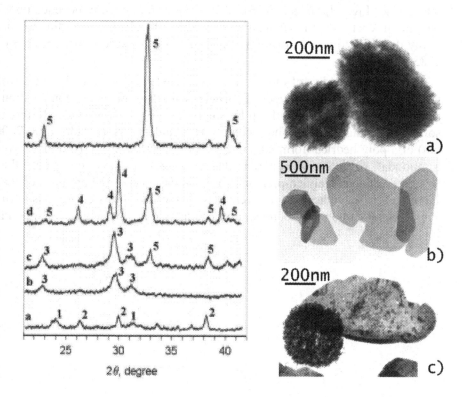

*Figure 1. XRD patterns of the compound $La_{0.7}Sr_{0.3}MnO_3$ being heated at (a) 300°C, (b) 400°C, (c) 600°C, (d) 700°C, (e) 850°C for 1 hour. The reflexes of the forming phases are shown by numbers: (1) $MnCO_3$, (2) $La_2O(CO_3)_2$. * H_2O, (3) La_2CO_5, (4) La_2O_3, (5) $La_{0.7}Sr_{0.3}MnO_3$.*

*Figure 2. The morphology of parent phases at different stages of the $La_{0.7}Sr_{0.3}MnO_3$ synthesis: (a) $MnCO_3$ and (b) $La_2O(CO_3)_2$ * H_2O at 300°C; (c) $La_{0.7}Sr_{0.3}MnO_3$ and La_2CO_5 at 500°C.*

The washed, filtered and dried mass of a precursor is a mixture of amorphous metal carbonates. As a result of calcination at 300°C, two crystal phases are observed on the powder X-ray patterns, i.e. $MnCO_3$ and $La_2O(CO_3)_2 \cdot xH_2O$ (Fig.1a). After calcination at 400°C $MnCO_3$ is destroyed because of decarbonisation. Reflexes of only one La_2CO_5 phase (Fig.1b) were fixed because dehydration and partial de-carbonization of the La component develop at heating within 300-400°C. The first signs of the perovskite $La_{0.7}Sr_{0.3}MnO_3$ phase make their appearance at 500°C. At 500-700°C, an increase of the manganite phase quantity takes place via a La_2CO_5 decrease (Fig.1c). La_2CO_5 looses CO_2 molecules at 700°C and transforms into La_2O_3 (Fig.1,d). This increase of the manganite quantity depends upon the rate of reacting with a lanthan oxide within 700-850°C. According to our X-ray data, the formation of a $La_{0.7}Sr_{0.3}MnO_3$ mono-phase comes to a close at 850°C (Fig.1e).

The TEM study of the obtained powders has found out two morphologically different types of particles with the same phase composition (Fig.2). The first type is built up by small (40 nm) round particles forming soft spherical aggregates. The second type includes large particles (100-200 nm) of a poly-hedrical form combined in poly-crystal forms. To define why a bimodal particle size distribution emerges, we have compared the structures of separately co-precipitated metal carbonates and products of the $La_{0.7}Sr_{0.3}MnO_3$ synthesis at different stages. Particle forms characteristic of source metal carbonates obtained after calcination at 300°C are presented in Fig.3.

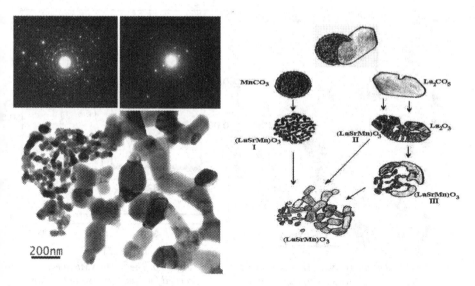

Figure 3. The microstructure of nanopowder La$_{0.7}$Sr$_{0.3}$MnO$_3$ with two morphologically different types of particles sintered at 850°C.

Figure 4. The scheme of phase transformations when obtaining La$_{0.7}$Sr$_{0.3}$MnO$_3$.

A manganese carbonate has the form of spherical globules while a lanthan carbonate has the form of plane crystals. Analysis of the synthesis products shows that there is a memory of the form in the course of phase transformations. The scheme of the transformations is presented in Fig.4.

We shall call MnCO$_3$ the first parent phase because fine dispersed La$_{0.7}$Sr$_{0.3}$MnO$_3$ (the 1st generation) arises within soft globules immediately after its decomposition. La$_2$CO$_5$ becomes the second parent phase initiating two phases, i.e. La$_2$O$_3$ and La$_{0.7}$Sr$_{0.3}$MnO$_3$ in the course of decomposition (the 2nd generation). Later La$_2$O$_3$ becomes a parent too and forms the 3rd generation of La$_{0.7}$Sr$_{0.3}$MnO$_3$. Generations 2 and 3 have inherited the size and form of La$_2$CO$_5$ particles. Thus, a bimodal distribution is determined by the formation of particles of two types based on a manganese carbonate and a lanthan carbonate.

An understanding of the decomposition processes and phase transformation allows us to obtain a mono-dispersed LSM nanopowder (Fig.5).

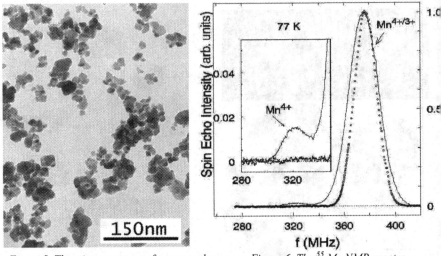

Figure 5. The microstructure of nanopowder $La_{0.7}Sr_{0.3}MnO_3$ with mono size particles synthesized at 620°C.

Figure 6. The ^{55} Mn NMR spectra recorded for nano-size (solid line) and large size (o) samples at 77 K. Inset: the same spectra on an enlarged scale, the NMR spectra at about 320 MHz shown in detail.

A lanthanum-strontium manganite ($La_{1-x}Sr_xMnO_3$) crystallizes into the perovskite structure. With Sr^{2+} in place of La^{3+} ions, $La_{1-x}Sr_xMnO_3$ transforms from anti-ferromagnetic insulating state to a ferromagnetic one within a certain range of concentrations at $x \approx 0.3$. Early theories tried to explain the ferromagnetic metallic state in terms of the double-exchange model. The presence of both trivalent and tetravalent Mn ions placed in the center of an oxygen octahedron causes a transfer of an electron from Mn^{3+} to Mn^{4+} through an intermediate oxygen.

The double exchange mechanism is sensitive to a Mn-O-Mn bound state, any structural disorder near a grain boundary (oxygen non-stoichiometry, vacancies, stress, etc.) modifies this exchange and leads to a spin disorder. Then, an interface between neighboring grains plays the part of a potential barrier. Nuclear magnetic resonance (NMR) is effective to investigate magnetic properties of manganites, because it can determine locally different charge states and their dynamics.

The ^{55}Mn NMR spectra of both nanosized and ceramic LSM samples with the same phase composition are presented in Fig.6.

The dominant line at $f_{res} \approx 377$ MHz is typical of mixed valence metallic-like manganites and corresponds to a manganese ion state with an intermediate valence $Mn^{3+/4+}$. Besides this frequency for nanosized samples, an additional line at $f_{res} \approx 315\text{-}332$ MHz was observed. This line is responsible for the Mn^{4+} ions presence on a sharp interface layer (~2nm) of a nanogranule.

Our results on a nuclear spin dynamics provide a direct evidence of previous statements that a grain boundary in magneto-resistive manganites is not a magnetically and electrically sharp interface, but should be considered as a transition region of several mono-layers different from a grain inner part by magnetic and structural orders.

REFERENCES

[1] Kakihana M. Sol-gel preparation of high temperature superconducting oxides. J Sol-Gel Sci. Tecnol., 1996, v.6, p.7-55,

[2] Vazques-Vazques C., Blanco M.C., Lopez-Quitela M.A. et al. Characterization of particles prepered by the sol-gel route. *J. Mater. Chem*, 1998, v.8, p. 991-1000,

[3] Marques R.F.C., Jafelicci M., Jr., Paiva-Santos C.O. et al. Nanoparticle synthesis of $La_{1-x}Sr_xMnO_3$ (0.1, 0.2 and 0.3) perovskites. IEEE Trans. on magnetics, 2002, v.38, p. 2892-2894.

[4] Konstantinova T. E., Danilenko I. A., Pilipenko N. P. et al. In Proceedings of the International Symposium "Solid Oxide Fuel Cells VIII," 2003, v. 2003–07, p.153, in Inglish.

[5] Doroshkevich A., Danilenko I. A.,. Konstantinova T. E. et al. Formation of monocrystal particles in $ZrO_2\text{-}3mol\%Y_2O_3$ system. Phys. Techn High Pressure, 2002, v.12, p. 38, in Russian.

PERSPECTIVE MATERIALS FOR APPLICATION IN FUEL-CELL TECHNOLOGIES

M. KUZMENKO, N. PORYADCHENKO

Frantsevych Institute for Problems of Materials Science, NASU, Kyiv, Ukraine

Abstract: In this work the investigations of heat resistance of titanium alloys elaborated were made. The tests performed in air in a temperature range of 600-1000 °C (time exposure up to 30 hours) allowed kinetic curves of oxidation for selected alloys to be obtained. It was shown on the basis of thermogravimetric, metallographic and X-ray micro-spectrum analysis that the alloys retain capability of working in oxidizing medium up to 800 °C. Level of their heat resistance being 0.1 mg/(cm^2·h) is 3 times more than of that for conventional titanium (BT1-0) and almost 5 times of pure titanium.

Key words: Heat Resistance/Titanium Alloys Microturbines/Titanium In Situ Composites/ Oxidation

1. INTRODUCTION

Requirements for a number of properties are established for metallic materials to be used as elements of high temperature fuel-cell – microturbines hybrids. Increasing heat and corrosion resistance are of the most importance among these. Titanium alloys can be considered as suitable materials offering low specific weight, high corrosion resistance and manufacturability.

The titanium alloys are not heat resisting materials being inferior to stainless steel in this respect. Recently titanium-based alloys alloyed with silicon, aluminium, zirconium (elements which considerably enhance heat resistance of technical metals Fe, Co, Ni) were elaborated in IPMS of NASU

N. Sammes et al. (eds.), Full Cell Technologies: State and Perspectives, 253-258.
© 2005 Springer. Printed in the Netherlands.

[1]. In this work the investigations of heat resistance and some mechanical properties of titanium alloys elaborated were made.

2. EXPERIMENTAL PROCEDURE

The alloys were smelted in the plasma-arc furnace. The compositions of alloys were defined by alloying of a BT1-0-basic alloy with silicon (1-6 wt. %), aluminium (3 wt. %) and zirconium (~5 wt. %). The oxidation was carried out in the apparatus "Derivatograph" with continuous recorder of weight change in temperature range 1073-1273 K during 6 hours. Long-term heat resistance tests were made in the electro-resistance furnace with duration up to 30 hours at a temperature of 1073 K. The specimens with sizes 10 mm x 4 mm x 3 mm were used for this purpose. Heat resistance of alloys was estimated by mean value of mass change on unit of sample surface (Δm/S, mg/cm^2). Comparison of heat resistance of different alloys was carried out using oxidation rate parameter (V, mg/(cm^2·h)).

The experimental results presented in this paper were obtained by traditional mechanistic procedures such as thermobalance, optic metallography, electron microprobe (SEM) and X-ray-analysis (XRD).

3. RESULT AND DISCUSSION

The researches showed that the considerable increase of mass of the explored alloys at oxidation took place at temperatures above 900 °C. At the temperatures of 800 and 900 °C difference in heat-resistant is insignificant. The sharp oxidation resistance decrease is observed at a temperature of 1000 °C. So, heat-resistance for alloy with 2 wt.% Si and 3 wt.% Al at temperatures of 800, 900 and 1000 °C equals to 0.1, 0.22 and 0.56 mg/(cm^2·h), and for the unalloyed alloy BT1-0 to 0.28, 0.55 and 1.41 mg/(cm^2·h), respectively.

Influence of alloying elements was analyzed based on results of oxidization kinetics. The data obtained are presented in Table 1.

It follows from these data that addition of silicon in the basic titanium alloy rises its resistance to oxidation. Especially it becomes apparent at silicon content of 2 wt. % and more. Further its increase does not change heat-resistance practically. Data of tests obtained by different methods for 5 hours exposure coincide practically. The results of short-time tests are confirmed at the increase of exposure time to 30 hours: the best characteristics have alloy with 2 wt. % Si as well as that with additional aluminium alloying. If at the short-time tests insignificant dependence is revealed at increasing silicon content up to 6 wt. % and addition of 3 wt. % Al, this is not revealed at the long-term exposure when resistance to oxidation gets even worse. Confirmed also is the negative influence of zirconium, which lowers influence of silicon and aluminium at the simultaneous alloying of titanium with these elements. However the oxidization rate of the complex alloyed alloy remains to be 2 times as below compared with that of unalloyed one (Table 1).

Table 1. Oxidation rate (mg/(cm^2·h)) of titanium alloys at 800 °C in different periods of exposure at oxid ation

Composition of alloy, % Si	Duration of exposure, h			
	5	10	20	30
0	0,25	0,155	0,11	0,16
2	0,15	0,1	0,06	0,05
4	0,12	0,13	0,085	0,07
6	0,10	0,15	0,09	0,07
2+3% Al	0,095	0,09	0,05	0,05
4+3% Al	0,11	0,13	0,09	0,08
6+3% Al	0,105	0,1	0,06	0,06
4+3%Al+5%Zr	0,20	0,14	0,1	0,08

At the same time, at the additional alloying with aluminum the influence of aluminium at different content of silicon and at different temperatures is ambiguous. So, addition of aluminium in the alloy with 4 wt. % Si causes some worsening heat-resistance at the temperature of 900 °C, an additional alloying with zirconium making it worse yet (Fig. 1).

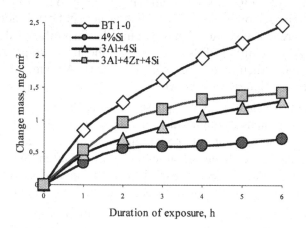

Figure 1. Kinetic curves of oxidation of titanium alloys at 900 °C, 6 h.

Only at the temperature of 800 °C titanium alloys with aluminium and silicon have the best heat-resistance, than alloys without aluminium and with silicon only. A considerable difference in influence of alloying on oxidization rate is observed at a temperature above 800 °C. It seems to be due to both the rise of temperature and to structural changes occurring in the alloy upon transition of titanium in β-phase at the temperatures of 900 and 1000 °C.

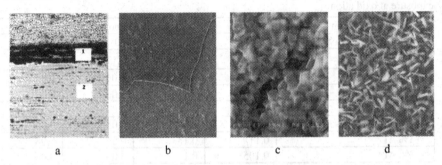

a   b   c   d

Figure 2. Microstructure (a) and morphology (b, c, d) of scale on the surface of alloys after oxidation at 800 °C: a – BT1-0, 6 h, x400; b - BT1-0, 30 h (general view); c - BT1-0, 30 h (external layer); d – alloy with 6% Si+3% Al; 1 – scale; 2 – matrix of metal.

Except for the basic alloy a scale with satisfactory adhesion appears on the surface of all alloys samples during oxidization. The scale is multilayer in the basic alloy (Fig. 2a) and begins to peel off after the 10-hour test. The scale on this layer is fine-grained and homogeneous (Fig. 2b) with cracks in an external layer (Fig. 2c). Upon alloying of alloy BT1-0 with ≥2 wt. % Si the scale is formed on its surface which does not split off. It does not split off after exposure for 30-hours. There is only the insignificant deterioration of adhesion. After addition of 2-6 wt. % Si in the alloy the prolonged crystals appear in the structure of the scale (Fig. 2d). Amount of these crystals increases at addition of 3 wt. % Al.

According to data of X-ray phase analysis the scale appearing on the surface of specimens during oxidation consists of oxide TiO_2 (rutile) mainly and of negligible amounts of α-SiO_2 and γ-Al_2O_3. One can see that during oxidation there is the change of stoichiometry of titanium oxide TiO_2 (rutile). In works [2, 3] such effect is explained by formation of rutile layer with complex structure, and also by competitive diffusion of titanium and oxygen.

Metallographic investigations and determination of microhardness testify that upon titanium oxidization the dissolution of oxygen in the surface layer of metal occurs simultaneously with formation of oxide film. The results of measuring of microhardness of surface layer, metallic matrix and depth of gas saturation show that the layer of metal adjacent to the scale has the increased hardness. Microhardness of surface layer is considerably more than hardness of alloy matrix. Metal hardness decreases gradually when going in direction of specimen middle. This can be associated with large solubility of oxygen in titanium [4] as well as with directions of diffusive streams [5] in oxide film.

It follows from analysis that the hardness of surface layer rises considerably after oxidation in all alloys under study. Hardness rises in a less measure in the alloys alloyed with silicon; the depth of gas saturation is also less. The additional alloying of alloys with 3 wt. % Al does not change the depth of gas saturation and hardness of surface layer essentially. With the increase of exposure time to 30 h the depth of gas saturation changes in a considerably greater degree than hardness of surface layer.

In most cases [2,6] influence of alloying elements on oxidation resistance of titanium is explained by formation of protective film with a structure and phase composition that differs from film formed on the unalloyed titanium. In our researches the phase composition of scale consisted mainly of oxide TiO_2 (rutile) for all alloys. The composition of scale did not practically depend on alloying and duration of exposure. Therefore the reason of rising heat-resistant in the alloys is presumably due to influence of their structure and phase composition on the scale formation.

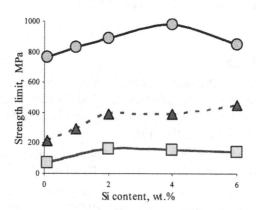

Figure 3. Strength of cast alloys of the Ti-Si system after the tests at: o – 20 °C ; Δ– 600 °C ;
□ – 750 °C.

Effect of silicon additions on strength of Ti-Si cast alloys after tests at the temperatures of 20, 600 and 750 °C is shown in Fig. 3. One can see that growth of silicon content from 0 to 4 wt. % results in the rising strength at room temperature. At test temperatures of 600 – 750 °C the strength limit comes to saturation already at 2 wt. % Si. That is, with growth of silicon content above 2 wt. % rise of the strength at high temperatures does not take place. It allows making a conclusion about inexpedience of alloying of titanic matrix with silicon in binary alloys of the Ti-Si system in amount more than 2 wt. %. At the subsequent rise of silicon content plasticity at room temperature substantially decreases. This is undesirable effect because it is necessary to provide sufficient reserve of plasticity taking into account the subsequent alloying of matrix with such elements as Al and Zr.

4. CONCLUSION

The investigated result of oxidation behavior in still air at 800 °C of BT1-0 based titanium alloys containing silicon, aluminium and zirconium shown that at oxidation the process of the scale formation and process of the gas saturation take place together. The parabolic law describes the common oxidation rate without dependence from duration exposure. The main influence on heat resistance makes silicon with content to 2 wt. %. This effect is connected with breaking the diffusion process in metal and scale.

The conducted researches also confirmed the similar dependence of oxidation and heat resistance on composition of alloys [3] and showed that the elaborated alloys are capable to work in an oxidizing environment up to 800 °C. Complex alloyed alloy has also high level heat resistivity. His ultimate tensile strength at 700 °C is 650 MPa.

REFERENCES

[1] Firstov S.A., Podrezov N.N., Kuzmenko N.N., Danilenko N.I. Izouchenie vliyaniya plastichescoy deformatsii on mehanichescie svoystva эvtectichescih splavov sistemы Ti-Al-Si-Zr // FTVD.- 2002.- t.12 №3.- P.48-56. (in Russion)

[2] Lazarev Э.M., Cornilova Z.I., Fedorchouc N.M. Ocislenie titanovih splavov M.: Science 1985.- 144c. (in Russion)

[3] Voytovich R.F., Golovko Э.I. Visocotemperatournoe ocislenie titan of I ego splavov. Kiev: Naykova dymka.-1970.- 317 p. (in Russion)

[4] Eremenko V.N. Titan of I ego splavi.Kiev: AN SSSR.- 1960.- 500 p. (in Russion)

[5] Revyacin A.V. C voprosou o cinetice ocisleniya titan //Titan of I ego splavi.- Vip. 8. Metallourgiya titan Izd-vo AN SSSR.- 1962.-P. 175-190. (in Russion)

[6] Mazyr V.J., Kapustnikova S.V., Blochina O.A. Researching High temperature gas corrosion in new titanium alloys eutectic type //3rd Int. Symp. "Corros. Resist Alloys Cracow", 20-22, June, 1996: Conf. Pap.- Cracow, [1996].- P. 141-148.

ELECTROCONDUCTIVITY OF THE OXIDE INTERLANTHANOIDS WITH PEROVSKITE STRUCTURE

V.V. LASHNEVA, V.A. DUBOK

Frantsevych Institute for Problems of Materials Science, National Academy of Science of Ukraine, Krzhyzhanovs'koho Str., Kyiv-142, 03680 Ukraine

Abstract: Results of measurements of electroconductivity of the oxide interlanthanoids with perovskite structure are presented. The studied sintered ceramic samples with X-ray pattern of perovskite have composition $LaYO_3$, $LaErO_3$, $LaYbO_3$, $GdScO_3$, $ScYO_3$ pure and doped by oxides of calcium CaO and hafnium HfO_2 with concentration up to 0,5 mol. %.

It was established that the electrical conductivity of the studied materials was determined by a non stoichiometric disordering and could be regulated in considerable extent by doping and changing equilibrium partial pressure of oxygen in gas phase.

Key words: Fuel Cell/Perovskite/Oxide Interlanthanoid/Electrical Conductivity Ceramics

1. INTRODUCTION

Due to necessity to develop a new generation of ceramic materials for fuel cells the oxides with perovskite structure are intensively studying. This kind of crystal structure inherents to ceramics with unique electrical, catalytic, magnetic and other properties. Among the given class of materials considerable interest presents perovskites composed of two or more rare earth oxides (REO) that can be called oxide interlanthanoids (OIL). Most of such compounds are characterized by melting point above 2000 °C, high

N. Sammes et al. (eds.), Full Cell Technologies: State and Perspectives, 259-264.
© *2005 Springer. Printed in the Netherlands.*

thermodynamic and corrosion resistance in aggressive chemical environments.

The typical representatives of this kind of chemical compounds are oxides $LaYO_3$, $LaErO_3$, $LaYbO_3$, $GdScO_3$, $ScYO_3$ that were synthesized and studied in this developments.

This study presents results of measurements of electrical conductivity of high density ceramics (porosity about 2 %) made of high purity OIL perovskite mentioned above as well as the OIL doped with oxides of calcium CaO and hafnium HfO2 in quantity up to 0,5 mol. %.

2. RESULTS AND DISCUSSION

The electrical conductivity was studied in temperature range 1000 - 1600 °C in equilibrium with gas phase with partial pressures of oxygen from 2.10^4 down to 1.10^{-10} Pa. To synthesize the OIL the chemicals were used with contents of the main substances more than 99,95 %, main impurities - other rare earth oxides. All samples were monophase with crystal structure of perovskite of rhombic syngony, the $GdFeO_3$ kind. Electrical conductivity was measured at temperatures up to 1650 °C with two burning platinum electrodes using direct or alternating current with frequency 1500 Hz. The partial pressure of oxygen P_{O2} in gas was controlled by ceramic solid electrolyte pump and measuring device based on ZrO_2 (8 mol. % Y_2O_3).

Results of measurement of electroconductance vs. temperature in air for synthesized OIL ceramics, pure and doped with CaO and HfO$_2$, are presented at fig. 1-3 accordingly. In Arrhenius coordinates lg ρ - 1/T in temperature range 1000 - 1600 °C these results can be approximated by a straight lines or consist of 2 straight-line parts, which declinations were used to calculate the activation energies of electrical conductivity, listed in the table.

As can be seen from presented results, the highest values of electrical resistance and its activation energies were found for $LaYO_3$. Doping by small quantities of heterovalent impurities considerably changes electrical conductivity and activation energy of electrical conductivity for all synthesized OIL. For example, for $GdScO_3$ the electrical conductivity in air at temperature 1000 °C alters by doping within the limits of 4 orders of magnitude, and for $LaYbO_3$ - approximately within an order of magnitude. At the same time if doping by CaO increases electrical conductivity of OIL and reduces activation energy of electrical conductivity, doping by HfO$_2$ changes these properties in opposite direction. The action of the CaO addition is the most effective at the contents up to 0,5 mol. %, and HfO$_2$ - up to 0,02 mol. %.

The variation of the specific resistance of the OIL vs. partial pressure of oxygen obeys the same regularities. In fig. 4 the typical for these materials isotherms are presented: lg $\rho=f(P_{O2})$ for $GdScO_3$, $GdScO_3+0,5$ mol.% CaO and $GdScO_3+0,1$ mol.% HfO_2. These curves are usually extrapolated by the formula $\rho=$const $P_{O2}{}^{\alpha}$. Thus at $P_{O2}>10^2$ Pa the resistance increases with decreasing partial pressure of oxygen. For this part of the curves $\alpha \approx 0,20$. In the P_{O2} interval 10^2 - 10^{-1} Pa the value of α monotonically decreases, and at $P_{O2}<10^{-1}$ Pa $\alpha \approx 0$, i.e. the electrical conductivity practically does not depend on partial pressure of oxygen.

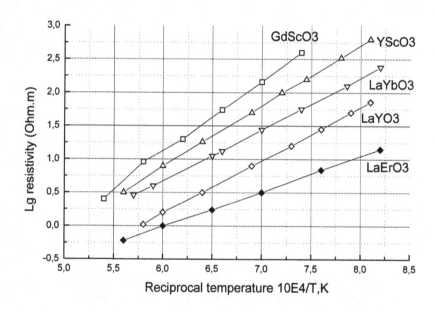

Figure 1. Resistivity vs temperature for oxide interlanthanoids $YScO_3$, $LaErO_3$, $GdScO_3$, $LaYbO_3$, $LaYO_3$

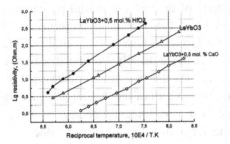

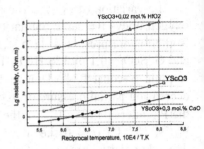

Figure 2. Resistivity vs temperature for
oxide interlanthanoid LaYbO₃, pure and
doped with CaO and HfO₂

Figure 3. Resistivity vs temperature for
oxide interlanthanoid YScO₃, pure and
doped with CaO and HfO₂

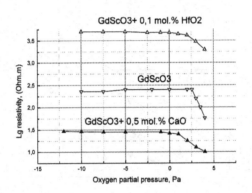

Figure 4. Resistivity of oxide interlanthanoid GdScO3, pure and doped with CaO and HfO2 at
temperature 1100°C vs. equilibrium oxygen partial pressure

Composition	Temperature range, °C	Activation energy of electroconductivity, eV
$LaYO_3$	1000-1600	1,6
$LaErO_3$	1000-1600	1,1
$YscO_3$	1000-1600	1,7
$YscO_3$ (0,1 мол.% CaO)	900-1330 / 1330-1600	1,6 / 1,1
$YscO_3$ (0,5 мол.% CaO)	900-1330 / 1330-1600	1,6 / 1,1
$YscO_3$ (0,02мол.%HfO_2)	900-1200 / 1200-1600	2,0 / 2,4
$YscO_3$ (0,2 мол.% HfO_2)	900-1250 / 1250-1600	1,8 / 2,2
$LaYbO_3$	1000-1600	1,5
$LaYbO_3$(0,5 мол. %CaO)	1000-1600	1,5
$LaYbO_3$(0,5 мол.%HfO_2)	1000-1600	2,0
$GdScO_3$	1000-1600	2,1
$GdScO_3$(0,02 мол. %HfO_2)	900-1100 / 1100-1600	1,8 / 2,3
$GdScO_3$(0,05 мол.% HfO_2)	900-1300 / 1300-1600	1,8 / 2,4
$GdScO_3$(0,5 мол.% HfO_2)	900-1300 / 1300-1600	2,1 / 2,8
$GdScO_3$(0,1 мол.% CaO)	900-1350 / 1350-1600	1,6 / 1,1
$GdScO_3$(0,5 мол.% CaO)	*900-1450 / 1450-1600*	*2,5 / 1,6*

Table 2. Activation energy (eV) of electroconductivity of oxyde interlanthanoids.

The measurements of ion transport numbers made by method of EMF and by sign of thermo-electric force reveal, that in all studied materials at $P_{O2} > 10^2$ Pa and in air the electronic conductivity of p-type (hole conductiviy) predominates, and at $P_{O2} < 10\text{-}1$ Pa the electrical conductivity is predominantly ionic. The transport number of ions t_i alters depending on oxygen partial pressure, temperature and doping from 0,9 down to 0,1.

Found values of α and other observed regularities of electrical conductivity variations for the OIL coincide mainly with the studied earlier for simple REO, including doped oxides. These results can be explained within the framework of theoretical model of equilibrium between electronic and ionic defects in crystals, which application for the analysis of defect structure of REO is reviewed in [1].

The application of this theory for REO can be extended on OIL, as the structures of valence electronic shells of their cations are identical.

In accordance with these regularities the electrical conductivity of oxide interlanthanoids is determined by interaction of stoichiometric and impurity defects. Variation of P_{O2} in gas environment which is in equilibrium with the oxide alters the kind of compensation of charge mismatch of impurity defects - from compensation only by ionic defects - cationic and anionic vacancies, interstitial ions up to compensation only by electrons or by holes contributing to the relative component of electroconductivity.

The doping impurities can increase and lower electrical conductivity and activation energy of electrical conductivity of OIL, as well as displace the area of independence of electrical conductivity from oxygen partial pressure towards larger or smaller values of P_{O2}.

REFERENCES

[1] New materials made of oxides and synthetic fluorsilicates / Ed. S.G.Tresvjatsky – Kiev: Naukiva Dumka, 1982.- 204pp.

[2] Huang Kegin, Feng Man, Goodenough B. Synthesis and Electrical Properties of Dense Ce0,9Gd0,1O1,95 Ceramics // J.Amer. Ceram. Soc. -1997. -V.81. - № 2. - P.357-362.

[3] Rare earth compounds. Zirconates, hafnates, tantalates, antimonates P.A.Arsenjev, V.B.Glushkova, A.A.Evdokimov et al. - Moskow: Nauka, 1985. -261 p.

ELECTROCONDUCTIVITY OF THE BULK AND GRAIN BOUNDARIES IN THE 0.8CeO$_2$-0.2Gd$_2$O$_3$ ELECTROLYTE PREPARED FROM NANOPOWDERS

A.S. LIPILIN[2], V.V. IVANOV[2], S.N. SHKERIN[1], A.V. NIKONOV[2], V.R. KHRUSTOV[2]

[1]Institute of High-Temperature Electrochemistry, Ural Branch RAS, Ekaterinburg
[2]Institute of Electrophysics, Ural Branch RAS, Ekaterinburg

Abstract: The impedance spectroscopy method was used to study conductivity of a ceramic oxygen-conduction electrolyte of the formula Ce$_{0.8}$Gd$_{0.2}$O$_{1.9}$, which was prepared by radial magnetic pulse compaction of nanopowders and their subsequent sintering.

Key words: Solid Oxygen-Conduction Electrolyte/Doped Cerium Oxide/Nanopowders

As compared to materials produced by traditional ceramic technologies using micrometer-sized powders, ceramics prepared from nanosized powders usually have different characteristics: a dense ceramic with a submicron structure is synthesized from nanopowders at relatively low temperatures (1100-1300 °C) and has a higher strength and a larger electroconductivity.

The initial powder of the solid oxygen ion electrolyte of the formula Ce$_{0.8}$Gd$_{0.2}$O$_{1.9}$ was synthesized by laser evaporation of a target at the Institute of Electrophysics, Ural Branch RAS [1]. The average size of particles was d$_g$ = 9.4 nm. More than 95% particles were 3-20 nm in size. The X-ray diffraction analysis confirmed that the powder structure had a cubic

N. Sammes et al. (eds.), Full Cell Technologies: State and Perspectives, 265-270.
© 2005 Springer. Printed in the Netherlands.

symmetry with the unit cell parameter of 0.5424 nm and crystallites about 19 nm in size.

Ceramic samples in the form of tubes were prepared by the method of radial magnetic pulse compaction of initial nanopowders and their subsequent sintering. The samples were compacted using pressure pulses with the amplitude of 240 MPa, which provided blanks whose density accounted for 50% of the theoretical density. The blanks were sintered in air by heating to 1300 °C at a rate of 2 K/min and then cooling at the same rate. The tube samples were about 13 mm in diameter and had walls nearly 0.7 mm thick.

The structure of the blanks was examined in a LEO 982 scanning electron microscope and the fracture of sintered ceramics was analyzed in a SOLVER-47 atomic force microscope. A microphotograph of the fracture of a compacted blank (Fig. 1a) clearly demonstrates a uniform distribution of particles of the initial weakly agglomerated nanopowder and its good compaction by the radial magnetic pulse method. Agglomerates were absent after compaction. Sintering of the ceramic produced a dense uniform structure (Fig. 1b), which was characterized by a small number of nanosized pores and an average size of crystallites equal to 100-200 nm.

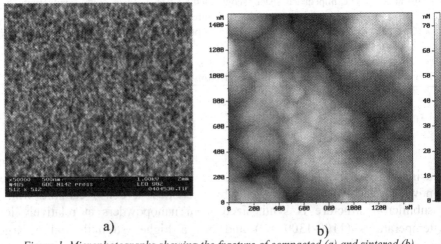

a)   b)

Figure 1. Microphotographs showing the fracture of compacted (a) and sintered (b) ceramics (the relief height regime of SOLVER-47)

A fragment of a tube with the wall 0.5 mm thick was cut for conductivity measurements. Large opposite surfaces had silver electrodes, which were fused at 700 °C for nearly 18 hours. Measurements were made using the two-probe impedance spectroscopy method on a Im6 instrument (Zahner-Elektrik) in air at frequencies from 10^{-1} to $8 \cdot 10^5$ Hz over the temperature interval of 220 to 700 °C with steps of 20-50 degrees.

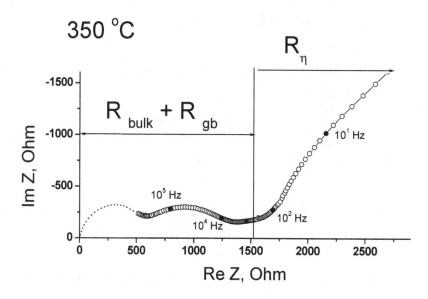

Figure 2. Example of the impedance spectrum of a Ag/electrolyte/Ag cell at 350 °C.
Separation of the resistance of the electrolyte bulk R_{bulk}, resistance of grain boundaries R_{gb},
and polarization resistance of the electrodes $R_η$ is shown

Fig. 2 presents typical hodograph curves of the cell impedance. The impedance spectroscopy measurements allowed separating contributions from electrode processes ($R_η$), resistance of the grain bulk (R_{bulk}), and resistance of grain boundaries (R_{gb}) in a polycrystal. The experimental data were processed using a program [2,3], which was written by Bernard Boukamp.

When the temperature increased, it was impossible to reliably separate the resistance of the electrolyte bulk R_{bulk} and the resistance of grain boundaries R_{gb} in the total resistance of the electrolyte. However, reliability of separation of the electrolyte resistance and the polarization resistance of the electrodes $R_η$ was beyond doubt.

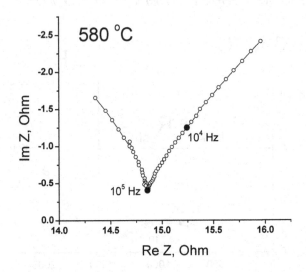

Figure 3. Example of the impedance spectrum of a Ag/electrolyte/Ag cell at 580 ℃

Fig. 4 presents temperature dependences of the total specific resistance of some electrolytes based on cerium oxide. It is seen that the total electroconductivity of the ceramic under study is comparable with best conductivity values, which were obtained earlier for flat samples prepared by magnetic pulse compaction (curves 6-7 in Fig. 4). Only the ceramic of the formula $Ce_{10}Gd$, which was synthesized by Steele, had better characteristics (3, Fig. 4). His ceramic usually had a negligibly small resistance of grain boundaries.

It is commonly assumed that the main reason for the decrease in the total electroconductivity of ceramics is grain boundaries and impurities, primarily Si, localized at boundaries. Notice that although the share of grain boundaries in the material, which was synthesized from nanopowders, was very large, their contribution to the decrease in the total electroconductivity of the ceramic was insignificant.

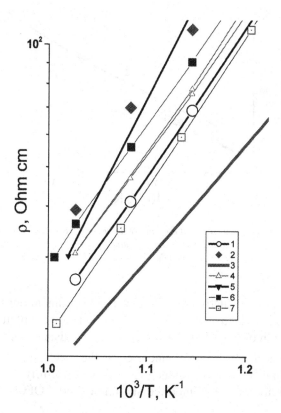

Figure 4. Temperature dependences of the total specific resistance of the electrolyte as found by different researchers: 1 – our study, Ce$_{20}$Gd; 2 – Milliken [4], Ce$_{20}$Sm; 3 – Steele [6], Ce$_{10}$Gd; 4 – Gorelov [7], Ce$_{10}$Sm, 5 – Zhan [5], Ce$_{20}$Sm; 6, 7 – Gorelov [7], Ce$_{20}$Gd

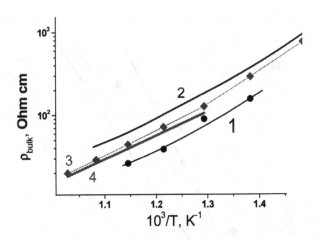

Figure 5. Temperature dependences of the bulk resistance of electrolyte grains as found by different researchers: 1 – our study, $Ce_{20}Gd$; 2 – Zhan [5], $Ce_{20}Sm$; 3 – Gorelov [7], $Ce_{10}Sm$; 4 – Steele [4], $Ce_{10}Gd$

The bulk electroconductivity of the electrolyte grains in our material was higher than in other electrolytes (Fig. 5). Steele's data were plotted assuming a zero contribution of grain boundaries to the total resistance of the ceramic.

After the loss due to grain boundaries in the electrolyte is minimized, probably we shall be able to produce an efficient electrolyte with a higher electroconductivity for use in intermediate-temperature SOFC's.

REFERENCES

[1] Kotov Yu.A., Osipov V.V., Samatov O.M. et al. // ZhTF, 2004, v. 74, No. 3, p. 72..

[2] Boukamp B.A. // Solid State Ionics. 1986. V. 18/19. P. 136.

[3] Boukamp B.A. // Solid State Ionics. 1986. V. 20. P. 31.

[4] Milliken C., Guruswamy S., Khandkar A. // J. Electrochem. Soc. 1999. V. 146. P. 872.

[5] Zhan Z.,Wen T.-L., Tu H., Lu Z.-Y. // J. Electrochem. Soc. 2001. V. 148. A. 427.

[6] Steele B.C.H., Hori K.M., Uchino S. // Solid State Ionics. 2000. V. 135. P. 445.

[7] Gorelov V.P. Unpublished data. By courtesy of the author.

A TWO PHASE MODEL FOR ELECTROCHEMICAL SYSTEMS

M. D. MAT[1], K. ALDAS[2], T.N. VEZIROĞLU[3]

[1]Mechanical Eng. Dept. Nigde University, 51100 Nigde, Turkey
Tel: +90 388 2250115, Fax: +90 388 2250112, mdmat@nigde.edu.tr
[2]Mechanical Eng. Dept. at Aksaray, Nigde University, 68100, Turkey
[3]Mechanical Engineering Dept., University of Miami, FL, USA

Abstract: Two phase flow is encountered in many electrochemical systems and play vital role on system efficiency, species transport, velocity distribution etc. A two phase flow model which accounts specific nature of liquid and gaseous phase is developed. The model applied to water electrolysis in an electrochemical cell. Transport equations are solved numerically for both phases with allowance for inter – phase mass and momentum Exchange. Liquid and gaseous phase distributions velocities, current density distribution are calculated under various working conditions.

It is found that gas layer accumulation on the electrode surface decreases the active reaction area and adversely affects the reaction rate.

Key words: Two-Phase Flow/Water Electrolysis/Current Distribution

1. INTRODUCTION

Gas bubbles are characteristics of the electrochemical cells sometimes being a desired product such as in hydrogen production or off product as in electroplating etc. Since density of the gas phase is much smaller than the liquid phase, a significant flow is generated as bubbles rises in electrochemical cells. Although this flow is very important in determining the mass transfer, gas and ion distribution it is often neglected or simplified assumptions are employed studies in the literature

N. Sammes et al. (eds.), Full Cell Technologies: State and Perspectives, 271-277.
© 2005 *Springer. Printed in the Netherlands.*

There is a similarity between gas – electrolyte flow and steam – water flow which is extensively studied in boiling heat transfer. There fore, the main purpose of this study is to incorporate the large knowledge accumulated on boiling heat transfer into electrochemical systems. Vogt [1] analyzed similarities between boiling and gas evolving cells and pointed out that void fraction and bubble sizes in electrochemical cell are relatively small. Vogt showed that to account for effects of gas phase, pressure drop should be multiplied by a factor $(1-f)^n$ where $n = 1$ in laminar flow and 0,25 in turbulent flow.

Understanding of gas-liquid flow in electrochemical systems is very important for system optimization, enhance mass transport and thus gas release efficiency. There are relatively little theoretical studies available in the literature which considers process as a two-phase flow problem. Zeigler and Evans[2] applied the drift – flux model of Ishii[3] to electrochemical cell and obtained velocity field, bubble distribution, mass transfer rate. Instead of treating the bubbles as a second phase, they obtained bubble distribution from concentration equation. Dahlkild [4] developed an extensive mathematical model for gas evolving electrochemical cells and performed a boundary layer analysis near a vertical electrode.

Wedin and Dahlkild [5] further improved the mixture two – phase flow model by incorporating empirical relations for shear induced hydrodynamic diffusion and migration of particles and applied the model to developing flow in a channel. Wedin and Dahlkild [5] found that the bubble thickness in front of the electrodes increases significantly when bubble size increased, however, with the larger bubble diameter, the slip velocity decreases. The performance of the model is then tested by comparing numerical results with data of Boissonneau and Byrne [6] and a good agreement is obtained.

Although this study considerable improved the understanding effects of gas phase on electrochemical cell still suffers inherent limitation of the mixture mathematical method employed. Specifically the motion of individual phases cannot be analyzed with two methods. Therefore the main motivation of this study is to apply a two-phase flow model that has no such limitation and investigate the electrolysis process in detail.

2. MATHEMATICAL MODEL

Consideration is given to hydrogen evolution in an electrochemical which have cell shown in Fig. 1. The cell consists of two electrodes namely cathode and anode 40 mm height, electrolyte and a membrane. The width of the channel between electrodes was set to 3 mm. The electrolyte is dilute solution of Na_2SO_4 and perfectly dissolved in the water. The hydrogen gas evolved at the cathode and oxygen at the anode after passing a current between two electrodes. Configuration and dimensions of the system considered mirrors the experimental set up of Boissonneau and Byrne [6] for direct comparison of the estimated results with experimental data.

To represent the flow behavior and heat transfer in the system a two phase mixture of liquid and gas is considered. The phases are assumed to share space in proportion to their existence probabilities such that their volume fractions sums to unity in the flow field. This can be expressed mathematically as:

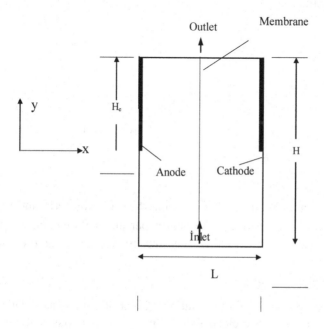

Figure 1. Schematic sketch of the system considered

$$\alpha_L + \alpha_G = 1 \tag{1}$$

where α_L and α_G are the volume fraction of liquid and gas respectively. The zone averaged quantities are obtained through solution of separate transport equations for each phase.

Within this framework the governing equations for boiling two phase flow can be expressed in Cartesian coordinate as follows:

Mass Conservation

$$\frac{\partial}{\partial x}(\rho_i \alpha_i u_i) + \frac{\partial}{\partial y}(\rho_i \alpha_i v_i) = M_{i-int} \tag{2}$$

where subscripts, i and j represent the phases and take the value of L,G in this problem. Subscripts L and G refer to liquid and gas phases, respectively, in this and subsequent formulations. The term on the right of the equation represents mass diffusion between two phases at electrolyte-gas interface.

x- momentum

$$\frac{\partial}{\partial x}(\rho_i \alpha_i u_i^2) + \frac{\partial}{\partial y}(\rho_i \alpha_i u_i v_i) = -\alpha_i \frac{\partial P}{\partial x} + F_r(u_j - u_i) + \frac{\partial}{\partial x}\left(\alpha_i \mu \frac{\partial u_i}{\partial x}\right) + \frac{\partial}{\partial y}\left(\alpha_i \mu \frac{\partial u_i}{\partial y}\right) \tag{3}$$

y- momentum

$$\frac{\partial}{\partial x}(\rho_i \alpha_i vu) + \frac{\partial}{\partial y}(\rho_i \alpha_i v_i^2) = -\alpha_i \frac{\partial P}{\partial x} + F_r(v_j - v_i) + \frac{\partial}{\partial x}\left(\alpha_i \mu \frac{\partial v_i}{\partial x}\right)$$
$$+ \frac{\partial}{\partial y}\left(\alpha_i \mu \frac{\partial v_i}{\partial y}\right) + F_b \tag{4}$$

F_r in both momentum equations is interface friction term and represents momentum exchange between the phases per unit volume and $F_b=\rho g$ is the buoyancy force where g being the gravity vector. μ is the molecular viscosity.

Ionic Species Transport

The liquid phase is dilute solution of Na_2SO_4 in water. This can be characterized as a binary electrolyte with no homogenous reactions taking place in the electrolyte fluid. In the alkaline solution overall homogenous reaction for hydrogen evolution at the cathode can be characterized as;

$$4H_2O + 4\bar{e} \rightarrow 2H_2 (gas) + 4OH^- \tag{5}$$

The anodic oxidation of hydroxide ions can be characterized as;

$$4OH^- \rightarrow O_2 (gas) + 4e^- + 2H_2O \tag{6}$$

The species transport in dilute solution can be calculated using Planck-Nernst law expressed as;

$$N_i = C_i U - D_i \nabla C_i - \frac{Z_i FD_i}{RT} C_i \nabla \phi \tag{7}$$

where C_i, D_i, Z_i are concentration diffusivity and charge number of species i respectively. ϕ represent the electric potential. U is the velocity vector for the solvent. The first, second and third term in Eq. 7 represent the convective, diffusive and migration contribution of mass flux respectively.

The current density i in the electrolyte can be calculated by Faraday's law

$$i = F (Z_1 N_1 + Z_2 N_2)$$ (8)

The current conservation can be expressed as;

$$\frac{\partial i}{\partial x} + \frac{\partial i}{\partial y} = 0$$ (9)

Numerical Technique

The nonlinear transport equations presented in previous section describe the species transport and gas hydrogen evolution in the electrolysis process. The equations are solved by an iterative; finite-domain solution procedure embodied is the PHOENICS computer code [7]. The multi-phase system of equations is solved by the inter-phase Slip Algorithm (IPSA) [7], involving the use of Partial Elimination Algorithm (PEA) to accelerate convergence of the volume fraction and scalar equations. The main advantage of the PHOENICS is to allow user incorporates source terms and transport equation which is not included in the main program with an appropriate coding. Therefore, electrochemical reaction, species conservation electric potential equations and source terms are added in this study.

3. RESULTS

Computed hydrogen and oxygen gas distribution is presented in Fig. 2. Hydrogen gas evolved at the cathode while oxygen gas formed at the anode sides. Gas fraction increases towards to top of the cell mainly because of accumulation of gas released at lower sections of the electrodes. Hydrogen gas fraction at the cathode side is considerable higher than that of oxygen gas at the anode. The difference between gas fraction of both gases result of rate electrochemical reaction. It is known that hydrogen release is twice as large as oxygen. It is also seen that both gases confined a region at the vicinity of the electrodes. The void fraction towards to the center of the channel vanishes. This result may be attributed to effect of convective flow which swept produced bubble away from the reaction zone.

Effect of current density on the gas release is shown in Fig. 3. It is seen that gas release rate increases at higher current densities as expected. At higher current densities gas penetrate at lateral direction mainly because of increase in lateral gas velocities. It is seen that the hydrogen release rate is not proportional to increase in the current density. This is result of accumulation of ion concentration of gas on the electrodes which adversely affects the chemical reaction rate.

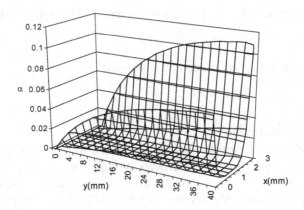

Figure 2. Estimated void fraction distribution along the electrode (i=2000A/m²).

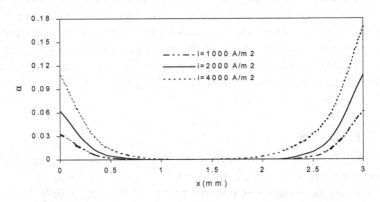

Figure 3. Effect of current density on the gas release (v_{in}= 0.08 m/s,y=H_e/2).

4. CONCLUSIONS

Gas evolution in a vertical electrochemical cell is investigated numerically with a two-phase flow model. The mathematical model involves solution of separate transport equation for gas and liquid phase. Therefore gas distribution is calculated from the first principle instead of employing empirical or semi-empirical relations.

It is found that gas release rate significantly affected from the current density. At higher current densities gas release increases however, when all conditions are held same, hydrogen release adversely affected from gas accumulated on the electrodes at high current densities. Therefore for an efficient electrolysis process gas released should be removed from the reaction sites to increase surface areas available for the reaction.

REFERENCES

[1] Voght H. Gas Evolving Electrodes. In Comprehensive Treatise of Electrochemistry. Plenum Press, New York 1983; 6:445.

[2] Ziegler D and Evans JW. Mathematical modeling of electrolyte circulation in cell with planar vertical electrodes. J. Electrochem. Soc. 1986; 103(3):567-576.

[3] Ishii M. Thermofluid dynamic theory of two-phase flow. Eyrolles,1975

[4] Dahlkild A A. Modeling The Two-Phase Flow and Current Distribution Along a Vertical Gas-Evolving Electrode. J. Fluid Mech 2001; 428:249

[5] Wedin R , Dahlkild A. A numerical and analytical hydrodynamic two-phase study of an industrial gas-lift chlorate reactor. ASME 1999; PVP-397(1):125-136.

[6] Boissonneau P, Byrne P. An Experimental investigation bubbles-induced free convection in small electrochemical Cell. J.Appl Electrochem 2000;30: 767-775.

[7] Rosten H, Spalding, DB. Phoenics Manual, CHAM, TR/100, London, 1986.

EFFECT OF IRON OXIDE ON STRUCTURE OF Y-STABILIZED ZIRCONIA CERAMIC

E.V. PASHKOVA, A.G. BELOUS, O.I. V'YUNOV, V.P. IVANITSKII

V.I.Vernadskii Institute of General & Inorganic Chemistry, National Academy of Science of Ukraine; 32/34, Palladina Ave., Kyiv-142, 03680, Ukraine, belous@ionc.kar.net

Abstract: The aim of present work is the investigation of structural peculiarities of zirconium oxide stabilized by combined dopant depending on chemical composition, synthesis conditions and heat treatment. It has been shown that solubility of iron in zirconium oxide increases with yttrium content. Increasing Y/Fe ratio in zirconia ceramics doped with the same total amount of doping oxides stabilizes the structure and inhibits low-temperature degradation. Nonequivalent sites of iron (III) in precipitated samples have been identified. Decrease in coordination number of iron ions in comparison with that of host cations stabilizes the structure and inhibits its degradation due to the increase in Me-O binding energy. It has been found that precipitated powders are composed of nanoparticles with the size of 10-20 nm.

Keywords: Stabilized Zirconia/Solid State Fuel Cell/Thermal Stability/Combined Dopant/ Mössbauer Spectra/Easy-Breaking Aggregates

1. INTRODUCTION

Fully stabilized zirconium dioxide is widely used as oxygen conductor for solid state fuel cell. One of the basic requirements to this material is the thermal stability of the structure. The most effective stabilizer for zirconium oxide is yttrium oxide. However, the structure of yttrium-stabilized ZrO_2 is susceptible to low-temperature degradation, which is caused by tetragonal-monoclinic transformation [1]. Therefore, the search for new stabilizers and modification of them are a topical problem. The choice of iron oxide as the third component of the system ZrO_2-Y_2O_3-Fe_2O_3 is dictated both by steric factors and by the possibility to lower the crystallization and sintering temperatures of ZrO_2-based solid solutions.

N. Sammes et al. (eds.), Full Cell Technologies: State and Perspectives, 279-285.
© 2005 *Springer. Printed in the Netherlands.*

The aim of the work is to investigate the structure features of zirconium dioxide stabilized by a complex dopant (Y_2O_3 and Fe_2O_3) as a function of the chemical composition, precipitation, heat treatment, and ageing conditions.

2. EXPERIMENTAL PROCEDURE

Samples for investigation were prepared by calcining $ZrO(OH)_2$-$Y(OH)_3$-FeOOH hydroxides precipitated from concentrated solutions of $ZrOCl_2$, $Y(NO_3)_3$, and $Fe(NO_3)_3$ with an ammonia solution by two methods: coprecipitation (CPH) and sequential precipitation (SPH). Using the SPH method, $Y(OH)_3$ was precipitated on coprecipitated $ZrO(OH)_2$ and FeOOH. Heat treatments were performed in a chamber furnace at 970-1470 K. Compositions corresponding to the formula $[1-(x+y)]ZrO_2 \cdot xY_2O_3 \cdot yFe_2O_3$, where x = 0, 0.01, 0.015, 0.02, 0.03 and x+y = 0.03 (*series I*); x = y = 0.02, 0.025, 0.03, 0.04 and x+y>0.03 (*series II*). The samples were investigated just after heat treatment and after storing them under atmospheric conditions for three years. The X-ray investigations were carried out on a DRON 4-07 diffractometer (Co K_{α} radiation). Mössbauer spectra of samples were recorded at room temperature on an electrodynamic spectrometer operating in the constant-acceleration mode with a ^{57}Co γ-quantum source in Rh matrix. The calibrations of the velocity scale in the magnetic and paramagnetic measurement ranges were performed by means of α-Fe and sodium nitroprusside respectively. The spectra were processed using the least-squares method. The micrographs were taken on a JEOL JEM 100 CX II electron microscope.

3. RESULTS AND DISCUSSION

Fig. 1 shows the temperature dependence of the content of *monoclinic* (*m*) modification of the samples from series I for different hydroxide precipitation methods, and different x:y ratios in as-heat-treated samples and after storage for 3 years. The optimal stabilization of zirconium dioxide for the compositions under study is observed at the ratio Y_2O_3:Fe_2O_3=0.02:0.01. The degree of stabilization at this ratio is higher than for the samples in which only Y_2O_3 was used as stabilizer (curves 3, 4). In the temperature range 1170-1570 K there is practically no stabilization of the high-temperature both *cubic* (*c*) and *tetragonal* (*t*) modifications of ZrO_2 in the case of complete substitution of Fe_2O_3 for Y_2O_3 (curve 1). The stabilizing effect of Fe_2O_3 on the structure of ZrO_2 at the above temperatures manifests itself only when Fe_2O_3 is present together with Y_2O_3 (curves 2-4). Fe_2O_3 also has a wholesome effect on the phase stability of ZrO_2 in time. During

storage, zirconium dioxide powders undergo the transformations
t-$ZrO_2 \rightarrow m$-ZrO_2 [2]. The results given in Fig. 1 indicate that the process of low-temperature degradation of zirconium dioxide in the ternary system
$0.97ZrO_2 \cdot xY_2O_3 \cdot yFe_2O_3$ (curves 3-3$'$) is slower than in the binary system
$0.97ZrO_2$-$0.03Y_2O_3$ (curves 4-4$'$).

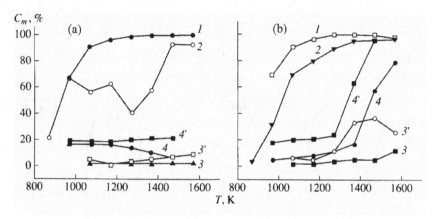

Figure 1. Content of monoclinic ZrO_2 modification (C_m) of $0.97ZrO_2 \cdot xY_2O_3 \cdot yFe_2O_3$
samples (x+y = 0.03) vs heat treatment temperature: (a) CPH, (b) SPH; x = 0 (1);
0.01 (2); 0.02 (3); 0.03 (4) - as-heat-treated samples; x = 0.02 (3'); 0.03 (4') - after storage
for 3 years.

Method of hydroxide precipitation affects greatly the degree of ZrO_2
stabilization, especially for samples after thermal treatment at $T > 1270$ K.
SPH technique results in the much lower degree of ZrO_2 stabilization
(Fig. 1b) as compared with CPH (Fig. 1a) for all $x : y$ ratios (curve 2-4)
excepting ratio $x : y = 0.02 : 0.01$ (curve 3). Soft, readily destructible
aggregates are formed after heat treatment of sequentially precipitated
hydroxides (in contrast to coprecipitated hydroxides) [2]. This make it
possible to produce fine ZrO_2 powders with a particle size of 10-20 nm
(Fig. 2) without disaggregation and/or milling. Therefore, in the case of
SPH it is very important to determine the composition of the powders, in
which both the high stabilization degree and the friability are retained after
heat treatment. The transformation t-$ZrO_2 \rightarrow m$-ZrO_2 during the ageing of
the samples obtained by coprecipitation (Fig. 1a, curves 3, 3$'$ and 4, 4$'$) is far
slower than in the sequentially precipitated samples (Fig. 1b, curves 3, 3$'$ and
4, 4$'$).

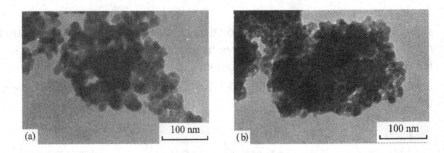

Figure 2. Micrographs of $Zr_{0.886}Y_{0.057}Fe_{0.057}O_{2-\delta}$ samples heat-treated at 870 K: (a) CPH, (b) SPH.

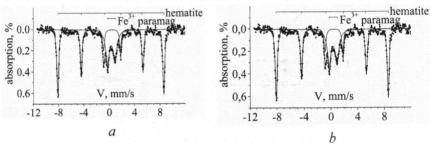

a   *b*

Figure 3. Mössbauer spectra of $Zr_{0.886}Y_{0.057}Fe_{0.057}O_{2-\delta}$ samples obtained in the magnetic measurement range: (a) CPH, (b) SPH.

Since for the composition $0.94ZrO_2 \cdot 0.03Y_2O_3 \cdot 0.03Fe_2O_3$ ($Zr_{0.886}Y_{0.057}Fe_{0.057}O_{2-\delta}$) complete stabilization of zirconium dioxide is observed at 1470 K for both methods of hydroxide precipitation (CPH and SPH), the above samples were chosen for the investigation of the local environment of Fe^{3+} ions by Mössbauer spectroscopy. X-ray investigations showed zirconium dioxide (CPH and SPH) to have a CaF_2-type structure with tetragonal distortion of lattice (space group P4/nmc (137)). The absence of other modifications (c-ZrO_2 and m-ZrO_2), evident from X-ray analysis, will make it possible to exclude the ambiguity of the interpretation of Mössbauer spectra.

The Mössbauer spectra of samples prepared by CPH and SPH, which were obtained in the magnetic range, are shown in Fig. 3. The spectra are represented by a superposition of a Zeeman splitting sextet and a quadrupole splitting doublet corresponding to high-spin Fe^{3+} ions in octahedral coordination [3]. The asymmetric character of absorption lines could be due to the presence of several nonequivalent sites of Fe^{3+} ions, whereas high values of absorption line full width at half-maxima (FWHM) could be ascribed to the inhomogeneity of the cation environment of Fe^{3+} ions.

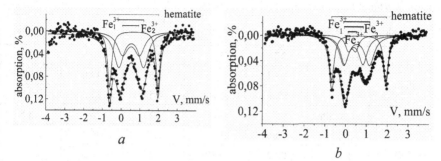

Figure 4. Mössbauer spectra of $Zr_{0.886}Y_{0.057}Fe_{0.057}O_{2-\delta}$ samples obtained in the paramagnetic measurement range: (a) CPH, (b) SPH.

To reduce experimental errors and to explain the character of the asymmetry of the doublet lines, we examined spectra in the paramagnetic measurement range, in which the resolvability of spectra is over three times higher than that in the magnetic range. This allowed the precise determination of the components of the Fe^{3+} spectra (Fig. 4). The parameters of the resolved components are listed in the Table. The number of resolved doublets makes it possible to identify two and three nonequivalent sites of Fe^{3+} ions in the solid solutions $Zr_{0.886}Y_{0.057}Fe_{0.057}O_{2-\delta}$ obtained by CPH and SPH respectively. The Mössbauer spectra parameters of samples obtained by SPH do not practically change after storage for three years in air (Table, samples 2 and 2*). This denotes the absence or considerable suppression of the low-temperature degradation of the structure of stabilized zirconium dioxide in the ZrO_2-Y_2O_3-Fe_2O_3 system in comparison with the binary ZrO_2-Y_2O_3 system.

Table. Parameters of Mössbauer spectra of $Zr_{0.886}Y_{0.057}Fe_{0.057}O_{2-\delta}$ samples obtained in the paramagnetic measurement range.

Sample	Phase, ion	IS, mm/s	QS, mm/s	FWHM, mm/s	S, %	S_o, %
1 (CPH)	Hematite	0.76	2.60	0.27	30.5	72.5
	Fe_1^{3+}	0.64	1.33	0.54	42.0	16.6
	Fe_2^{3+}	0.63	0.84	0.53	27.5	10.9
2 (SPH)	Hematite	0.75	2.60	0.25	26.6	68.5
	Fe_1^{3+}	0.66	1.21	0.55	36.5	15.7
	Fe_2^{3+}	0.52	0.92	0.46	26.2	11.2
	Fe_3^{3+}	0.43	0.40	0.37	10.7	4.6
2* (SPH, after three years)	Hematite	0.75	2.61	0.29	29.8	71.8
	Fe_1^{3+}	0.66	1.18	0.52	34.0	13.7
	Fe_2^{3+}	0.49	0.94	0.62	26.4	10.6
	Fe_3^{3+}	0.40	0.44	0.42	9.8	3.9

Note: H_{eff} = effective magnetic field; IS = isomer shift relative to α-Fe, QS = quadrupole splitting; FWHM = absorption line full width at half-maxima; S = relative area of component; S_o = areas of components reduced with respect to the six-line spectrum of hematite. IS's have been reduced with respect to sodium nitroprusside. Measurement error IS, QS and FWHM: $\pm$ 0.03 mm/s, S: $\leq$10 %.

When interpreting Mössbauer spectra, the value of quadrupole splitting (QS) can be used as a measure of the distortion of coordination polyhedra, and deviation of their symmetry from cubic symmetry. The value of isomer shift (IS) can be associated with the coordination of Fe^{3+} ions.

Taking into account the results of [4], where clusters were found using high-resolution electron microscopy, the presence of several doublets with different areas in the Mössbauer spectra of $Zr_{0.886}Y_{0.057}Fe_{0.057}O_{2-\delta}$ samples can be attributed to the cluster character of iron distribution in the solid solution ZrO_2-Y_2O_3-Fe_2O_3. The doublets due to Fe_1^{3+}, Fe_2^{3+}, Fe_3^{3+} can be attributed to the presence of Fe^{3+} sites with the coordination number of six (IS = 0.63-0.66; QS = 0.84-1.33), five (IS = 0.49 - 0.52; QS = 0.92-0.94) and four (IS = 0.40-0.43; QS = 0.40-0.43) respectively in $Zr_{0.886}Y_{0.057}Fe_{0.057}O_{2-\delta}$. It is evident that the change in the local environment of cations and Me–O bond energy corresponding to the decrease in the coordination number (CN) of Fe^{3+} ions as compared with that of Zr^{4+} in the basic structure (c, t-ZrO_2, CN 8; m-ZrO_2, CN 7) stabilizes high-temperature zirconium dioxide modifications and considerably suppresses (depending on the Zr/Y/Re ratio) low-temperature degradation of the structure.

4. CONCLUSION

The polymorphic composition of ZrO_2-based materials has been determined in the series of samples, which correspond to the formula $[1-(x+y)]ZrO_2 \cdot xY_2O_3 \cdot yFe_2O_3$, in the temperature range 620-1570 K. It has been found that at the same molar ratio ZrO_2 : doping oxides, the degree of ZrO_2 stabilization increases, and the low-temperature degradation process is suppressed by the partial substitution of Fe^{3+} for Y^{3+}. Nonequivalent sites of Fe^{3+} ions have been identified: two sites with octahedral coordination (CPH), and three sites with octa-, penta- and tetrahedral coordination (SPH). The possibility of cluster distribution of Fe^{3+} ions as well as the dependence of the number of vacancies on synthesis conditions have been shown.

REFERENCES

[1] Lepisto T.T., Mantyla T.A. A model for structural degradation of Y-TZP ceramics in humid atmosphere. Ceram. Eng. and Sci. Prod., 1989, v.10, № 7-8, p.1362.

[2] Makarenko A.N., Belous A.G., Pashkova Y.V. Structure Formation and Degradation Partially Stabilized Zirconium Dioxide, J. Europ. Ceram. Soc., 1999, v.19, p. 945-947.

[3] Shirane G., Cox D.E., Ruby S.L. Mössbauer study of isomer shift, quadrupole interaction and hyperfine field in several oxides containing Fe^{57}, Phys. Rev., 1962, v.125, № 4, p.1158-1165.

[4] Olkhovik G.A., Naumov I.I., Velikokhatnyi O.I., Aparov N.N., Electronic structure and optical properties of yttrium-stabilized ZrO_2, Inorg. Mater., 1993, v.29, № 5, p.636-640.

STRUCTURE, THERMAL EXPANSION AND PHASE TRANSITION IN $La_{0.92}Sr_{0.08}Ga_{0.92}Ti_{0.08}O_3$ SINGLE CRYSTAL

YE. PIVAK[1], L. VASYLECHKO[1], A. SENYSHYN[1], M. BERKOWSKI[2], M. KNAPP[3]

[1] *"L'viv Polytechnic" National University, Lviv Ukraine*
[2] *Institute of Physics Polish Academy of Sciences, Warsaw, Poland*
[3] *Darmstadt University of Technology, Germany*

Abstract: The structure of mixed $La_{0.92}Sr_{0.08}Ga_{0.92}Ti_{0.08}O_3$ (LSGT-8) single crystal, obtained by Czochralsky method, have been investigated in a wide temperature range of 12–1203 K by means of *in situ* high resolution powder diffraction technique using synchrotron radiation as well as DTA/DSC analysis. Both, diffraction and calorimetric data indicate a first-order phase transition from orthorhombic (space group *Pbnm*) to rhombohedral (space group *R-3c*) structure near the room temperature ($T_c \approx 303$ K). Lattice parameters and atomic coordinates of low- and high-temperature phases have been refined in the whole temperature range investigated. It was established that the critical temperature of the phase transition in the system $(1-x)LaGaO_3$-$(x)SrTiO_3$ in the concentration range of $x = 0 \div 0.08$ falls down linearly according to the empirical relation: $T_c \approx 419(9) - 15.4(17)*x$.

Keywords: $LaGaO_3$-Based Perovskites/Thermal Expansion/Phase Transitions

1. INTRODUCTION

$LaGaO_3$-based compounds, doped with alcaline-earth cations and Mg show very high oxygen-ion conductivity and have a great importance for applications in high-temperature electrochemical devices, such as solid oxide fuel cells [1,2]. Perovskite-like $LaGaO_3$ lattice tolerates a substitution for

N. Sammes et al. (eds.), Full Cell Technologies: State and Perspectives, 287-293.
© *2005 Springer. Printed in the Netherlands.*

both A- and B-cation sites in a wide scale, thus giving a possibility to improve their transport and other properties.

In the present work we report the results of the study of thermal behaviour of $La_{0.92}Sr_{0.08}Ga_{0.92}Ti_{0.08}O_3$ (LSTG-8) structure in a wide temperature range of 12–1203 K. Structural changes, occurred at the phase transition and the thermal evolution of the lattice parameters are analysed and discussed.

2. EXPERIMENTAL

Single crystal of LSGT-8 was grown by Czochralsky method. Details of the growth procedure are described in [3]. High-resolution powder diffraction (HRPD) experiments using synchrotron radiation have been performed at the powder diffractometer at beamline B2, HASYLAB, DESY. The low temperature (LT) behaviour of the LSGT-8 structure has been studied in the temperature range 12–300 K by using He closed-cycle cryostat and NaI scintillation counter. Full diffraction patterns were collected at 12 K (2θ - range 16°–76°, $\Delta 2\theta$ = 0.006°, λ= 1.11952 Å) and 303 K (2θ - range 16°–67°, $\Delta 2\theta$ = 0.004°, λ= 1.12435 Å). In order to obtain the thermal behaviour of the lattice parameters, the measurements of the thermal evolution of 12 characteristic reflections were carried out in the temperature range 12–300 K. High temperature (HT) structure investigations in the temperature range 300 – 1250 K were performed applying HRPD technique and synchrotron radiation by using a newly developed on-site readable image plate detector (*OBI*) [4]. A 0.3 mm quartz capillary was filled with powdered LSGT-8 specimen and mounted inside a STOE furnace in Debye-Scherrer geometry, equipped with a Eurotherm temperature controller and a capillary spinner. Full patterns were collected over a 2θ - range of 15° to 86° in the temperature range 300 – 1200 K with the temperature step of 100 K. The wavelength was determined to be 1.1414 Å based on 6 reflections of Si standard NIST640b.

Full profile Rietveld method was used for the refinement of the lattice parameters, positional and displacement parameters of atoms. Data evaluation was performed by using the WinCSD program package [5].

DTA/DSC examination of the LSGT-8 crystal was carried out using STA 409 calorimeter (NETZSCH, Selb, SiC-furnace, heating rate 10.0 K/min) in a flowing Ar atmosphere.

3. RESULTS AND DISCUSSION

X-ray phase analysis of the sample performed at room temperature (RT) ensured a perovskite-type structure of the crystal; reflections of other phases were not revealed. Deformation and partial splitting of the reflections at the diffraction pattern indicated the deviation from ideal perovskite structure. However, the attempts to refine the structure in one of the possible space groups (e.g. *Pbnm, R-3c, P2/c, Imma*) were unsuccessful. Therefore, the HRPD technique and synchrotron radiation were applied for the structural characterization of LSGT-8 crystal. Extremely high beam collimation (typical FWHM values are 0.01 - 0.02 degrees) and good signal-to-noise ratio allowed us to detect the line splitting, which could not be observed in conventional X-ray pattern. Indexing of the diffraction maxima has shown, that some of them could be indexed in orthorhombic (Or) lattice, whereas another ones - in rhombohedral cell. Such simultaneous existence of two perovskite-like phases in the La$_{0.92}$Sr$_{0.08}$Ga$_{0.92}$Ti$_{0.08}$O$_3$ sample could be explained by two possible reasons: (1) heterogeneity of a chemical composition of the sample and accordingly presence of two types of domains with slightly different cation ratio; (2) existence of the structural phase transition in the crystal near RT. *In situ* HRPD experiments, performed in the temperatures range of 270 - 320 K, clearly prove the homogeneity of a chemical composition of LSGT-8 and showed the existence of structural phase transition at temperature ~ 303 K (fig. 1). On heating from RT the magnitudes of the Or peaks decrease and disappear completely above 308 K. Only reflections of Rh phase are presented at the patterns above 308 K. In contrast, the cooling of the sample led to the decrease of the intensities of Rh peaks with their further disappearing below 286 K.

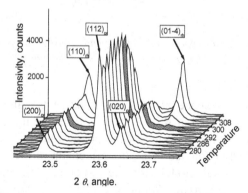

*Figure 1. Temperature evolution of reflections (200)+(112)+(020) of **Or** phase and (110)+(01-4) reflections of **Rh** one in LSGT-8 structure at heating from 277 K to 315 K.*

In the temperatures range of 292 - 315 K the co-existence of both **Or** and **Rh** phases is observed. As it was recently reported [6, 7], the simultaneous co-existence of LT and HT phases in some temperature regions of the *Pbnm* - *R-3c* phase transition is typical for La$_{1-x}$RE$_x$GaO$_3$ solid solutions.

DTA/DSC examination of the crystal showed an endothermic signal on heating indicating a phase transformation at 306 K. The transition was found to be reversible and the corresponding exothermal peak has been observed on cooling at 297 K. Both *in situ* diffraction data and DTA/DSC one revealed that phase transition occurs with small hysteresis in 5-8 K. The decrease of the temperature of the *Pbnm* - *R-3c* phase transition from 421 K in LaGaO₃ to 303 K in LSGT-8 is explained in terms of the reduction of the perovskite structure deformation, caused by partial substitution of Sr and Ti for the La and Ga sites, respectively. It was established, that the critical temperature of the *Pbnm* - *R-3c* phase transition in the (1-*x*)LaGaO₃ – *x*SrTiO₃ pseudo-binary system falls down linearly according to the empirical relation: $T_c \approx 419(9) - 15.4(17)*x$ in the concentration range of $x = 0÷0.08$ (fig. 2).

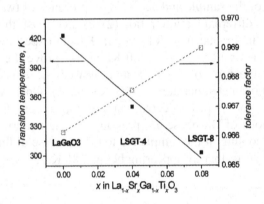

Figure 2. Concentration dependencies of the phase transition temperature and tolerance factors in the (1-x)LaGaO₃–xSrTiO₃ pseudo-binary system.

Simultaneous presence of the reflections of LT and HT phases at the HRPD pattern collected at RT allowed us to refine the crystal structures of both phases just in the the point of the phase transformation. Reflections of HT Rh phase were excluded from the refinement of the LT Or structure, and *vice versa*. As initial models for the structure refinement of Or and Rh structures in space groups *Pbnm* and *R-3c*, the parameters of atoms in LaGaO₃ structure at 300 K and 673 K were chosen [8]. Graphical results of the Rietveld refinement of LT and HT modifications of LSGT-8 structure at 303 K are presented in fig. 3.

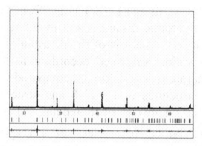 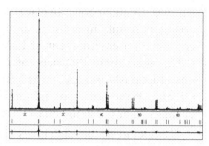

Figure 3. Graphical results of the Rietveld refinement of Or and Rh modifications of LSGT-8 structure at 303 K (HRPD, λ=1.1197 A).

In situ powder diffraction experiments performed in a wide temperature range of 12 – 1200 K show, that LSGT-8 structure is Rh in the temperature range 12- 303 K, and HT structure remains Or at least up to 1220 K. Refined structural parameters of LT and HT phases of LSGT-8 at 12 K, 303 K and 1203 K are summarized in table 1.

Table 1. Refined structural parameters of LT and HT phases of LSGT-8

		Pbnm		R-3c	
		12 K	303 K	303 K	1203 K
a, Å		5.51865(4)	5.52769(3)	5.52430(5)	5.5665(1)
b, Å		5.48558(3)	5.49167(3)		
c, Å		7.76107(5)	7.76806(5)	13.3709(1)	13.5600(3)
V, Å³		234.951(4)	235.809(4)	353.38(1)	363.87(2)
x		-0.0047(3)	-0.0028(5)	0	0
y	La/Sr,	0.0157(2)	0.0128(3)	La/Sr, 0	0
z	4c	¼	¼	6a ¼	¼
B(is/eq)		0.38(3)	0.79(5)	0.90(8)	1.39(7)
x		½	½	0	0
y	Ga/Ti,	0	0	Ga/Ti, 0	0
z	4b	0	0	6b 0	0
B(is/eq)		0.52(7)	0.80(10)	0.84(14)	0.96(13)
x		0.60(2)	0.066(3)	0.562(3)	
y	O1,	0.490(2)	0.497(4)	O, 0	0
z	4c	¼	¼	18e ¼	¼
B(is/eq)		0.4(3)	2.6(7)	2.8(8)	2.3(9)
x		-0.261(3)	-0.273(3)		
y	O2,	0.264(3)	0.272(4)		
z	8d	0.0418(10)	0.033(2)		
B(is/eq)		0.6(2)	2.6(5)		

The main structural changes occurred in LSGT-8 at the phase transition become apparent in anion sublattice and result in the different character of the octahedra tilting in LT and HT structures. In terms of Glaser notations

[9] the tilt system is changed from $a^- a^- c^+$ in LT Or structure to $a^- a^- a^-$ in HT Rh one. Simultaneous displacement of oxygen and La(Sr) atoms from their positions results in redistribution of interatomic distances and in change of coordination environment of La(Sr) cations. Analysis of the distribution of R-Ga interatomic distances and their ratio, performed according to [10] unambiguously proves that HT Rh structure is less deformed compared with Or one. Peculiarities of LT and HT structures of LSGT-8 and structural changes occurred at the phase transformation are described in detail in [11].

Temperature dependencies of the lattice parameters of LSTGO-8, normalized to the perovskite-like cell, have nonlinear and anisotropic character (fig. 4). A small decrease in the cell volume ($\Delta V = -0.09$ %) occurs at the phase transition.

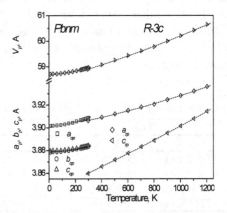

Figure 4. Temperature behaviour of the lattice parameters of the LSTG-8.

ACKNOWLEDGMENTS

The work was supported in parts by WTZ (UKR 01/12), Ukrainian Ministry of Education and Science (Projects "Cation" and M/85-2003), Polish Committee for Scientific Research (Grant N 7T08A 00520) and ICDD Grant-in-aid program.

REFERENCES

[1]   T. Ishihara, H. Matsuda, Y. Takita, J. Em. Chem. Soc. 116 (1994) 3801.

[2]   M. Feng, J. B. Goodenough, Eur. J. Solid State Inorg. 31 (1994) 633.

[3]   S. M. Kaszmarek, Cryst. Res. Technol 36 (2001) 8.

[4] M. Knapp, V. Joco, C. Bähtz, H.H. Brecht, A. Berghëuser, H. Ehrenberg, H. von Seggern, H. Fuess, Nucl. Instrum. Meth. A 521 (2004) 565.

[5] L.G. Akselrud, P.Yu. Zavalij, Yu. Grin, V.K. Pecharsky, B. Baumgartner and E. Woelfel, Materials Science Forum 133-136, (1993) 335.

[6] L. Vasylechko, R. Niewa, H. Borrman, M. Knapp, D. Savytskii, A. Matkovskii, U. Bismayer, M. Berkowski, Solid State Ionics 143 (2001) 219.

[7] L. Vasylechko, D.Savytskii, A. Matkovskii, M. Berkowski, M. Knapp, U. Bismayer, J. Alloys Compds 328(1-2) (2001) 264.

[8] C.J. Howard, B.J. Kennedy, J. Phys.: Condens. Matter 11 (1999) 3229.

[9] A.M. Glazer, Acta Crystalogr. B28 (1972) 3384.

[10] L. Vasylechko, A. Matkovskii, D.Savytskii, A.Suchocki, F. Wallrafen, J. Alloys Compds 292(1-2) (1999) 57.

[11] A.T. Senyshyn, Ye.V. Pivak, L.O. Vasylechko, A.O. Matkowskii, M. Berkowsky Visnyk NU "Lvivska Politekhnika". Elektronika. 482 (2003) 13 (in Ukraine).

INVESTIGATION OF HYDROGEN INFLUENCE ON STRUCTURE OF AMORPHOUS ALLOYS Mg-Cu-Y AND Mg-Ni-Y BY NEW METHOD OF DIFFRACTION FROM ATOMIC COORDINATE SPHERES

O.P. RACHEK, A. GEBERT*, M.P. SAVYAK, I.V. UVAROVA

Frantcevych Institute for Problems of Materials Science, National Academy of Science of Ukraine; Zirconia Ukraine Ltd. ,3 Krzhyzhanivs'koho Str., Kyiv-142, 03680 Ukraine.

Leibniz-Institute for Solid State and Material Research (IFW) Dresden, P.O. Box 270016, Dresden, Germany, gebert@ifw-dresden.

Abstract: This work is based on using Ehrenfest's formula in processing X-ray patterns from amorphous alloys. The radius of the first coordinate sphere (CSR) and the content of the corresponding amorphous phase in an alloy were calculated from the position and intensity of the first peak in the pattern. A decrease in the peak intensity down to its complete disappearance under hydrogen charging was attributed to the appearance and growth of a certain amorphous phase, which may contain a set of various CSR and therefore reflection from the first coordinate sphere is not detectable. We named this phase as "very disordered". For Ni-containing alloys, CSR did not change under hydrogen charging and was equal to the average CSR in the corresponding crystalline alloy with a body centered cubic (BCC) structure. For Cu-containing alloys, CSR changed from the value of the average CSR in the corresponding BCC crystalline alloy to that in the face-centered cubic (FCC) alloy. Under hydrogen charging the FCC phase disappeared completely, thus the amount of the very disordered phase reached 100 %. With further hydrogen charging two amorphous phases precipitated .

Keywords: Amorphous Alloy/Charging/Diffraction/Peak Intensity/Coordinate Sphere

N. Sammes et al. (eds.), Full Cell Technologies: State and Perspectives, 295-300.
© 2005 Springer. Printed in the Netherlands.

Mg and its intermetallic combinations in alloys are known to be efficient hydrogen sorbents [1-3]. To increase the capacity and to decrease the temperature of H sorption and desorption, the metastable state of material may be used. Amorphous alloys are considered to be promising materials for this purpose [4].

In this study $Mg_{65}Cu_{25}Y_{10}$, $Mg_{50}Ni_{30}Y_{20}$ and $Mg_{63}Ni_{30}Y_7$ glassy ribbons with a width of 5 mm and thickness of 30 μm were prepared by single-rolled melt spinning in an Ar atmosphere. Ribbon samples 20 mm long were electrochemically charged with H at room temperature in 0.1 N NaOH electrolyte at a cathodic current density of 1 mA/cm^2. The H content was determined using the hot extraction method. The alloys were investigated by XRD (Philips 1050 diffractometer, Co Kα radiation), TEM (Philips CM 20FEG microscope) and SEM in the Leibniz-Institute for Solid State and Material Research (Germany) [5].

For calculation of coordinate spheres radii (CSR) Ehrenfest's formula [6] was used :

$$2r \times \sin t = 1.23 \times f \qquad (1)$$

where r is the CSR, t is the reflected X-ray angle, f is the X-ray wave length (f=0.17902 nm for Co Kα radiation). This method of calculation was first used by Yu.I. Sozin for diamond and other crystalline materials [7]. A.F. Skryshevskiy [8] established two conditions under which formula (1) is valid: amorphous material or liquid must consist of one element and the material structure is almost close packed. In our case, the first condition is not satisfied. That is why this research was carried out with some assumption: in the alloys studied the "X-ray sizes" of different atoms are approximately identical except for H, which is not "visible"for X-rays. In the first approximation the second condition is satisfied.

In Figs.1, 2 XRD patterns from these alloys are shown. As seen, the diffuse maximum shifts depending on the amount of the charged hydrogen. The H-free alloys have an amorphous structure: diffraction from them is characterized by diffuse peaks, related to the reflection angles from the first coordinate sphere. There are also narrow small peaks from Y_2O_3, which form during melt spinning. The diffuse peaks are nonsymmetrical. This is connected with the existence of two almost close- packed structures: BCC and FCC.

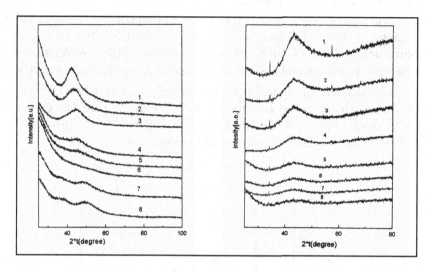

Figure 1. XRD patterns from the alloy
$Mg_{65}Cu_{25}Y_{10}$ *with H, at. %: 1) 0 ; 2) 11.8; 3)*
31.4; 4) 43.9; 5) 48.7; 6) 51.9; 7) 59.4; 8)
62.7.

Figure 2. XRD patterns from the alloy
$Mg_{50}Ni_{30}Y_{20}$ *with H, at. %: 1) 0; 2) 7.9; 3)*
16.6; 4) 26.4; 5) 36.5; 6) 40.2; 7) 42.9; 8)
47.

To compare CSR for BCC and FCC structures in the corresponding crystalline alloys which are characteristic for Mg, Cu, Ni, Y, the average CSR was calculated for each alloy. The results of this comparison for H-free alloys are presented in Table 1.

Table 1. The comparison of CSR values (nm x10) between the theoretical data for crystals with BCC and FCC structures and the experimental data for H-free amorphous alloys

Alloy	Theoretical (BCC)	Theoretical (FCC)	Experimental
$Mg_{65}Cu_{25}Y_{20}$	2.904		3.038
		2.370	---------
$Mg_{50}Ni_{30}Y_{20}$	3.037		2.990
		2.482	2.421
$Mg_{63}Ni_{30}Y7$	2.981		2.930
		2.434	2.504

In Refs. [9, 10] hydrogen adsorption was shown to cause a peak shift to the smaller angles, which corresponds to an increase in CSR. The diffraction peaks broadening was explained by hydrogen atom interstitial penetration in pores and defects in amorphous alloys. In this study the amorphous alloy $Mg_{65}Cu_{25}Y_{10}$ charging decreases the diffuse peak intensity and shifts it to the right. For the H-free alloys $Mg_{50}Ni_{30}Y_{20}$ and $Mg_{63}Ni_{30}Y_7$ XRD peaks are nonsymmetrical. This is connected with the presence of CSR sizes corresponding to the FCC structure. The hydrogen charging of these alloys decreases diffuse peak intensity too.

With using XRD data (Figs. 1, 2) and equation (1) we calculated a change in the amorphous phase content and CSR for BCC and FCC structures (taking as 100% the area under the diffuse peaks from H-free amorphous alloys), and CSR in alloys depending on the hydrogen content.

The amount of material with a set of various CSR increases with hydrogen charging and is equal to 100% minus the percentage of material with determined CSR. Let us name this material "very disordered". The results of the calculation are presented in Figs. 3, 4.

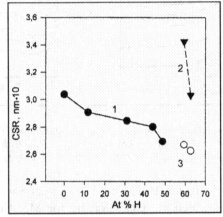

Figure 3. The change in CSR against the hydrogen content in the Cu-containing alloy with CSR corresponding to the structures: 1- BCC and FCC , 2- YH_3, 3- Cu_2Mg.

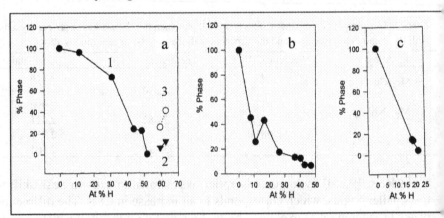

Figure 4. The change in the amorphous phase amount against the hydrogen content for alloys: a) Cu-containing alloys with CSR corresponding to the structures: 1-BCC, FCC, 2-YH3, 3- Cu_2Mg ; b) $Mg_{50}Ni_{30}Y_{20}$; c) $Mg_{63}Ni_{20}Y_7$.

For the Cu-containing alloy two diffraction peaks rise when the content of H is more than 53 at. %, which correspond to two formed amorphous phases. The CSR of one of them is close to that of the nanocrystalline precipitate Cu_2Mg and the CSR of the other phase is close to that of YH_3 (Fig. 3, 4).

When the H content in the alloy $Mg_{50}Ni_{30}Y_{20}$ increases, the CSR of the amorphous phase does not change but its amount decreases and this drop is more abrupt compared with the Cu-containing alloy. The alloy $Mg_{63}Ni_{30}Y_7$ is analogous to the Cu-containing alloy. Its CSR does not change with H charging and is close to that for the $Mg_{50}Ni_{30}Y_{20}$ alloy. The amorphous phase content decreases more quickly in $Mg_{63}Ni_{30}Y_7$ than in $Mg_{50}Ni_{30}Y_{20}$. Ni is a transition metal and has stronger covalent bonds than Cu and therefore it forms CSR sizes close to those in the FCC structure.

- Electrochemical charging with hydrogen increases disordering in Mg-based amorphous alloys. Herein a very disordered amorphous phase forms, which produce no reflection from the coordinate sphere.

- Under hydrogen saturation of Cu-containing alloys the values of CSR change from the average CSR in the corresponding BCC-structure alloys to those in FCC-structure alloys, with the amorphous phase content decreasing down to zero, that is, the very disordered phase content increasing up to 100 %. Under further saturation with hydrogen two amorphous phases form.

- The alloys $Mg_{65}Cu_{25}Y_{10}$, $Mg_{50}Ni_{30}Y_{20}$ and $Mg_{63}Ni_{30}Y_7$ can dissolve about 22, 44 and 60 at. % H, respectively. The less amount of dissolved hydrogen in Ni-containing alloys may be attributed to stronger covalent bonds.

- Using Ehrenfest's formula is efficient in studying amorphous material structures.

REFERENCES

[1] Stander S.M. Kinetics of decomposition of magnesium hydride. J.Inorg. Chem. – 1977. – 39,№2.- p. 221 – 223.

[2] Reilli J.J., Wiswall R.H. The reaction of hydrogen with alloys of magnesium and copper . Inorganic chemistry, - 1967, v.6, 2220 – 2223.

[3] Seiler A., Schlapbach L., von Waldrich Th., Shaltiel D., Stucki F. Surface analysis of Mg_2Ni-Mg, Mg_2Ni, Mg_2Cu. J. Less-Com. Met., 1980, v.73, p.193-199.

[4] Savyak M, Hirnyj S. , Bauer H.-D., Uhlemann M., Gebert A.. W., Eckert J., Schultz L.. Stability and electrochemical properties of $Mg_{65}Y_{10}Cu_{25}$ metallic glass. J of Alloys and Compounds. 2004, v. 364, p.229-237.

[5] Gebert A., Wolff U., John A., Eckert J. Scripta Mater. 2000, v.43, 279.

[6] Ehrenfest. Proc. Ams. Acad. 17, 1132 (1915).

[7] Sozin Yu.I. Diffraction from coordination spheres, Crystallography 1994, v. 39, №1, p. 10-18 in Russian.

[8] Skryshevsky A,F, Structural analysis of liquids and amorphous bodies, Vysshaya shkola, 1980. 327p, in Russian.

[9] Suzuki K.,Fudzyumori X Hasimoto. Amorphous metals, M.. Metallurgia . 1987, 328 p, in Russian.

[10] Amorphous metallic alloys. M., Metallurgia, 1987, 584p.,in Russian

THE YSZ ELECTROLYTE SURFACE LAYER: EXISTENCE, PROPERTIES, AND EFFECT ON ELECTRODE CHARACTERISTICS

S. N. SHKERIN

Institute of High-Temperature Electrochemistry of Russian Academy of Science, Ekatherineburg 620016, Rissia

Abstract: It was shown that surface of a perfect single crystal of YSZ had a polycrystalline layer with a different (low-symmetry) structure. Reasons for existence of the Surface Layer of the YSZ, the change of electrolyte properties, and effect on electrode characteristics are discussed in this review.

Key words: Oxide Electrolytes/Yttria Stabilized Zirconia/Surface Layer/Un-Uniform Properties Of Electrolyte/Solid Oxide Fuel Cells

Complex FCC oxides of the fluorite type represent oxygen-conduction solid electrolytes (SOE's). They comprise a typical class of materials for the manufacture of sensors of oxygen activity in complex gas mixtures, oxygen pumps, electrolyzers and high-temperature fuel elements. These materials are based on doped oxides of cerium and thorium, zirconium and hafnium, and bismuth oxide. Materials based on **zirconium oxide**, for example, yttrium stabilized zirconia (YSZ) are the most known and studied among them. This fact is explained both by their processibility and a wide spectrum of practical applications and by the possibility to conduct studies on single crystals, which have the commercial name "fianites" and are used in jewelry.

N. Sammes et al. (eds.), Full Cell Technologies: State and Perspectives, 301-306.
© *2005 Springer. Printed in the Netherlands.*

1. EXISTENCE

Single crystals, which we studied [1-3], were perfect: mosaic blocks were as large as 1.5 cm, while misorientation angles were small. However, diffraction patterns of samples of single crystals contained small reflections at angles smaller than the position of the <111> reflection, which is the lowest-index reflection among those allowed for the fluorite structure. Usually this fact is interpreted as the presence of more than one phase in these materials [4]. The examination by the method of small-angle X-ray topography proved [1,2] that two phases were unusual: the second phase formed a surface layer. If hard radiation and/or large incidence angles were used, X-ray patterns showed that the sample was a perfect single crystal. If soft radiation hit the crystal at a small angle [1,2], Debye rings were observed in addition to Laue spots of a single crystal. This observation proved unambiguously that the surface of a perfect single crystal had a polycrystalline layer with a different (low-symmetry) structure.

2. PROPERTIES

Whatever the reason for appearance of the surface layer, the change in the material structure should be reflected in **the change of its properties**. One may assert that properties of some surface layer do differ from properties of the crystal bulk:

- Conductivities of the electrolyte single crystal and the thin film differ not only by their values, but also by the shape of the temperature dependence [5]. Therefore, different specific values cannot be explained by erroneous measurements of the film thickness. Moreover, the conductivities themselves differ by more than one order of magnitude.

- The oxygen diffusion coefficient, which was determined by the method of isotope exchange for thin electrolyte layers [6],

$$D_0 \ [cm^2 s^{-1}] = 0.123 \exp(-160 \ [kJ \cdot mole^{-1}]/RT,$$

is 1.5 order of magnitude smaller than the diffusion coefficient in the bulk of the crystal [7]

$$D_0 \ [cm^2 s^{-1}] = 1 \ 10^{-3} \exp(-86 \pm 2 \ [kJ \cdot mole^{-1}]/RT).$$

Activation energies differ too. Recently this result was reproduced independently [8].

- The forbidden gap, which was measured by passage of light with different wavelengths through the single crystal, was 2.4-2.8 eV [9] depending on temperature. The forbidden gap, which was estimated by surface methods [10], was 5.2 eV and was independent of temperature.

3. EFFECT ON ELECTRODE CHARACTERISTICS

It was interesting to analyze how the surface layer, whose structure and properties differed from those of the SOE bulk, influenced electrode processes in cells with YSZ as the electrolyte. A set of experimental facts about the oxygen reaction in SOE systems was considered in [11]. The largest contribution to the development of ideas about electrode processes in SOE systems was made by research teams headed by S. Karpachev and M. Perfiliev (Sverdlovsk/Ekaterinburg, Russia) [12,13] and M. Kleitz (Grenoble, France) [14,15]. The overwhelming majority of results can be described in terms of the **base model** [11], which was developed by those two teams. However, there is a set of experimental facts, which cannot be described by this model in principle:

- The dependence of the measured resistance of the cell electrolyte on oxygen activity in the gas phase. This dependence, which was first revealed by E. Shouler [16], led to the notion "expansion of the three-phase region". However, if we estimate numerically the value of the electron conduction, which would provide the observed change of the measured resistance of the electrolyte, it becomes apparent that the observed dependences cannot be explained in terms of the base model. This expansion of the three-phase region was studied recently using model electrodes and the metal/electrolyte contact geometry was determined well [17-20].

- One more group of experimental facts is connected with the manifestation of "memory effects", the effect of treatment by current, by reducing gas, etc. [21-27]. For example, when a model mesh electrode is used [26], the polarization resistance of the electrode in the initial state depends on the orientation of the face of the electrolyte single crystal. It increases in the series of <100>, <110>, and <111> orientations. A similar series of changes of the oxygen exchange on the surface of the ZrO_2-Y_2O_3 (9.5 mole %) electrolyte depending on the orientation of the surface of the electrolyte single crystal was revealed recently [28] in experiments on optical absorption of light.

Treatment with current improved reproducibility of measurement results, eliminated the dependence on the orientation of the single crystal face [1, 29], and led to a considerable decrease in the polarization resistance. The decrease in the polarization resistance is commonly considered in terms of the base model [13] as a result of changes in the contact microgeometry under the action of the current. Special measures were taken in [26] to control the change of the electrode/electrolyte contact. Moreover, the experiments were performed at a temperature when the interpretation of the effect as a consequence of oxidation-reduction of platinum is unacceptable.

- The last most vivid fact, which cannot be explained in terms of the modern representations, is the electrolyte mass transfer [30, 31].

It is asserted [11] that addition of two provisions to the base model, the first and most important of which is existence of a layer with a different structure on the SOE surface, will allow explaining and describing the aforementioned three groups of experimental results.

4. REASONS FOR EXISTENCE OF THE SURFACE LAYER OF THE ELECTROLYTE

It was shown that "odd" forbidden small-angle reflections in the X-ray pattern appeared even from a fresh fracture of a single crystal at room temperature [1-3]. Therefore, the surface layer of the electrolyte cannot be attributed to contamination of the SOE surface with impurities, primarily silicon (see, for example, [32]). Impurities and, also, segregation of SOE components, e.g., yttrium, on the electrolyte surface undoubtedly influence properties of the SOE surface layer.

There is no consensus of opinion about the cause for appearance of the surface layer on FCC oxides having the structure of the fluorite type. We reason that solid electrolytes represent materials with a large concentration of defects. For example, the concentration of oxygen vacancies reaches 25% in the Bi_2O_3 δ-phase. This is the maximum possible concentration of defects in the structure of the fluorite type. Defects interact. The interaction energy is sufficiently large. Measures of the accumulated interaction energy are the characteristic size of the lattice coherence region and the defect concentration. In the case of a single crystal of the electrolyte based on zirconium dioxide with the characteristic size of the mosaic block equal to 1.5 centimeter, thickness of the surface layer was ~0.2-0.6 μm as measured by three different methods. Clearly, it will be different from thickness of the surface layer of a ceramic having the same composition with grains of the order of a micrometer. This layer on the bismuth oxide ceramic can be easily detected by X-ray methods [1, 2] even despite a small size of the coherence region.

REFERENCES

[1] *Shkerin S.N.* PhD Thesis, Sverdlovsk. Institute of Electrochemistry, Ural Scientific Center of the USSR Academy of Sciences. 1989. in Russian.

[2] *Shkerin S.N.* Proc. All-Union conference on electrochemistry. V. 2. Part 1. "Physicochemical methods for the study of the surface structure and composition". Sverdlovsk. 1991. p. 32. in Russian.

[3] *Shkerin S.N.* Izv. Akad. Nauk. Ser. Fiz. 2002. V. 66. p. 890. in Russian.

[4] *Suzuki Y., Kohzaki T.* // Solid State Ionics. 1993. V.59. P. 307.

[5] *van Hassel B.A., Burggraaf A.J.* // Solid State Ionics.1992. V.57.P.193.

[6] *Kurumchin E.Kh.* Doctoral Thesis. Ekaterinburg, Institute of High-Temperature Electrochemistry, Ural Branch RAS. 1997. in Russian.

[7] *Solmon H., Chaumont J., Dolin C., Monty C.* // Ceramic transactions. 1991. V. 24. P. 175.

[8] *de Ridder M., van Welzenis R., Brongersma H., Kreissig U.* // Solid State Ionics. 2003. V. 158. P. 67.

[9] *Park J., Blumental R.* // J. Amer.Ceram.Soc. 1988. V.71. p. 642.

[10] *Vohrer U., Wiemhofer H.-D., Gopel W., van Hassel B.A., Burggraaf A.J.* // Solid State Ionics. 1993. V.59. P. 141.

[11] *Shkerin S.N.* // Elektrokhimiya. In print.

[12] Electrochemistry of Solid Electrolytes / *V.N. Chebotin, M.V. Perfiliev.* – Moscow, Khimiya. 1977. 312 p. in Russian.

[13] High-Temperature Electrolysis of Gases / *M.V. Perfiliev, A.K. Demin, B.L. Kuzin, A.S. Lipilin.* – Moscow, Nauka. 1988. 232 p. in Russian.

[14] *Kleitz M., Fabry P., Schouler E.* Electrode polarization and electronic conductivity determination in solid electrolytes. /In "Fast ion transport in solids" Ed. by W. Van Gool. – Amsterdam, London: North-Holland Publ. Comp., 1973. – P. 439-451.

[15] *Kleitz M., KloidtT., Dessemond L.* /In: "High Temperature Electrochemical Behaviour of Fast Ion and Mixed Conductors". 14th RISO International Symposium on Materials Science. 1993. P.89.

[16] *Shouler E.* Thesis. Grenoble. 1979.

[17] *Shkerin S., Gormsen S., Mogensen M.* // Ionics. 2002.V. 8. P. 439.

[18] *Shkerin S., Primdahl S., Mogensen M.* // Ionics. 2003. V. 9. P 140.

[19] *Shkerin S., Gormsen S., Primdal S., Mogensen M.* // Elektrokhimiya. 2003. V. 39. P. 1183. in Russian. In English – www.maik.ru, Russian Journal of Electrochemistry.

[20] *Shkerin S., Gormsen S., Mogensen M.* // Elektrokhimiya. 2004. V. 49. P. 156. in Russian. In English – www.maik.ru, Russian Journal of Electrochemistry.

[21] *Sridhar S., Stancovski V., Pal U.* // Solid State Ionics. 1997. V. 100. P. 17.

[22] *Stancovski V., Sridhar S., Pal U.* // J. Electroceramics. 1999. V. 3. P. 279.

[23] *Sridhar S., Stancovski V., Pal U.* // J. Electrochem. Soc. 1997. V. 144. P. 2479.

[24] *Jacobsen T., Zachau-Christiansen B., Bay L., Jorgensen M.* // Electrochim. Acta. 2001. V. 46. P. 1019.

[25] *Jacobsen T., Bay L.* // Electrochim. Acta. 2002. V. 47. P. 2177.

[26] *Shkerin S.N.* // Elektrokhimiya. 2003. V. 39. P. 957. in Russian. In English – www.maik.ru, Russian Journal of Electrochemistry.

[27] *Shkerin S.N.* // Elektrokhimiya. 2004. V. 40. P. 576. In print. in Russian. In English – www.maik.ru, Russian Journal of Electrochemistry.

[28] *Sasaki K., Maier J.* // Solid State Ionics. 2003. V. 161. P. 145.

[29] *Shkerin S.N., Perfiliev M.V.* // Elektrokhimiya. 1990. V. 26. P. 1461. in Russian in Russian. In English – www.maik.ru, Russian Journal of Electrochemistry.

[30] *Bay L., Jacobsen T.* // Solid State Ionics. 1997. V. 93. P. 201. (Erratum: Solid State Ionics. 1997. V. 97. P. 159).

[31] *Bay L.* Ph.D. Thesis. Copenhaven. 1998.

[32] *Mogensen M., Jensen K., Jorgensen M., Primdahl S.* // Solid State Ionics. 2002. V. 150. P. 123.

SOLID ELECTROLYTES FROM STABILIZED ZIRCONIUM OXIDE THAT ARE DEVELOPED IN O.J.S.C. "UKRAINIAN RESEARCH INSTITUTE OF REFRACTORIES NAMED AFTER A.S. BEREZHNOY"

V.V. PRIMACHENKO, V.V. MARTYNENKO, I.G. SHULIK,
T.G. GAL'CHENKO, G.P. OREGHOVA

O.J.S.C. «The Ukrainian research institute of refractories named after A.S. Berezhnoy»
18, Gudanova Str., Kharkiv, 61024, Ukraine

Abstract: As a result of realized investigations of phase composition, structure and electrical conductivity of samples of ternary solid solutions $0,9$ $ZrO_2 + 0,1$ (Sc_2O_3, Y_2O_3) and $0,9$ $ZrO_2 + 0,1$ (Sc_2O_3, Lu_2O_3) the optimum compositions of solid electrolytes were set that are characterized with high electrical conductivity(æ_{1000} $o_c = 0,17\text{-}0,22$ Ohm^{-1} cm^{-1}) and ageing stability – $R\tau/R_o$ – $1,03\text{-}1,06$. The technology is developed and production with output of high-density electrical conductivity ceramics from stabilized zirconium oxide is organized in O.J.S.C. "Ukrainian research institute of refractories named after A.S. Berezhnoy". The application of this ceramics is developed when using it in high-temperature electrochemical devices.

Key words: Zirconium Ceramics/Solid Electrolytes/Electrical Conduction

N. Sammes et al. (eds.), Full Cell Technologies: State and Perspectives, 307-315.
© 2005 Springer. Printed in the Netherlands.

1. INTRODUCTION

The development of new areas of engineering is inseparably linked with application of materials that have high refractoriness, heat resistance, strength, corrosion stability. Among these materials the solid solutions of zirconium oxide that are widespread both in Ukraine and abroad take an individual position. In particular, they are used as solid electrolytes in a whole series of electrochemical devices including oxygen sensors, high-temperature fuel elements, oxygen pump and high-temperature electrolyzers for production of hydrogen and oxygen.

The unique property of solid solutions on the basis of zirconium oxide is oxygen-ionic conductivity and it is due to their crystal structure type. The solid solution of stabilized zirconium oxide has a cubic structure of fluorite type with anionic vacancies that leads to electrical conductivity abrupt increase at temperature increase of $> 600°$ C.

Among the most important demands that are made for electrolytes on the basis of zirconium oxide is combination of high electrical conductivity with ageing stability in the range of operating temperatures (900-1000° C).

In the present feature there are research results of influence of stabilized additions of scandium, yttrium or lutecium oxides and their amount on its phase composition, temperature and time dependence of electrical conductivity of zirconium oxide samples that are fired at 1900° C.

2. EXPERIMENTAL PART

Zirconium oxide containing 99,5% of ZrO_2 as well as yttrium, scandium and lutecium oxides containing the basic oxide of $> 99,99\%$ were used as initial materials for test operations.

Mixtures of the compositions N 1-5 ($Y_2O_3 + Sc_2O_3 = 10$ mol. %), N 6 (Sc_2O_3 — 10 mol.%) and N 7-11 ($Lu_2O_3 + Sc_2O_3 = 10$ mol.%) were prepared for test operations.

The 5% solution of polyvinyl alcohol was used to moisten mixtures. The actual batch humidity was 8,0-8,2% of mass %. The shaping of tablets was realized on the laboratory hydraulic press at pressure of 200 MPa. Molded samples after drying in natural conditions were fired at 1900 °C with isothermal curing during 1 hour in the laboratory electrical furnace with graphite heaters.

The phase composition of samples was determined by means of their rentgenography on the diffractometer in Cu-K_α emission (with Ni-filter). Parameters of a unit cell of cubic solid solutions of fluorite type were calculated according to the line profiles by scan technique on the diffractometer. The line (311) was used for calculation and characterization accuracy was equal to ±0,002 Å.

The electrical conductivity and its time variations were measured on samples in the form of flat-parallel tablets with diameter of 10 mm and thickness of ~3 mm (with ratio of the sample thickness to the area of platinum electrode [tablet diameter area] ~ 0,05 mm^{-1}) according to the procedure that is described in [1]. The samples (up to 11 pieces) were placed in series in a measuring cell from corundum tube. The conducting and collecting wires were taken to the first and last samples and current collecting wires were taken to intermediate samples. The tight contact between samples and current conducting and collecting wires was provided with special holders. The cell with samples was placed to the furnace with a platinum-rhodium heater so as to have samples in the working area. The furnace temperature was adjusted with a high-accuracy temperature regulator and it was measured with a thermal converter using a combined digital device. The alternating current with fixed frequency of 100 kHz was passed through the serially connected samples. The frequency was given with a generator of signals of special shapes. The standard frequency independent resistance was switched on with samples in series. The measurement of electrical conductivity is based on comparison of voltage drop at the sample and standard resistance. The voltage was measured with multi-purpose digital voltmeter. Samples of all compositions after 0,5-hour isothermal curing at 1500° C (constancy of electrical conductivity was obtained in 15 minutes) were cooled quickly till the given temperature. After that their electrical conductivity from temperature was measured in the temperature range of 400-1400° C with a step of 15-20° C. The time dependence of resistance was determined at 930° C and isothermal curing during 740 hours and it was characterized by an ageing factor (R_τ/R_0) that is by ratio of the electrolyte resistance at the moment of time τ to the initial resistance. The measuring error of electrical conductivity was ±2-3%.

3. RESULTS AND DISCUSSION

On the Fig. 1 there is a dependence of the cell parameters of cubic solid solutions $0,9\ ZrO_2 + 0,1\ (Sc_2O_3, Y_2O_3)$ and $0,9\ ZrO_2 + 0,1\ (Sc_2O_3, Lu_2O_3)$, that are investigated in the present scientific work, on the average ionic radius $r_k = \sum c_i r_i$ where c_i and r_i correspondingly are concentrations and radii of component cations that form a solid solution. Ionic radii R^{3+} of rare-earth elements Sc, Lu, Y and Zr are taken according to Templeton-Deben scale [2], because this scale of ionic radii is the most accurate ($R_{Y3+} = 0,905Å$, $R_{Sc3+} = 0,730\ Å$, $R_{Lu3+} = 0,848$) for oxides of rare-earth elements that have structure similar to fluorite type).

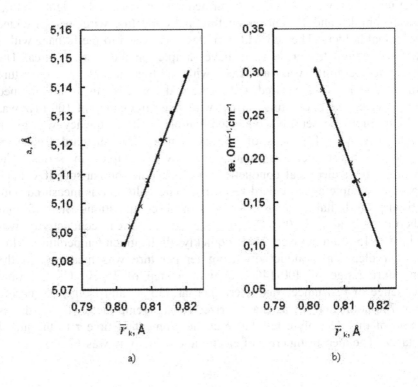

Figure 1. The dependence of the cell parameters a (a) and electrical conductivity χ at 1000^0C (b) on average ion radius of cation r_k samples of solid solutions $0,9\ ZrO2 + 0,1(Sc_2O_3, Y_2O_3)$ (•) and $0,9ZrO_2 + 0,1(Sc_2O_3, Lu_2O_3)$ (✗)

As it is clear from the Fig. 1a the cell parameters of cubic solid solutions of both sample series increase when increasing an average ionic radius of cation in the solution, making a united linear dependence that confirms a correct use of Templeton-Deben scale of ionic radii in this case.

On the Fig. 1b there is a dependence of electrical conductivity (at 1000 °C) on the average ionic radius of cation in solid solutions on the basis of ZrO_2. As it is clear from this figure, the values of electrical conductivity for studied triple solid solutions on the basis of ZrO_2 decrease linearly when increasing the average ionic radius (r_k) in the solid solution lying on a common straight line. The observed linear change of cubic cell parameters from the average ionic cation radius (r_k) in the solid solution on the basis of ZrO_2 (Fig. 1a) allows to build a graphics dependence of electrical conductivity of mentioned solid solutions on the cell parameters (Fig. 2a). As it is clear from the Figure 2a the values of electrical conductivity of studied solid solutions on the basis of ZrO_2 increase when decreasing parameters of their grid that corresponds to decreasing of average ionic cation radius (r_k) in the solid solution. That confirms a part of the ion size factor in diffusion processes in solid solutions of fluorite type [3].

The analysis of electrical conductivity isotherms of solid solutions samples in sections 0,9 ZrO_2·+ $(0,1 - x)Y_2O_3 + Sc_2O_3$ and 0,9 ZrO_2 + $(0,1 - x)Lu_2O_3 + Sc_2O_3$ (Fig. 26) shows that:

- samples from zirconium oxide (stabilized with 10% Sc_2O_3) are characterized with maximum electrical conductivity at all temperatures;

- samples with addition of Sc_2O_3 and Lu_2O_3 have higher indications of electrical conductivity at all studied temperatures and at all investigated ratios than samples with addition of Sc_2O_3 and Y_2O_3;

- difference in values of electrical conductivity for samples of both series decreases when approaching to the samples that contain addition of 10% Sc_2O_3 at fixed temperature as well as when increasing temperature at constant composition of samples and at the temperature of 1400° C the isotherms of samples practically coincide.

As with decreasing of the cation average ionic radius the electrical conductivity of solid solution samples on the basis of ZrO_2 increases and stability of solid solution deceases [4] and one of the most important demands that are made to the solid electrolytes is a combination of high electrical conductivity with the ageing stability (R_τ/R_0), the ageing of investigated solid electrolytes was studied in the present work.

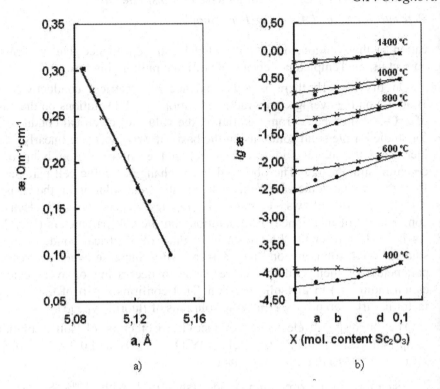

a)   b)

Figure 2. The dependence of electrical conductivity (χ) at 1000^0C on cell parameters of solid
solution (a) and on electrical conductivity isotherm (b) $0.9\,ZrO_2 + (0,1-x)\,Sc_2O_3 +xY_2O_3$ (•)
and $0.9\,ZrO_2 + (0,1-x)Sc_2O_3 + xLu_2O_3$ (×)

The study of time dependence of electrical conductivity (ageing) of solid
electrolytes shows that after long-duration curing (up to 740 hours) at 930 °C
(Fig.. 3) the ageing degree (R_τ/R_0) of all samples under study ranges 1,03-
1,06 and much lower than for samples of the composition 0,9 ZrO_2 ·+0,1
Sc_2O_3. This composition is characterized with the highest electrica
conductivity at the initial time period and its electrical resistance increases
by ~ 1,8 times in ageing process at the same conditions.

As a result of the realized research the optimum compositions of solid
electrolytes are set. They are characterized with high electrical conductivity
($æ_{1000\,0c} = 0,17-0,22\,Ohm^{-1}cm^{-1}$) and ageing stability – R_τ/R_0- 1,03–1,06.

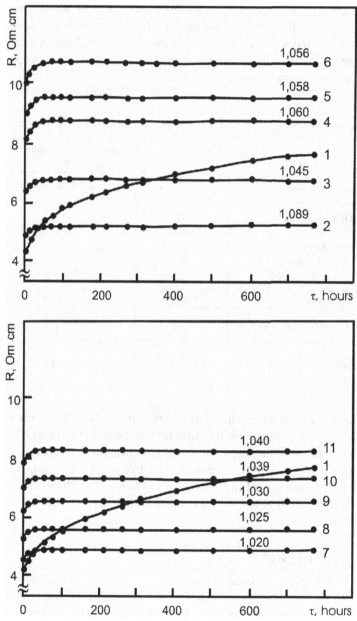

Figure 3. Temporal dependence of electrical resistance at temperature $930^{\circ}C$ for samples of solid solutions. Numerals above curves — extent of ageing (R_∞/R_0), numeral with curves — numbers of compositions.

O.J.S.C. "The Ukrainian research institute of refractories named after A.S. Berezhnoy" developed a technology using the obtained results and organized production with turnout of high-density electroconductive ceramics on the basis of zirconium oxide that is stabilized with yttrium oxide and combined with addition of scandium and yttrium oxides. Properties of these compositions are given in the table.

Table. Properties of high-density electroconductive ceramics from zirconium oxide

Denomination	Property indices of products from zirconium oxide that is stabilized with	
	Y_2O_3	$Y_2O_3 + Sc_2O_3$
Chemical composition:		
$ZrO_2 + HfO_2$, not less	82	85
Y_2O_3	15–17	6,5–7,5
Sc_2O_3	—	6–7
SiO_2, not more	0,5	0,5
Fe_2O_3, not more	0,2	0,2
Open porosity, %	0,0–0,1	0,0–0,1
Apparent density, g/cm^3, not less	5,7	5,6
Specific electrical resistance at 1000 °C, Ohm/cm, not more	12	6

The solid electrolytes on the basis of ZrO_2 that is stabilized with Y_2O_3 (or $Y_2O_3 + Sc_2O_3$) are produced in the form of discs, capillaries as well as covers and tubes with wall thickness of 0,3-3 mm and they are used in high-temperature electrolyzers for generation of hydrogen and oxygen, in gas-analyzers for analyses of neutral and oxidizing shielding mediums, in hygrometers of moisture and absolute humidity in nitrogen and air.

4. CONCLUSION

As a result of the realized research the optimum compositions of solid electrolytes are set. They are characterized with high electrical conductivity ($\ae_{1000^\circ c} = 0,17–0,22$ Ohm^{-1}cm^{-1}) and ageing stability – R_τ/R_0- 1,03–1,06.

O.J.S.C. "The Ukrainian research institute of refractories named after A.S. Berezhnoy" developed a technology and organized production with turnout of high-density electroconductive ceramics from stabilized zirconium oxide. The application of this ceramics was developed in high-temperature electrochemical devices.

REFERENCES

[1] Perfil'ev M.V., Inozemzev M.V., Vlasov A.N. // Electrochemistry, 1982, v.18, №9, p.1230-1236

[2] Templeton D.H., Dauten C.H. // Journ. Amer. Chem. Soc., 1954, v.76. p.5237

[3] Móbius H.H. // Z.Chem. 1962, Bd.2. p.100-102

[4] Vlasov A.N., Shulik I.G.// Electrochemistry, 1982, v.18, №9, p.909-913

NANOCRISTALLINE POWDERS SYSTEM ZrO$_2$ – Sc$_2$O$_3$ FOR SOLID-STATE COMBUSTION CELLS

V.G. VERESCHAK, N.V. NIKOLENKO, K.V. NOSOV

Ukrainian State Chemical Technology University, Gagarin Av., 8, Dniepropetrovsk 49005, Ukraine, E-mail: ughtu@dicht.dp.ua

Abstract: The physicochemical features of the formation of collateral precipitated Zr(IV) and Sc(III) hydrosols were found. It was shown, that the basic influence on formation of non-aggregative monodispersible oxides powders was made by physicochemical processes which proceed at dehydration of the oxyhydroxides. It was investigated the influence of organic solvents on processes of fracture of oxyhydroxides polymeric structure at its dehydration and formation of nanocrystalline powder. It was shown, that the molecules of alcohol, being adsorbed on a surface of particles, supersede water from volume of a precipitate, screening surface OH-groups and a reduce forces of capillary tension. It is found, that the special treatment of the coprecipitated Zr^{4+} and Sc^{3+} oxyhydroxides by the organic solvents lets to produce the oxide powders with primary particles about 100-200 A$^{\circ}$.

Key words: Nanocrystalline   Zirconium   Dioxides   Stabilized   By   Scandium(III)/ Oxyhydroxides/A Sol - Gel Method

The powders of zirconium dioxide stabilized by scandium(III) are of interest as materials of solid electrolytes for high-temperature combustion cells. The basic advantages of solid electrolytes on the basis of zirconium dioxide with Sc^{3+} stabilizer are higher conductance and thermostability in comparison with other stabilizers (Ca^{2+}, Mg^{2+}, Y^{3+}).

The tendency of use in combustion cells of film technologies and ceramics with thickness of devices less than 100 microns is caused the

N. Sammes et al. (eds.), Full Cell Technologies: State and Perspectives, 317-322.
© 2005 *Springer. Printed in the Netherlands.*

strong technological requirements to initial powdered materials, namely a high reactivity during moulding of fuel cells devices, a sintering ability at reduced temperatures, a thermostability at change of the thermal loadings. The solution of these problems is possible with use of nanocrystalline-powdered materials.

Nanocrystalline powders of stabilized by scandium(III) zirconium dioxides was produced by us a sol-gel method. It was investigated the optimal conditions and mechanism of a collateral deposition of Zr(IV) and Sc(III) cations in their chloride solutions. It is shown, that the essential influence on formation of scandium (III) solid solutions in a matrix of zirconium dioxide and degree of its stabilization has a requirements of preparation of initial solutions, receptions and requirements of their coprecipitation, dehydration and calcination.

The synthesis of nanodimensional powders of solid solutions Sc^{3+} in zirconium dioxide was carried out by methods a sol-gel technology with application of special processing methods of the coprecipitated Zr^{4+} and Sc^{3+} oxyhydroxides which prevented the processes of spontaneous aggregation of primary oxyhydroxides particles at removal of a dispersion medium and dehydration.

It is known [1], that the particles sizes of the oxyhydroxides precipitates are 20-40 nm. However at water removal in result of capillary forces the primary particles of a precipitate are approached to each other and therefore the spatial requirements for passing interparticle olation and oxolation processes are created that give uncontrollable formation of particles aggregates. At burning of such aggregates the processes of agglomeration of primary particles and formation rather large and inert powders are occurred.

For prevention of the polymerization processes in result of interaction between primary particles of hydrogels the wet precipitates were treated with organic solvents such as aliphatic alcohols. The interaction of the organic molecules with surface OH-groups of primary particles of a precipitate prevented from formation of agglomerates at subsequent deaquation.

It was supposed, that molecules of alcohol with relatively long hydrocarbonaceous radicals and one functional OH-group would be adsorbed on an oxyhydroxides surface and in result of its hydrophobization reduced forces of capillary tension at removal of a dispersion medium.

The removal of a dispersion water medium from hydrogels precipitates is possible by an azeotropic distillation too. This method of water elimination from gels is represented to us as the most technological because it allows to produce rather fine-crystalline powders of oxides.

The carried out examinations had shown that the burning of powders of zirconium oxyhydroxide anhydrous by azeotropic distilling in pentanol medium permits to obtain finely divided powders of zirconium oxide with a narrow enough particle size distribution.

We believe that obtaining of non-agglomerated powders of the stabilized zirconium dioxide after processing its hydrogels by organic solvents is a result of prevention of the interparticle polymerization processes. This conclusion was confirmed by methods of a thermogravimetric analysis and IR spectroscopy.

The derivatograms of hydrogels, which were dried without processing and with alcohol processing at room temperature, had shown their identity: on the differential heating curve at 150 °C the endothermic effect caused by evolving of water from oxyhydroxide is observed. However, in case of a hydrogels, which were processed by boiling in alcohol, in addition to process of heat absorption at 140 °C the exothermic effect is observed at 320 °C. According to a curve of mass diminution this effect is caused by process of burning of alcohol which molecules were adsorbed on a surface of zirconium oxyhydroxide particles.

In result of IR spectroscopic investigations it was founded that alcohol molecules can chemically interact with zirconium oxyhydroxides. In IR spectrum of the samples observed poorly intensive peaks of uptake in the field of frequencies 2850 – 2950 cm^{-1}, which are referred by us to valence vibrations of C–H and C–OH of amyl alcohol. On all visibility, the inappreciable amount of alcohol remained to bound in pores of hydrogel. Formation of hydrogen bridges of molecules of alcohol with surface OH-groups of oxyhydroxide does not find. For hydrogel treating by alcohol at boiling point, in the field of 2850 – 2950 cm^{-1} observed considerably more intensive peaks of uptakes caused by presence of a lot of amyl alcohol in an explored sample. In its spectrum also absolutely there is a strip uptake 1637 cm^{-1}, caused by straining oscillations of molecules of water. In the field of 1055 cm^{-1} the valence vibrations of a surface ether C–O are registered.

We offer the mechanism of interaction of alcohol with zirconium hydrosol, according to which the molecules of water and alcohol are capable to form hydrogen bridges with OH-groups of oxyhydroxide surface or to substitute hydroxyl ions in coordination sphere of Zr(IV). It was concluded, that interaction between alcohol molecules and oxyhydroxide (as and water molecules) was controlled by values of the effective charges of their interacted atoms (see Fig. 1). This interactions cannot be controlled by orbital overlap because the energy of the lowest unoccupied orbitals of alcohol molecules are much less than energies of a 2p-levels of the zirconium oxyhydroxide surface. The excess of water molecules prevents from process of OH$^-$ ions replacement in coordination sphere of a zirconium cation by alcohol. Such replacement for zirconium hydroxide will begin only at decreasing of water content in a dispersion medium or with increasing of temperature.

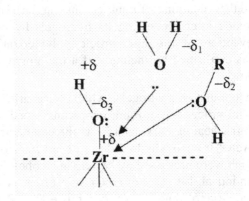

Figure 1. The scheme of interaction between alcohol molecules and a surface zirconium oxyhydroxide

On the basis of derivatograhpic examinations and analysis of infrared spectrums it was found that the surface ethers are generated and their formation is similar to the process of etherification:

$$\equiv Zr(OH)_n + nROH \rightarrow \equiv Zr(OR)_n + nH_2O.$$

The formation of surface ethers essentially changes the mechanism of formation of oxyhydroxides at their heat treatment. Arising at high-temperature processing the surface zirconium ethers prevent from formation of bonding between particles of oxyhydroxides, and at their approach to each other there is a steric repulsion of surface hydrocarbonaceous radicals.

The result of electron-microscopical examinations of zirconium dioxide stabilized by 10 % Sc_2O_3, which was produced by a sol - gel method with application of an alcoholic deaquation, are submitted in Fig. 2. It is found, that the special treatment of the coprecipitated Zr^{4+} and Sc^{3+} oxyhydroxides by the organic solvents lets to produce the oxide powders with primary particles about 100-200 A°. It was found that the agglomeration of powders and the size of primary particles grow with increasing of amount of stabilizing cations.

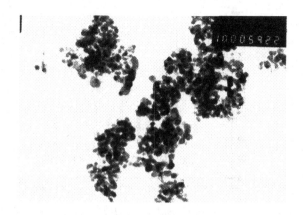

Figure 2. Electron-microscopical photo of powders of ZrO_2 + 10 %(mol) Sc_2O_3, obtained by sol-gel method with application of an alcoholic deaquation (100000x)

The phase analysis of zirconium dioxide, which was obtained with the various content of Sc_2O_3, has shown that the preparation of solid solutions with a fluorite phase is observed already at adding of 2 % (mol) Sc(III) in initial solutions. With increasing of scandium concentration in a solution the amount of the fluorite phase in synthesis products is incremented. For zirconium dioxide with more than 12 %(mol) Sc_2O_3 it was found the formation of phases with structure which most close to structure of fluorite (Fig. 3). It is necessary also to note that in comparison with solid phase method of synthesis the sol-gel procedure of synthesis allows to obtain the solid solutions of scandium in ZrO_2 at considerably lower temperatures (750-900 °C).

The carried out examinations allow to make a conclusion that the high-temperature processing of a gel by aliphatic alcohol is one of method of management by structure and dispersity of oxide powders.

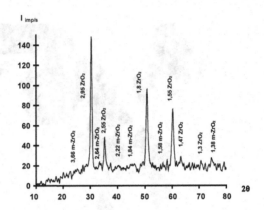

Figure 3. X-ray pattern of sample $ZrO_2+10\%$ (mol.)Sc_2O_3 Cu-K☐ radiation.

Thus, varying time, temperature and the kind of a hydrocarbonaceous alcohol radical, it is possible to controlled by terminating disperse structure of obtained zirconium oxides.

ACKNOWLEDGEMENTS

Author thanks the Science and Technology Center in Ukraine for their financial support of works directed on Research and Development in fields of zirconia ceramics.

REFERENCES

[1] Cleafild A., Vaughan P.A. The crustal of zirconyl chloride octahydrate and zirconyl bromide octahydrate .- Acta crystallogr., 1956, Vol. N 7, p. 555-558.

[2] Nikolenko N.V., Vereschak V.G., Grabchuk A.D. Adsorption of organic compounds by means of coordination and hydrogen communication. // Magazine of physical chemistry.-2000.-V. 74. - № 12. - P. 2230-2235.

STRUCTURAL PECULIARITIES AND PROPERTIES OF (La,Sr)(Mn,Me)O₃ (Me=Cu,Cr)

O.I. V'YUNOV,[1] A.G. BELOUS,[1] O.Z. YANCHEVSKII,[1] D.A. DURILIN,[1] A.I. TOVSTOLYTKIN,[2] V.O. GOLUB[2,3], D.Y. PODYALOVSKY[2]

[1] *V.I.Vernadskii Institute of General & Inorganic Chemistry, National Academy of Science of Ukraine; 32/34, Palladina Ave., Kyiv-142, 03680, Ukraine, belous@ionc.kar.net*

[2] *Institute of Magnetism, 36-b Vernadsky Blvd.,03142 Kyiv-142 , Ukraine*

[3] *AMRI, University of New Orleans, 2000 Lakeshore Dr., New Orleans, LA 70148, USA*

Abstract: Actual oxidation state of doped ions in $La_{0.7}Sr_{0.3}Mn_{1-x}Me_xO_{3\pm\delta}$ systems (Me = Cu, Cr; $0 \le x \le 0.15$) has been determined based on the analysis of structural, magnetic and resonance properties of these materials. It has been shown that for copper-substituted manganites there are two regions of concentrations with different magnetic properties. At $x \le 0.05$ they are homogeneous ferromagnetics with a small variation of magnetic parameters, while at $x > 0.05$ they become substantially inhomogeneous and both the saturation magnetization and Curie temperature significantly decrease with x increasing. The chromium-substituted compounds are homogeneous ferromagnetics in all investigated concentration range. These results can be explained only in supposition of divalent copper ions and trivalent chromium ones, which was verified and confirmed experimentally.

Keywords: Lanthanum Strontium Manganite/SOFC Cathode/Magnetic Sensors/Oxidation State/Rietveld Refinement/Magnetization/Ferromagnetic Resonance

1. INTRODUCTION

Lanthanum strontium manganites (LSM) are very attractive both from scientific and practical point of view as cathode materials for solid-state fuel cells as well as materials with colossal magnetoresistance for magnetic sensors [1]. Substitutions in manganese sublattice allow essentially changing the properties of LSM. On the one hand, such substitutions can modify the conductivity, phase stability, thermal expansion coefficient, chemical compatibility with YSZ and redox behavior of LSM cathode for SOFC. On the other hand, they can lead to a reduction in Zener energy of the double exchange and allow a controlling decrease of the operating magnetic fields.

323

N. Sammes et al. (eds.), Full Cell Technologies: State and Perspectives, 323-328.
© *2005 Springer. Printed in the Netherlands.*

The aim of the present work is to specify the charge compensation mechanisms occurring at substitutions in the manganese sublattice, basing on the analysis of the interrelation between structural, magnetic, and resonance properties of the $La_{0.7}Sr_{0.3}Mn_{1-x}Me_xO_{3\pm\delta}$ (Me=Cu, Cr) compounds.

2. EXPERIMENTAL PROCEDURE

The polycrystalline $La_{0.7}Sr_{0.3}Mn_{1-x}Me_xO_{3\pm\delta}$ (Me = Cu, Cr) samples were prepared using solid state sintering technique. The X-ray investigations were carried out on DRON 4-07 diffractometer (Co $K\alpha$ radiation). The structural parameters were refined by Rietveld full-profile. Shannon's system of ionic radii was used to analyze the crystal structure aspects of the substitutions. Magnetic properties were measured in Quantum Design MPMS-5S SQUID magnetometer. Ferromagnetic resonance (FMR) spectra were recorded using X-band RADIOPAN spectrometer on the samples of parallelepiped shape (1×1×5 mm), with the magnetic field along the long axis.

3. RESULTS AND DISCUSSION

The structure parameters of copper-containing manganites were in detail investigated in our previous work [2]. The analysis of experimental data allowed to reveal the most probable mechanisms of charge compensation during the substitutions in manganese sublattice and to calculate the percentage of Mn^{4+} ions, one of the important parameters governing magnetic properties of doped manganites [3]. It also allowed us to come to the conclusion about 2+ oxidation state of copper in $La_{0.7}Sr_{0.3}Mn_{1-x}Cu_xO_{3\pm\square}$ (LSMCu) compounds. The process of charge compensation can be described by *two* simultaneous reactions of substitution with the formation of oxygen vacancies:

$$2Mn^{3+}{\rightarrow}Mn^{4+}+Cu^{2+} \text{ and } 2Mn^{3+}{\rightarrow}2Cu^{2+}+V_O^{\cdot\cdot} . \qquad (I)$$

In this work, an analogous procedure was applied to the investigation of chromium-containing manganites. Fig. 1 shows the experimental dependence (points) of the unit cell volume and the interatomic Mn–O distance of $La_{0.7}Sr_{0.3}Mn_{1-x}Cr_xO_{3\pm\square}$ (LSMCr) on chromium content as well as theoretical curves calculated using the assumptions of different charge compensation mechanisms at substitution of Cr for Mn. In these calculations, we assumed that Mn is in oxidation states 3+ and 4+ only [1,3], while chromium can be in different states (2+, 3+, 4+, 5+, 6+). It is clear from Fig. 1, that experimental data can be fairly good fitted with theoretical curves (4) and (5), which correspond to the following mechanisms of charge compensation:

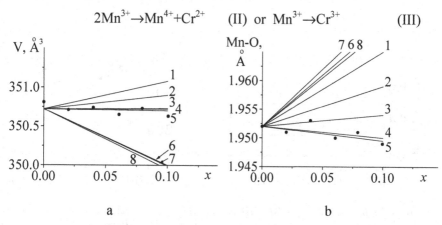

Figure 1. Experimental dependences of the unit cell volume (a) and the interatomic Mn–O distance (b) of $La_{0.7}Sr_{0.3}Mn_{1-x}Cr_xO_{3\pm\square}$ samples on the chromium content (points) and theoretical curves calculated in supposition of the different charge compensation mechanisms for substitution of Cr for Mn: $3Mn^{4+} \to 2Mn^{3+} + Cr^{6+}$ (1); $2Mn^{4+} \to Mn^{3+} + Cr^{5+}$ (2); $Mn^{4+} \to Cr^{4+}$ (3); $2Mn^{3+} \to Mn^{4+} + Cr^{2+}$ (4); $Mn^{3+} \to Cr^{3+}$ (5); $2Mn^{3+} \to 2Cr^{2+} + V_O^{\bullet\bullet}$ (6); $Mn^{4+} \to Cr^{2+} + V_O^{\bullet\bullet}$ (7); $2Mn^{4+} \to 2Cr^{3+} + V_O^{\bullet\bullet}$ (8).

To make a choice between these mechanisms the investigation of magnetic properties of the materials should be carried out. Keeping this goal in mind, we calculated the fraction of Mn^{4+} ions ($[Mn^{4+}]/[Mn_{total}]$) in $La_{0.7}Sr_{0.3}Mn_{1-x}Me_xO_{3\pm\delta}$ as a function of x, based on the Eqs. (I)-(III). It is known [1,3] that a homogeneous ferromagnetic phase can exist in Sr-containing manganites if the Mn^{4+} content is in the 0.18-0.50 range. Out of this range, the tendency to antiferromagnetic ordering predominates, which gives rise to antiferromagnetism or more complex types of magnetic ordering [3]. However it should be noted that the results of such analysis are valid only if the magnetically active ions are manganese ones only, while doped ions do not participate in magnetic interaction and stand as nonmagnetic impurity [1,2]. The situation when Me ions take part in magnetic interaction should be considered separately.

For LSMCu system, homogeneous ferromagnetism can exist only below $x = 0.05$ (see Fig. 2). At $x > 0.05$ more complex types of magnetic ordering should be expected [1,3]. The dependence of Curie temperature T_C on x (Fig. 3, curve 1) and ferromagnetic resonance data (Fig. 4a) are in good agreement with this statement. For the samples with $x = 0$ and 0.025, the resonance spectra consist of a single line with parameters that correspond to the ferromagnetic state of manganites and similar to those obtained in other works for $La_{0.7}Sr_{0.3}MnO_3$ samples [4]. Meanwhile the magnetic resonance spectra for the samples with $x > 0.050$ consist of two well-defined absorption lines, which correspond to two different magnetic phases.

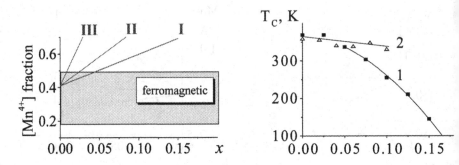

Figure 2. *Mn^{4+} fraction for LSMCu (I) and LSMCr (II, III) samples. The lines correspond to the charge compensation mechanisms described by Eqs (I) – (III), respectively. Hatched area shows the region of existence of a homogeneous ferromagnetic phase.*

Figure 3. *Curie temperature as a function of doped ions content (x) for LSMCu (1) and LSMCr (2) samples.*

Fig. 2 also demonstrates the magnetic diagram of chromium-containing compounds as a function of the Cr concentration calculated from Eqs. (II) and (III). The calculations were based on the supposition that chromium ions do not participate in the magnetic interaction. As can be seen from the figure, in both cases we should expect a magnetic inhomogeneity above $x \approx$ 0.03. However the experiments showed that up to $x = 0.10$ the compounds behaved as homogeneous ferromagnetics. This was supported by the facts that the shape of FMR spectra remained practically the same (Fig. 4b) and T_C only slightly changed with the variation of chromium content (Fig. 3, curve 2).

The evidence of weak influence of Cr substitution on magnetic properties of the manganites was obtained in several previous works [5,6]. Based on the results of magnetic measurements the authors [5] suggested that Cr ions in these compounds are in 3+ oxidation state and participate in double exchange, which is the base of manganite ferromagnetism. The results obtained in this work showed that only such an approach could explain all complex of structure and magneto-resonance properties of LSMCr system. As is seen from the above analysis, other mechanisms, including those connected with the appearance of chromium ions in the oxidation state other than 3+, cannot describe the behavior of chromium-substituted manganites. The most likely reason for the participation of chromium ions in magnetic interaction inside Mn-O-Cr chain is the identity of electronic configurations $(t_{2g}^{3}e_{g}^{0})$ of Mn^{4+} and Cr^{3+} ions [5].

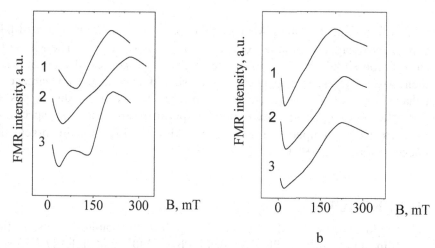

Figure 4. FMR spectra (T = 77 K) of: a) LSMCu samples with different copper content: x = 0.025 (1); 0.050 (2); 0.075 (3); b) LSMCr samples with different chromium content: x = 0 (1); 0.04 (2); 0.08 (3).

To confirm the existence of double exchange between Mn and Cr ions the following additional argument can be presented. If the ferromagnetic double exchange dominates in manganite system, the substitution of Cr for Mn should weakly affect the magnetism due to the participation of chromium ions in the double exchange. However, if the double exchange is not a dominant interaction, the substitution of Cr for Mn should lead to the enhancement of ferromagnetism. Such effects were observed in several works devoted to the investigation of Mn $\rightarrow$ Cr substitution on properties of weakly ferromagnetic manganites [7] and antiferromagnetic charge-ordered manganite compounds [8].

4. CONCLUSIONS

Based on the results of comprehensive analysis of dependences of structural, magnetic and resonance properties of $La_{0.7}Sr_{0.3}Mn_{1-x}Me_xO_{3\pm\delta}$ (Me = Cu, Cr) compounds on x the following conclusions have been made:

1. The mechanism of charge compensation in copper substituted manganites has to include participation of oxygen vacancies and is described by *two* simultaneous reaction of substitution of Cu for Mn: $2Mn^{3+}\rightarrow Mn^{4+}+Cu^{2+}$ and $2Mn^{3+}\rightarrow 2Cu^{2+}+V_O^{\cdot\cdot}$. Realization of such mechanism in $La_{0.7}Sr_{0.3}Mn_{1-x}Cu_xO_{3\pm\delta}$ leads to the transition from homogeneous ferromagnetism to inhomogeneous magnetism (two magnetic phases state) with x increasing.

2. Chromium ions in LSMCr system are in 3+ oxidation state and take part in ferromagnetic double exchange with manganese ions. The Mn $\rightarrow$ Cr substitution should weakly affect the magnetic properties of manganites if their basic state is ferromagnetism, but should lead to the appearance or enhancement of ferromagnetism for manganites with another types of magnetic ordering. It has been shown experimentally that FMR spectra as well as Curie temperature of $La_{0.7}Sr_{0.3}Mn_{1-x}Cr_xO_{3\pm\square}$ remain practically unchanged up to $x = 0.10$.

REFERENCES

[1] Haghiri-Gosnet A-M., Renard J.-P. CMR manganites: physics, thin films and devices, J. Phys. D: Appl. Phys., 2003, v.36, p.R127-R150.

[2] Belous A., V'yunov O., Yanchevskii O., Tovstolytkin A., Golub V. Oxidation states of copper ions and magnetic properties of $La_{0.7}Sr_{0.3}Mn_{1-x}Cu_xO_{3\pm\square}$ system, Proc. Intern. Conf. on Electroceramics and Their Applications, Cherbourg, France, May 31 – June 3, 2004.

[3] Dagotto E., Hotta T., Moreo A. Colossal magnetoresistant materials: the key role of phase separation, Physics Reports, 2001, v.344, p.1-153.

[4] Budak S., Ozdemir M., Aktas B. Temperature dependence of magnetic properties of $La_{0.67}Sr_{0.33}MnO_3$ compound by ferromagnetic resonance technique, Physica B, 2003, v.339, p.45-50.

[5] Sun Y., Tong W., Xu X., Zhang Y. Tuning colossal magnetoresistance response by Cr substitution in $La_{0.67}Sr_{0.33}MnO_3$, Appl. Phys. Lett., 2001, v.78, No 5, p.643 - 645.

[6] Ghosh K., Ogale S.B., Ramesh R., Greene R.L., Venkatesan T., Gapchup K.M., Bathe R., Patil S.I. Transition-element doping effects in $La_{0.7}Ca_{0.3}MnO_3$, Phys. Rev. B., 1999, v.59, No 1, p.533-537.

[7] Zhang L.W., Feng G., Liang H., Cao B.S., Meihong Z., Zhao Y.G. The magnetotransport properties of $LaMn_{1-x}Cr_xO_3$ manganites, J. Magn. Magn. Mater., 2000, v.219, p.236-240.

[8] Barnabe A., Maignan A., Hernieu M., Damay F., Martin C., Raveau B. Extension of colossal magnetoresistance properties to small A site cations by chromium doping in $La_{0.5}Ca_{0.5}MnO_3$ manganites, Appl. Phys. Lett., 1997, v.71, No 26, p.3907 - 3909.

SIMPLE METHOD FOR THE STUDY OF TRANSITION METAL VALENCE STATE IN MIXED-VALENCE COMPOUNDS

A.V. ZYRIN, T.N. BONDARENKO
Frantcevych Institute for Problems of Materials Science.Kyiv.Ukraine

Abstract: This paper concerns a simple rapid method of transition metal valence state determination in mixed valence materials. The X-ray common use diffractometers which are utilized in X-ray phase analysis may be used to that end. The X-ray absorption spectra of researched materials were obtained on X-ray diffractometer with the use of a copper X-ray tube brake radiation. The position of the absorption K-edge for a number of the pure transition metals and their individual oxides in which the metal had various valence states was measured. The valence state of the Mn, Ni, Co atoms in a number of doping La-manganites, nickelites and cobaltites was investigated using this method.

Key words: X-ray Absorption Spectra/3d-Transition Metal Oxides/SOFS-Cathodes Materials/Valence State/Doped Lanthanum Manganite

1. INTRODUCTION

The evaluation of elements' valences (charge state of an atom in compound with the ionic type of chemical bond) is especially needed for studying and designing such materials as mixed valence semiconductors based on 3d-transition metal oxides. The preliminary set electron or hole current carrier density in such materials can be created by applying the valence regulation method. Such electroconducting oxide materials are widely used as electrodes of fuel cells and other current sources, gas sensors, electric heating elements, thermistors etc.

N. Sammes et al. (eds.), Full Cell Technologies: State and Perspectives, 329-334.
© *2005 Springer. Printed in the Netherlands.*

Mainly such materials are presented by doped rare earth manganites, cobaltites, nickelites, chromites. One from methods permitting to determine the atoms' valence is the X-ray spectroscopy method.

1.1. X-RAY ABSORPTION SPECTRA

X-ray spectroscopy is widely used for the research of materials electronic structure and for the estimation of atoms charge state. X-ray absorption spectroscopy may be applied for analyzing the chemical environment of an element in an unknown material.

When X-rays are absorbed in matter, the energy of the X-rays is converted into the kinetic energy of photoelectrons, Auger electrons, secondary electrons, or fluorescent X-rays.

The amount of energy absorbed by a matter is usually estimated by a transmission method, but can also be estimated by measuring these secondary phenomena, such as photoelectrons, Auger electrons, secondary electrons, fluorescent X-rays, thermal radiation, and drain electric currents.

At the transmission method the X-ray intensity of wavelength λ before (I_0) and after (I) the transmission of a thin film of thickness d is expressed by $I(\lambda)=I_o(\lambda)exp[-\mu_i(\lambda)\rho_i d]$, where $\mu_i(\lambda)$ and ρ_i are the mass absorption coefficient and mass density, respectively, of the i-th element in the thin film and their dimensions are $[cm^2 \cdot g^{-1}]$ and $[g \cdot cm^{-3}]$, respectively. The mass absorption coefficient μ of a specimen which contains n kinds of elements is expressed by $\mu=\mu_1(\lambda)W_1+\mu_2(\lambda)W_2+...+\mu_n(\lambda)W_n$, where $W_1, W_2, ...W_n$ are the weight fractions of element 1, 2,...n in the specimen. The wavelength dependence of the absorption coefficient $\mu(\lambda)$ is clarified when $log \, \mu(\lambda)$ is plotted against $log \, \lambda$.

The plot of the mass absorption coefficients of matter against the incident X-ray energy or wavelength is called an X-ray absorption spectrum (XAS), where we find some jumps at particular X-ray energy, corresponding to K-, L_I-, L_{II}-, L_{III}- etc. electron shell binding energies.

The jump is called the absorption edge, and the wavelength is highly correlated with the atomic number similarly to Moseley's law in the X-ray emission spectra. The Moseley's law in emission spectra is expressed equation: $1/\lambda=K^2(Z-s)^2$, where λ is the X-ray wavelength, Z is the atomic number, and K and s are constants for a spectral series.

The absorption coefficient is crudely proportional to $Z^4\lambda^3$ except for the edge jump.

The absorption edge energy or wavelength can be found in the literature [1,2]. The value of the absorption edge energy is close to the electron binding energy, which is used in electron spectroscopy, ESCA (electron spectroscopy for chemical analysis) or XPS (X-ray photoelectron spectroscopy).

The comparison of X-ray K-emission lines energy with absorption K-edge energy shows it proximity to $K\beta_5$-line energy (Table).

Table. K-lines and K-edge energy of the some 3d-metals (eV)

Element	$K\alpha_1$	$K\alpha_2$	$K\beta_{1,3}$	$K\beta_5$	K-edge
Cr	5405.5	5414.7	5946.7	5986.9	5988.8
Mn	5887.7	5898.8	6490.5	6532.2	6537.6
Fe	6390.8	6403.8	7058.0	7108.1	7111.2
Co	6915.3	6930.3	7649.4	7705.9	7709.5
Ni	7460.9	7478.2	8264.7	8328.6	8331.7

A change of the absorption edge energy position is called the chemical shift when the element valence changes.

Therefore the measurement of a chemical shift value on a designing of new materials helps to evaluate of their doping efficiency. It is applied to increase of the electronic (hole) current carriers density.

The variety of authors researched the dependency of the emission and absorption spectra chemical shift from valence of an element in compounds. So Kunzl [3] has found that chemical shift of K- and L_{III}- absorption edge of the third and fourth periods metals in oxides is to approximately in direct ratio valence of the atom. Then chemical shift of emission $K\beta_5$ lines as of [4] for elements from Na to Cl in compounds with oxygen and fluorine is connected with valence by square-law dependency of the type: $\delta E \sim w^2$, where δE is energy shift, and w - valence.

Many contributors have determined the transitional metals valence states in oxides by X-ray spectroscopy methods. The sophisticated unique specialized equipment fitted with high-vacuum systems and significant time expenditure is necessary for its implementation.

For example, in [5] for this purpose authors have carried out the research of the X-ray absorption spectra of the Ti doped complicated oxides in CoO-MnO-O_2 system with the direct and return spinels structures. All these compounds spectra have been compared with those of simple oxides containing ions Mn in various valence states. These spectra were registered during 6-12 hours by vacuum spectrograph.

The simplified accelerated X-ray spectroscopy method is presented in this work. The X-ray common use diffractometers which are utilized in X-ray phase analysis may be used to that end. The duration of the date recording did not exceed several minutes in our experiments.

2. INSTRUMENTATION AND EXPERIMENTAL RESULTS

The method allows determining the position of X-ray absorption edges of an element in matter. The shift of absorption edge in a compound in relation to its position in pure element makes it possible to evaluate the valence of an element. The described method was applied for researching some of them.

The X-ray absorption spectra of the researched materials were obtained on X-ray diffractometer (DRON) with the use of a copper X-ray tube brake radiation at 30 kV anode voltage. The monochromator (quartz of different cutting, single-crystals of mica, tourmaline etc.) were installed instead of samples for XRD-analysis on a goniometer. A thin layer of a researched material was then set in place of the filter cutting Kβ-radiation. The registration of X-ray intensity reflected by crystal-analyzer and passed through an absorber was done by the scintillator counter during standard θ-2θ rotation of counter and stage with the monochromator.

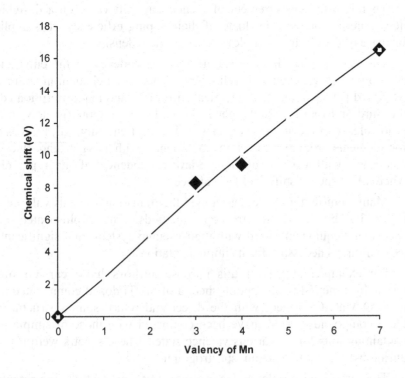

Figure 1. Relation between chemical shift of the absorption Mn K-edge XAS spectra and Mn valence

The position of the absorption K-edge for a number of the pure transition metals and their individual oxides in which the metal had various valence states was measured. For illustration, Fig. 1 shows the example of a chemical shift of the absorption edge for Mn compounds: Mn, Mn_2O_3, MnO_2 , $KMnO_4$.

Fig. 2 shows typical examples for the Mn K-edge of mixed-valence calcium doped lanthanum manganite ($La_{0.6}Ca_{0.4}MnO_3$). Quasi-chemical formula of this compound can look as:

$$La^{3+}_{(1-x)}Ca^{2+}_x Mn^{3+}_{(1-x)} Mn^{4+}_x O_3.$$

The chemical shift of the K-edge components relatively K-edge for metal coinsides reasonably well with those as that observed for Mn^{3+} and Mn^{4+} in Fig. 1. That yields a rough estimate of the mixed-valence state.

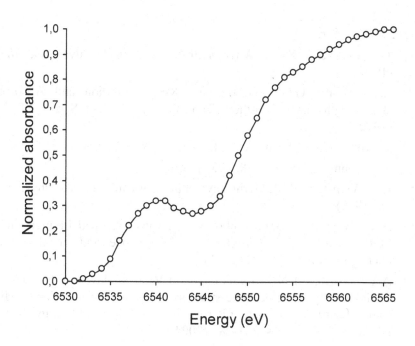

Figure 2. Mn K-edge of calcium doped lanthanum manganite ($La_{0.6}Ca_{0.4}MnO_3$)

The spectrum represented in Fig. 2 has been recorded with the flat quarz crystal (2d = 0.236013 nm) as a monochromator. A scanning rate has been 0.25deg.$2\theta \cdot$ min^{-1}, the value of an integrator time constant refers to 16 s.

Earlier we have investigated by the X-ray emission spectroscopy method the electron structure of another cathode materials − of doping La-nickelites

[6, 7]. In this work we have studied the 3d-metals valence state in these compounds and doping La-cobaltites with the use of the calibrated dependences like for Mn.

3. CONCLUSION

The simple method of shortcut determination of a transitional metals charge state in mixed-valence compounds which are applied as SOFC cathode materials is tested. The standard X-ray diffractometers which are present in the majority material-sciency laboratories are used for this purpose. The offered method is applied at the research of 3d-metals valency in some lanthanum manganites, nickelites and cobaltites. As the measurement standards the metals and their oxides were used.

REFERENCES

[1] J.A. Bearden, 'X-ray Wavelengths', Rev. Mod. Phys., 39, 78-124 (1967).

[2] E.W. White, G.G. Johnson, Jr, 'X-ray Emission and Absorption Wavelengths and Two-theta Tables', ASTM Data Series DS 37A, (1970).

[3] V. Kunzl, Coll. Chechoslov. Chem. Com., 4, 213 (1932).

[4] N.G. Johnson, Nature, 138, 1056 (1936).

[5] E.E. Weinstein, R.M. Ovrutskaya et al., Russian Sol. State Phys., 5, 29-35 (1963)

[6] A.V. Zyrin, T.N. Bondarenko, V.N. Uvarov, 'Doped La-nickelites as SOFCs cathodes', Hydrogen material science and metal hydrides chemistry, Kiev, 772-773 (2001).

[7] A.V. Zyrin, T.N. Bondarenko, I.V. Urubkov, V.N. Uvarov, 'Conductivity and electronic structure of lanthanum nickelites', Mixed Ionic Electronic Conducting Perovskites for Advanced Energy Systems, NATO Science Series, 281-287 (2004).

PEROVSKITE RELATED MATERIALS FOR CATHODE APPLICATIONS

S. J. SKINNER, R. SAYERS, R.W. GRIMES

Department of Materials, Imperial College London, Prince Consort Road, London, SW7 2BP, UK.

Abstract: K_2NiF_4 type oxides are promising candidates for solid oxide fuel cell cathode applications. Processes such as oxide ion diffusion and surface exchange are of critical importance in determining likely performance in a device and understanding of these necessary in order to develop and optimise new materials. The current status of the development of K_2NiF_4 oxides for cathode applications is reviewed and suggestions for the direction of further work made.

Key words: SOFC/K_2NiF_4/Cathode Performance/Oxygen Ion Diffusion/Compatibility

1. INTRODUCTION

Increasing interest in the development of alternative forms of power generation have led to the relatively rapid development of devices such as solid oxide fuel cells to the point where commercial devices are undergoing long term trials. Whilst this progress is encouraging there are still a number of significant issues related to the materials used in these devices. In particular the cathode kinetics have been identified as limiting the electrochemical performance of the device. Combined with the demand for fast ion conducting electrolytes this has resulted in the production of high

N. Sammes et al. (eds.), Full Cell Technologies: State and Perspectives, 335-347.
© *2005 Springer. Printed in the Netherlands.*

operating temperature devices (> 800 °C) and consequently rigorous demands on the interconnects and seals.

Several strategies have been adopted in an effort to overcome these issues including the development of new electrolytes and alternative interconnects, lowering operating temperature and improvement of the processing of materials, but the kinetics of the cathode function are still found to limit the performance of the device.

In terms of cathode development the state-of-the-art materials are perovskite (ABO_3) based mixed conducting ceramics [1] which are compositionally tailored to provide maximum mixed conductivity, stability and also to match the thermal expansion coefficient of the electrolyte of choice. These materials have been optimised over a number of years and to date few alternatives have been suggested.

More recently efforts have been directed towards the development of perovskite related cathode materials, particularly those of the K_2NiF_4 structure type [2,3], Fig. 1. These oxides consist of alternating layers of ABO_3 and AO (rock salt) and have been demonstrated to accommodate a significant oxygen non-stoichiometry.

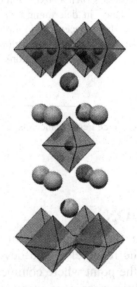

Figure 1. Materials of the K_2NiF_4-type structure have shown great promise as mixed ionic/electronic conductors, for example $La_2NiO_{4+\delta}$

Of interest for SOFC cathodes are those materials that exhibit oxygen hyperstoichiometry, located on interstitial sites. In these materials oxygen mobility has been reported and in the case of $La_2NiO_{4+\delta}$ at temperatures as low as 450 °C [4-8]. These oxides also have a significant total electronic conductivity and are therefore exciting new materials for SOFC cathodes. However, considerable further research is required and the most recent developments regarding the perovskite related cathodes are reviewed here.

2. DIFFUSION AND SURFACE EXCHANGE

In order to gain a direct measurement of the oxide ion diffusion and surface exchange coefficients in ceramic materials an isotopic exchange and SIMS analysis technique has to be used. This technique is detailed elsewhere [9] and has been successfully used for a number of SOFC component materials, most notably with electrolytes [10,11] and cathodes [12].

Diffusion coefficient and surface exchange coefficient measurements have been reported for the K_2NiF_4 type oxide materials by a number of authors [4-8, 13-19] and have been complemented by electrochemical permeation measurements [20-27] all of which demonstrate the fast oxide ion conduction of hyperstoichiometric K_2NiF_4 type oxides. Early reports also demonstrate the relatively poor oxide ion mobility in those materials found to be hypostoichiometric [28,29]. Initial reports of the fast oxide ion conduction in $La_2NiO_{4+\delta}$ [4, 6-8] have generated a number of further studies [13-19] regarding the optimization of composition and determination of the effects of anisotropy on the conduction properties of these materials. Each of these features will be discussed in more detail below.

The oxide ion diffusivity of La_2NiO_{4+d} over the temperature range 640 – 840 °C is shown in Fig. 2 and is compared with the data for the most common perovskite oxide ion mixed conductors.

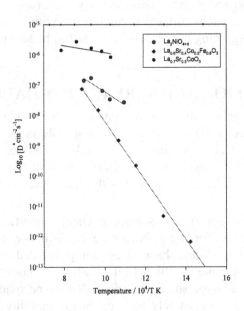

Figure 2. Comparison of the diffusion coefficients of cathode materials

From this it is evident that the diffusivity of La_2NiO_{4+d} was competitive with current mixed conducting perovskites with diffusion coefficients of the order of 10^{-7} cm^2s^{-1}. Evidently the identification of a new mixed conducting ceramic warranted further optimization and hence substitution of both A and B sites was investigated [4,5,7,8,14,18]. Introducing a divalent cation to either the A or B site was expected to reduce the hyperstoichiometry and it was therefore unsurprising to find that the presence of Sr lowered the oxygen content and, as the excess oxide species were viewed as the mobile species, the diffusivity [7]. The effects of a number of dopants are illustrated in Table 1. Most notably the incorporation of cobalt was shown to enhance the low temperature diffusivity of these materials [5, 18] and these materials may prove to be attractive candidates for lower temperature cathode materials (< 750 $°C$).

Table 1 - *Comparison of diffusion data for doped K_2NiF_4 type cathode materials.* * *data recorded at 725 & 800 °C,* †- *data only available for 700 °C*

Composition	Self diffusion coefficient, D* /cm^{2}s^{-1}		
	700 °C	800 °C	Reference
$La_2NiO_{4+\delta}$	3.4 x 10^{-8}	1.7 x 10^{-7}	7
$La_{1.9}Sr_{0.1}NiO_{4+\delta}$	1.0 x 10^{-8}	1.3 x 10^{-8}	7
$La_2Cu_{0.5}Ni_{0.5}O_{4+\delta}$	7.5 x 10^{-8}	3.6 x 10^{-7}	14
$La_2Ni_{0.9}Fe_{0.1}O_{4+\delta}$	3.3 x 10^{-8}	1.1 x 10^{-7}	8
$La_2Ni_{0.9}Co_{0.1}O_{4+\delta}$*	2.5 x 10^{-8}	5.0 x 10^{-8}	4,5
$La_2Ni_{0.8}Co_{0.2}O_{4+\delta}$*	4.0 x 10^{-8}	6.3 x 10^{-8}	4,5
$La_2Ni_{0.5}Co_{0.5}O_{4+\delta}$†	1.1 x 10^{-7}	--------	18

Further, introduction of copper on the B site [14] has been shown to have little effect on the diffusion coefficient, with values of the order of 10^{-9}-10^{-6} cm^{2}s^{-1} over a temperature range of 500-900 °C. Consideration has also been given to the likely use of Pr and Nd nickelates [6,13] as cathodes and diffusion data has been recorded from these K$_2$NiF$_4$ type materials. Indeed the oxygen excess content and diffusion behaviour indicated that both the Nd$_2$NiO$_{4+\delta}$ and Pr$_2$NiO$_{4+\delta}$ compositions would be better candidates than the La based materials. However subsequent investigation highlighted the decomposition of the Pr$_2$NiO$_{4+\delta}$ composition to a Pr$_4$Ni$_3$O$_{10-\delta}$ Ruddlesden-Popper phase [13]. Evidently these phases all show promising diffusion behaviour, but considerable work remains in identifying the optimum material for a cathode application.

Whilst the mobility of the oxide species is significant it is also essential to understand the surface exchange phenomena and the limiting effect this can have on the cathode performance, particularly at lower temperatures. Hence it is important to examine the surface exchange coefficient and the dependence of this with temperature. For a number of the compositions previously discussed it has been reported that the surface exchange coefficients are in the 10^{-8} – 10^{-6} cm s^{-1} range, showing a temperature dependence consistent with activation energies of the order of 0.8 eV. However, one composition, La$_2$Ni$_{0.5}$Co$_{0.5}$O$_{4+\delta}$, has been shown, Fig. 3, to deviate from this behaviour, with virtually temperature independent exchange behaviour at temperatures below ~600 °C. As there is a continued drive to lower the operating temperature of SOFCs the kinetics of diffusion and surface exchange at temperatures as low as 500 °C have become increasingly important and it is therefore of considerable significance that this class of materials has demonstrated the ability to maintain fast surface exchange at these lower temperatures.

As a further complication the data previously discussed was reported for bulk materials and takes no account of the anisotropic structure of the K_2NiF_4 materials. It has been well documented that the oxygen interstitial in these types of oxides is located in the rocksalt layers and that the most likely conduction path for interstitial mobility is through the *ab* plane [30-32]. Hence it is likely that the attractive diffusion performance reported for the nickelate based materials could be further improved through the orientation of the cathode layer. In order to characterize the anisotropy of both diffusion and surface exchange, Bassat et al [19] have investigated isotopic exchanges

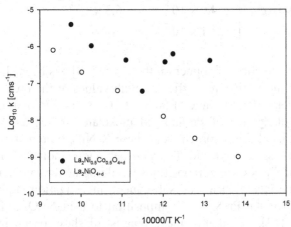

Figure 3. Comparison of surface exchange
behaviour for two perovskite related cathodes

in single crystal samples, finding that the *ab* plane diffusion coefficient is significantly higher than that for the *c* direction. However, a remarkably low activation energy for diffusion of 0.22 eV in the *c* direction was reported which is in contrast to the expected values suggested by atomistic simulation [32]. This surprising result was attributed to the mobility of $O_i^{''}$ in the *ab* plane and of $O_i^{'}$ in the *c* direction.

Significant further data on likely K_2NiF_4 type oxide candidates for SOFC cathodes can also be extracted from the work on oxygen permeation by several authors [20-27]. Many of the results from permeation measurements are concerned with the parent $La_2NiO_{4+\delta}$ composition doped on the B site with either Co, Fe or Cu. The data obtained from these measurements indicated that the permeation through the dense membranes was limited by surface exchange processes and that in order to maximize the flux of oxygen through the dense ceramics an active surface layer of Pt or praseodymium oxide was required. It is of interest to note that samples containing Fe gave the maximum oxygen permeability. Furthermore the $La_2Ni_{0.8}Cu_{0.2}O_{4+\delta}$ material exhibited higher oxygen permeability than the current perovskite cathode material, $La(Sr)Fe(Co)O_{3-\delta}$ [23], again indicating that the perovskite related materials, after suitable development, could have a significant performance advantage over current materials.

3. CATHODE PERFORMANCE

Whilst the kinetic parameters give an indication of candidate materials for cathodes, modelling using, for example, the Adler Lane Steele (ALS) model [33] can, based on the diffusion and surface exchange coefficients, predict more accurately cathode performance. However this model is not infallible and it is absolutely essential to test new materials in, ideally, single cell SOFCs. This of course may not be possible and an alternative would be to characterize the cathode performance using the polarization or area specific resistance (ASR) from symmetrical single cells of the cathode material deposited on a common electrolyte such as $La_{1-x}Sr_xGa_{1-y}Mg_yO_{3-\delta}$ (LSGM) or $Ce_{1-x}Gd_xO_{2-\delta}$ (CGO). To date no single cell SOFC based on a K_2NiF_4 type cathode has been prepared and tested, although a number of studies have investigated the cathode performance from ASR data. From the data obtained by isotopic exchange measurements it was calculated [5] that the ASR of a $La_2NiO_{4+\delta}$ cathode would be of the order of 0.25 Ωcm^2 at 800 °C but from experimental data this was found not to be the case, with ASR values orders of magnitude higher than expected [5]. Later work identified that the adhesion of the cathode to the electrolyte material was an issue with these compositions [3] and that sintering on CGO was problematic, leading to performance three orders of magnitude worse than on a LSGM electrolyte, Fig. 4. Furthermore Munnings et al [34] identified a significant effect on the cathode performance of the LSGM surface related to the depletion of Ga from the surface during processing of the electrolyte. Bassat et al [2] have also investigated the performance of these materials as higher temperature cathodes on yttria stabilized zirconia (YSZ) electrolytes, finding that the most competitive composition was the $Pr_2NiO_{4+\delta}$ material. It is, however, clear from all of these studies that considerable development is required in compositional testing, electrochemical testing and, critically, investigation of the cathode-electrolyte interface. This will enable rapid adoption of this alternative intermediate temperature cathode material.

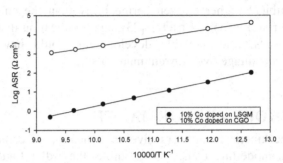

*Figure 4. ASR data for La₂NiO₄₊d cathodes on
LSGM and CGO electrolytes [3]*

4. ATOMISTIC SIMULATIONS

Atomistic simulations have been used to investigate the most favourable migration mechanisms for oxygen mobility in the K_2NiF_4 type oxides, and this technique has also now been used in the investigation of the structural transitions on varying the oxygen stoichiometry of these materials. The preliminary work on this system was performed by Minervini et al [31] who examined in detail the oxidation and oxide ion incorporation mechanisms of $La_2NiO_{4+\delta}$ using energy minimization techniques. From this work it was reported that the favoured defect compensation mechanism for oxygen incorporation was either:

$$\tfrac{1}{2}O_2 + Ni_{Ni}^x \rightarrow O_i' + Ni_{Ni}^{\bullet} \tag{1}$$

$$\text{or } \tfrac{1}{2}O_2 + 2Ni_{Ni}^x \rightarrow O_i'' + 2Ni_{Ni}^{\bullet} \tag{2}$$

This obviously means that the charge compensation involved only nickel holes. On increasing the oxygen excess concentration the compensating defects were then found to be associated with the NiO planes rather than isolated on individual nickel sites, meaning that there were two possibilities for the mobile oxygen interstitial species – O^- or O^{2-}. Further work was then performed on the migration mechanism, finding that as anticipated the activation energy for migration in the *ab* plane was relatively low at 0.29 eV. This activation energy was found for the O^- interstitial, whereas for the O^{2-} interstitial a significantly higher activation energy of 0.88 eV was reported. When considering the *c* direction, activation energies for both species were significantly higher at 2.90 and 3.15 eV respectively. Subsequent experimental work by Bassat et al [19], as discussed earlier, has confirmed the activation energy for the *ab* plane migration, but suggests a much lower activation energy of 0.22 eV in the *c*-axis. Obviously this raises questions regarding the migration mechanism in the perovskite layer which have yet to be resolved.

Based on the success of these results atomistic simulation has also been used to investigate the structural transitions occurring in the $La_2NiO_{4+\delta}$ materials as interstitial concentration is varied. A number of compensation mechanisms have been modeled and the defect energies calculated for each of these previously reported orthorhombic structures and are detailed in Table 2.

Table 2 - Defective lattice energies of oxygen excess compensation mechanism for isolated defects

Mechanism	Isolated Defect Energy (eV)	
	64 (Cmca)	69 (Fmmm)
$3La_2O_3 \rightarrow 2La_{Ni}^{\bullet} + O_i'' + 4La_{La}^x + 8O_O^x$	+11.3246	+16.1136
$3La_2O_3 + \frac{1}{2}O_2 \rightarrow 2La_{Ni}^{\bullet} + 2O_i' + 4La_{La}^x + 8O_O^x$	+4.8747	+9.8204
$\frac{1}{2}O_2 + O_O^x \rightarrow 2O_i' + V_{O(8e)}^{\bullet\bullet}$	-1.9034	-1.5739
$\frac{1}{2}O_2 + O_O^x \rightarrow 2O_i' + V_{O(8f/i)}^{\bullet\bullet}$	-1.2696	-0.3022
$\frac{1}{2}O_2 + O_O^x \rightarrow O_i' + O_{O(8e)}^{\bullet}$	-4.3551	-3.8097
$\frac{1}{2}O_2 + O_O^x \rightarrow O_i' + O_{O(8f/i)}^{\bullet}$	-3.4527	-2.1816
$\frac{1}{2}O_2 + O_O^x \rightarrow O_i'' + 2O_{O(8e)}^{\bullet}$	-2.2605	-1.3262
$\frac{1}{2}O_2 + O_O^x \rightarrow O_i'' + 2O_{O(8f/i)}^{\bullet}$	-0.4557	1.9300
$\frac{1}{2}O_2 + Ni_{Ni}^x \rightarrow O_i' + Ni_{Ni}^{\bullet}$	-6.5125	-6.5858
$\frac{1}{2}O_2 + 2Ni_{Ni}^x \rightarrow O_i'' + 2Ni_{Ni}^{\bullet}$	-6.5751	-6.8784
$\frac{1}{2}O_2 + Ni_{Ni}^x + O_O^x \rightarrow O_i'' + Ni_{Ni}^{\bullet} + O_{O(8e)}^{\bullet}$	-4.4178	-4.1023

On examination of these results it is apparent that the most favoured compensation mechanisms are:

$$\tfrac{1}{2}O_2 + Ni^x_{Ni} \rightarrow O'_i + Ni^{\bullet}_{Ni} \tag{1}$$

and $$\tfrac{1}{2}O_2 + 2Ni^x_{Ni} \rightarrow O''_i + 2Ni^{\bullet}_{Ni} \tag{2}$$

with isolated defect energies of -6.5125 and -6.5751 eV for mechanism (1) and (2) respectively for space group Cmca. On increasing interstitial content and therefore transforming to spacegroup Fmmm it is immediately apparent that mechanism (2) is slightly more favourable with a defect energy of -6.8754 eV. These results also indicate that there is a small driving force for the structural transition from Cmca to Fmmm symmetry. Of course further modeling is required to investigate the effect of defect clustering to confirm the results obtained from isolated defects and to also extend the modeling to more complex systems where, for instance, the B site consists of two elements.

5. FURTHER WORK

Whilst several authors are working on the perovskite related oxides as possible cathode materials, there remains a considerable amount of scope for further development of these materials. One of the critical issues is the cathode structure and it should be noted that to date all studies have been carried out on cathodes with poorly characterized morphologies. Also, a relatively small number of lanthanide based materials have been investigated with few attempts to investigate solid solutions series. As this was found to be essential in the perovskite cathodes with compositions such as $La_{1-x}Sr_xCo_{1-y}Fe_yO_{3-\delta}$ taking many years to optimize, there is a rich vein of potential new cathodes awaiting discovery. With the advent of high throughput screening and atomistic simulations the rapid identification of new materials will be possible. Furthermore, it is conceivable that higher members of the Ruddlesden Popper series would have attractive conduction properties and these materials have yet to be investigated.

There is also the possibility that these compositions will find use in related electrochemical applications such as oxygen separators and reactors for syngas production, and indeed there are already developments in this field [35, 36]. These applications are all likely to require careful control of the K_2NiF_4 type oxide film and some recent progress has been made on depositing $La_2NiO_{4+\delta}$ films [37] on YSZ substrates; however the anisotropic nature of the material suggest that investigation of the deposition of oriented films would be a useful step forward. Hence, after many years of investigation into the low temperature characteristics of K_2NiF_4 type oxides it may be that the high temperature properties are significantly more attractive and applicable to a wide variety of industries.

6. CONCLUSIONS

Perovskite related cathode materials have been shown to possess levels of ionic and electronic conductivity comparable to existing perovskite cathode materials. The development of these new cathodes is in the very early stages and significant research is required before these can compete with the established $La_{1-x}Sr_xCoO_{3-\delta}$ (LSC) and $La_{1-x}Sr_xMnO_{3-\delta}$ (LSM) based cathodes. However, one of the main advantages offered by these materials is the prospect of fast surface exchange and ionic diffusion at temperatures considerably lower than the established candidate materials thus enabling development of low temperature SOFC devices.

REFERENCES

[1]   S.J. Skinner, Int. J. Inorg. Mater., 3 2001 113-121

[2]   [J.M. Bassat, E. Boehm, J.C. Grenier, F. Mauvy, P. Dordor, and M. Pouchard, 5th ESOFC, Lucerne, 2002, Ed. J. Huijsmans, pp. 586-593

[3]   S.J. Skinner, C.N. Munnings, G. Amow, P.Whitfield and I. Davidson, SOFC VIII, Ed. M. Dokiya and S.C. Singhal, Electrochem. Soc. Proc. Vol. 2003-07 pp 552-560

[4]   C.K.M. Shaw and J.A. Kilner, 4th ESOFC, Lucerne, 2000, Ed. A.J. McEvoy, pp. 611-620

[5]   C.K.M. Shaw, PhD Thesis, University of London, 2001

[6]   E. Boehm, J.M. Bassat, F. Mauvy, P. Dordor, J.C. Grenier and M. Pouchard, 4th ESOFC, Lucerne, 2000, Ed. A.J. McEvoy, pp. 717-724

[7]   S.J. Skinner and J. A. Kilner, Solid State Ionics, 135 2000 709-712

[8]   S.J. Skinner and J. A. Kilner, Ionics, 5 1999 171-174

[9]   J.A. Kilner, Mater. Sci. Forum, 7 1986 205-222

[10] T. Ishihara, J.A. Kilner, M. Honda, N. Sakai, H. Yokokawa and Y. Takita, Solid State Ionics, 115 1998 593-600

[11] P.S. Manning, J.D. Sirman and J.A. Kilner, Solid State Ionics, 93 1996 125-132

[12] J.A. Kilner, Bol. Soc. Esp. Ceram. V. 37 1998 245-255

[13] P. Odier, C. Allancon and J.M. Bassat, J. Solid State Chem., 153 2000 381-385

[14] F. Mauvy, J.M. Bassat, E. Boehm, P. Dordor, J.C. Grenier and J.P. Loup, J. Eur. Ceram. Soc., 24 2004 1265-1269

[15] F. Mauvy, J.M. Bassat, E. Boehm, P. Dordor and J.P. Loup, Solid State Ionics, 158 2003 395-407

[16] J.A. Kilner and C.K.M. Shaw, Solid State Ionics 154-155 2002 523-527

[17] F. Mauvy, J.M. Bassat, E. Boehm, J.P. Manuad P. Dordor and J.C. Grenier, Solid State Ionics, 158 2003 17-28

[18] C.N. Munnings, S.J. Skinner, G. Amow, P.Whitfield and I. Davidson, NATO ARW, 2003, Kiev, Eds. N. Sammes and N. Orlovskaya, In Press

[19] J.M. Bassat, P. Odier, A. Villesuzanne, C. Marin and M. Pouchard, Solid State Ionics, 167 2004 341-347

[20] V.V. Kharton, A.P. Viskup, E.N. Naumovich and F.M.B. Marques, J. Mater. Chem., 9 1999 2623-2629

[21] D.M. Bochkov, V.V. Kharton, A.V. Kovalevsky, A.P. Viskup and E.N. Naumovich, Solid State Ionics, 120 1999 281-288

[22] V.V. Vashook, I.I. Yushkevich, L.V. Kokhanovsky, L.V. Makhnach, S.P. Tolochko, I.F. Kononyuk, H. Ullmann and H. Altenburg, Solid State Ionics, 119 1999 23-30

[23] V.V. Kharton, E.V. Tsipis, A.A. Yaremchenko and J.R. Frade, Solid State Ionics 166 2004 327-337

[24] A.A. Yaremchenko, V.V. Kharton, M.V. Patrakeev and J.R. Frade, J. Mater. Chem., 13 2003 1136-1144

[25] V.V. Kharton, A.A. Yaremchenko, A.L. Shaula, M.V. Patrakeev, E.N. Naumovich, D.I. Loginovich, J.R. Frade and F.M.B. Marques, J. Solid State Chem., 177 2004 26-37

[26] V.V. Kharton, A.P. Viskup, A.V. Kovalevsky, E.N. Naumovich and F.M.B. Marques, Solid State Ionics, 143 2001 337-353

[27] V.V. Kharton, A. A. Yaremchenko, E.V. Tsipis and J.R. Frade, SOFC VIII, Ed. M. Dokiya and S.C. Singhal, Electrochem. Soc. Proc. Vol. 2003-07 pp 561-570

[28] H. Kanai, T. Hashimoto, H. Tagawa and J. Mizusaki, Solid State Ionics 99 1997 193-199

[29] E.J. Opila, H.L. Tuller, B.J. Wuensch and J. Maier, J. Am. Ceram. Soc., 76 1993 2363-2369

[30] S.J. Skinner, Solid State Sciences, 5 2003 419-426

[31] J. Jorgensen, B. Dabrowski, S. Pei, D. Richards and D. Hinks, Phys. Rev. B 40 1989 2187-

[32] L. Minervini, R.W. Grimes, J.A. Kilner and K.E. Sickafus, J. Mater. Chem., 10 2000 2349-2354

[33] S.B. Adler, J.A. Lane and B.C.H. Steele, J. Electrochem. Soc., 143 1996 3554-3564

[34] C.N. Munnings and S.J. Skinner, J. Fuel Cell Sci & Tech., submitted.

[35] D.C. Zhu, X.Y. Xu, S.J. Feng, W. Liu and C.S. Chen, Cat. Today, 82 2003 151-156

[36] L. Borovskikh, G. Mazo and E. Kemnitz. Solid State Sci., 5 2003 409-417

[37] M-L. Fontaine, C. Laberty-Robert, F. Ansart and P. Tailhades, J. Solid State Chem., 177 2004 1471-1479

PURE HYDROGEN FOR FUEL CELLS: A NOVEL DESIGN OF COMPACT ENDOTHERMIC CATALYTIC STEAM REFORMER

A. TOMASI, L. MUTRI, F. FERRARI

ITC-Irst, via Sommarie 18, 38050 Trento, Italy; Scandiuzzi S.r.l., viale Dante 78 , Pergine 38057 Trento, Italy; Mechanical Department, Univ. of Trento, via Mesiano 77, 38050 Italy

Abstract: The design and the development of a new concept steam reformer for the production of hydrogen from light hydrocarbon gas feedstock with high efficiency and improved life service are presented. Design incorporates a radiant section and a convection section within a single cylindrical furnace vessel. Thank of this new geometry, the flue gas exits from the steam reformer furnace at about 560 °C instead of the 980 °C released through a conventional design. The reformer consists of a tube regenerative heat exchanger design, so that the final process gas temperature exiting from the reformer tube is only at 530 °C instead of 825 °C. Moreover, in the new conception, the steam reformer doesn't produce excess steam, with an improved efficiency. A radiant burner is used instead of a burner flame; the distance between the burner and the reformer tubes is less than the one used in a conventional steam reformer, allowing a more compact design. Consequently, the maximum reformer tube wall temperature is low, as the reforming pressure. This system is well adapted with the small-scale generation of gases for fuel-cells applications in the range of 1 to 50 kW. Comparing the characteristics of conventional steam reforming technologies, the experimental results obtained with the described system show a low cost and highly energy efficiency.

Key words: Hydrogen/Fuel Cells/Steam Reformer

N. Sammes et al. (eds.), Full Cell Technologies: State and Perspectives, 349-354.
© *2005 Springer. Printed in the Netherlands.*

1. INTRODUCTION

Nowadays, in order to reduce emissions of pollutants and to diversify primary energy supply, Hydrogen and Fuel Cells offers the greatest potential benefits [1-2]. The development of a hydrogen energy infrastructure is often seen as a formidable technical and economic barrier to the use of hydrogen as an energy carrier, but approximately the 95 % of hydrogen production comes from carbonaceous raw materials, and only a fraction of this hydrogen is currently used for energy purposes [3-4]. The main part of the production is used in a chemical feedstock for petrochemical, food, electronics and metallurgical processing industries. However, following the wave of interest in sustainable energy development, hydrogen's share in the energy market is increasing with the implementation of fuel cell systems and the growing demand for zero-emission fuels [5]. The cost of building a sufficient distribution infrastructure and transporting hydrogen over large distances are the harder economic barriers to the implementation in the application of the hydrogen-based technologies. Moreover, large-scale centralized production depends on the increasing of market volumes to compensate the capital expenditures required to build greater capacity. Thus, distributed production via smaller reformer systems is viewed as an attractive near- to mid-term option for supplying hydrogen, particularly for vehicles and regions where low cost natural gas is readily available [6]. Distributed hydrogen production via small-scale reforming at refueling stations could be an attractive near-to min-term option for supplying hydrogen to vehicles, especially in regions with low-cost natural gas production.

Results of collaboration among different industries and research institutes to produce a new concept of small-scale reformer for distributed hydrogen production are presented in this work.

2. STATE-OF-THE-ART

Hydrogen can be produced from different feedstocks using several processes [7]. All these processes, except, of course, dissociation of water, generate synthesis gas as the intermediate product. An overview of these processes is reported in Fig. 1.

From a design point of view, a conventional steam reformer typically consists of a large reformer furnace that contains long vertical tubes that are filled with catalyst [8]. The reformer tubes are supported with counterweights and are heated from burner flames that radiate within the furnace cavity or radiant box. The radiant box is sufficiently large to provide a relatively uniform radiant heat flux.

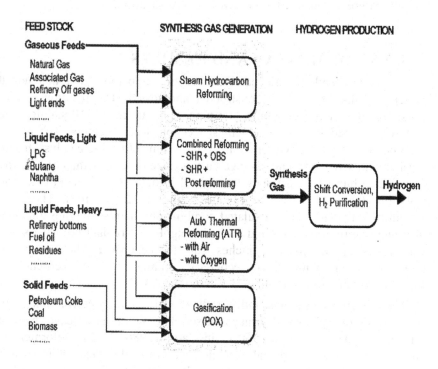

FEED STOCK   SYNTHESIS GAS GENERATION   HYDROGEN PRODUCTION

Gaseous Feeds

Natural Gas
Associated Gas
Refinery Off gases   Steam Hydrocarbon
Light ends   Reforming
....

Liquid Feeds, Light   Combined Reforming
LPG   - SHR + OBS
Butane   - SHR +   Synthesis
Naphtha   Post reforming   Gas   Shift Conversion,   Hydrogen
....   H₂ Purification

Liquid Feeds, Heavy   Auto Thermal
Refinery bottoms   Reforming (ATR)
Fuel oil   - with Air
Residues   - with Oxygen
....

Solid Feeds
Petroleum Coke   Gasification
Coal   (POX)
Biomass
....

Figure 1

In a conventional steam reformer, the outlet gas exiting from the radiant box is typically at about 980 °C. A separate convection section is required to recover additional heat from the flue gas. In a conventional steam reformer, the process gas exits from the steam reformer tubes at about 825 °C, and the excess (export) of steam doesn't supply the optimal energetic efficiency. Moreover, in this kind of systems the reformer tubes are made from expensive high temperature alloys and endothermic heat is commonly supplied by the combustion of carbonaceous fuel and oxidant in a diffusion or turbulent flame burner that radiates to the refractory wall of a combustion chamber, thereby heating them to incandescence, and providing a radiant source for heat transfer to a tubular reaction chamber. Uniform radiation to the surfaces of the tubular reaction chamber is essential because excessive local overheating of the tube surface can produce a mechanical failure. In large scale commercial steam reformers, poor distribution of heat within the furnace chamber is minimized by providing large spacing between the individual reactor tubes, the furnace wall and the burner flames. However, for small-scale catalytic reaction apparatus that is uniquely compact, such as for the production of hydrogen for small fuel cells applications, special design features are needed to prevent tube overheating.

3. EXPERIMENTAL AND RESULTS

A compact endothermic catalytic reaction apparatus that matches the preferred embodiment was constructed and tested. The reaction chamber consisted of 1-inch schedule to 40n pipe, constructed of 310 stainless steal that was formed in a arrangement spaced on 3-inch centers. The reaction chamber was a packet with a commercial steam reforming catalyst that was crushed and screened to an average particle size of approximately 0.25 inch. The radiant burner consisted of 4-inch by 1.5-inch outer diameter cylindrical assembly that had an active radiant angle of 120 degrees. The burner assembly was placed in a insulated combustion chamber having dimension of 6-inch internal diameter and 10-inch height. The radiant burner assembly was spaced approximately 4 inches from the centerline. The new concept steam reformer, with other secondary apparatus, short, vertical, self-supporting steam reformer tubes are used (Fig. 2).

The design incorporates both a radiant section and a convection section within a single cylindrical furnace vessel. The outlet gases leave the steam reformer furnace at about 560°C instead of 980°C. The reformer tube consists of a regenerative heat exchanger design so that the final process gas temperature exiting from the reformer tube is only 530°C instead of 825°C. Moreover in the new design the steam reformer does not produce excess steam and this allows more efficiency. Because a radiant burner is used instead of a burner flame, the distance between the burner and the reformer tubes is less than for a conventional steam reformer, thus allowing a more compact design. Because the maximum reformer tube wall temperature is low and the reforming pressure is low, the reformer tubes are constructed from relatively inexpensive, thin-wall materials.

In conclusion, compared to conventional steam reforming technology, our new-concept steam reformer lends itself to short vertical heights that are suitable for packaging into transportable containers with minimum field erection. Therefore, the system is uniquely compact, low cost, and highly energy efficient.

4. DISCUSSION

To reiterate, the new concept hydrogen generator represents a state-of-the-art integration of several innovative features which enhance overall system efficiency, reliability and performance relative to competitors currently present in the market. Some of these features are:

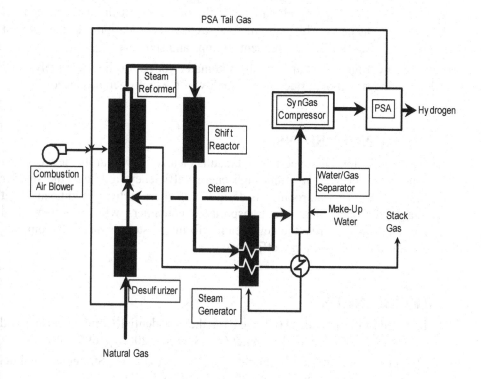

Figure 2. Process Flow Diagram

- a fully automated package that can be easily transported and installed on-site with minimum field installation cost. The unit is designed for fail-safe operation and includes multiple safety features.

- Proprietary reformer system, compact, highly efficient, reliable and easy to control. The reformer combines innovative heat recovery features that reduces process combustion gas temperature ad about 550 °C at the exit of the unit.

- The reformer uses an unique design of burner arrangement that is able to combust natural gas, propane, hydrogen-rich gas (such as Pressure Swing adsorption, PSA and tail gas) or combinations of these gases. The burner has a rapid response and can ramp up from low-load to high-load in seconds.

- The combination of the burner system with the overall system control allows for smooth and reliable transition from start-up fuel to the combustion of PSA gas.

- The hydrogen generator incorporates novel start-up features that allow for rapid start up of the unit without the use of electrical heaters. This system also avoids degradation of the catalyst resulting from frequent startups and shutdowns.

The hydrogen generator can be automated for controlling relatively rapid load changes without affecting the quality of the hydrogen produced

5. CONCLUSIONS

In conclusion, the experimental results clearly show that the new concept steam reformer system offers high energy efficiency, although the smaller volume of its reaction rooms. Its design, with short vertical heights, makes it suitable for packaging into transportable containers with minimum field erection. In few words, the new steam reformer system is uniquely compact, low cost and highly energy efficient.

REFERENCES

[1] A. de Groot et al. "Hydrogen for the Residential Combined Heat and Power", Proceed. of 2002 Fuel Cell Seminar, 2002, p. 695-699.

[2] G.Cacciola, et al. "Technology up Date and New Strategies on Fuel cells", J. of Power Sourcer, 2001, v. 100, p. 67-79.

[3] V. Recupero et al. "Hydrogen Generator via Catalytic Oxidation of Methane for Fuel Cells", J. of Power Sources, 1998, Vol. 100, p. 208-214.

[4] M.Krumpelt et al. "Catalytic Autothermal Reforming for Fuel Cell Systems", Proceed. of 2000 Fuel Cell Seminar, 2000, p. 542-546.

[5] M.V. Kantak et al. "Catalytic Partial Oxidation Reformer Development", Proceed. of 2002 Fuel Cell Seminar, 2002, p. 583-586.

[6] H.K. Geyer et. al."GCtool for Fuel Cells System Design and Analysis: User Documentation", Argonne National Laboratory, 1998, Report ANL-98/8.

[7] R. Bonrup et al. "Fuels and Fuels Impurity Effects and Fuel Processing Catalyst" Proceed. of 2000 Fuel Cell Seminar, 2000, p. 288-291

[8] R. Peters et al. "Kinetics of methane steam reforming", Proceed. of 2000 Fuel Cell seminar, 2000, p. 305-309

MICROSTRUCTURAL CHARACTERIZATION OF La-Cr-O THIN FILM DEPOSITED BY RF MAGNETRON SPUTTERING ON THE STAINLESS STEEL INTERCONNECT MATERIALS FOR SOFC APPLICATION

N. ORLOVSKAYA[1], A. NICHOLLS[2], S. YARMOLENKO[3], J. SANKAR[3], C. JOHNSON[4], R. GEMMEN[4]

Department of Materials Science and Engineering, Drexel University, Philadelphia, PA 19104, USA

Research Resources Center, University of Illinois at Chicago, Chicago, IL 60607, USA

Center for Advanced Materials and Smart Structures, North Carolina A&T State University, Greensboro, NC 27516, USA

National Energy Technology Laboratory, Department of Energy, Morgantown, WV 26507, USA

Abstract: $LaCrO_3$ based perovskites thin films are a prospective material as a protective coating against oxidation and corrosion of metallic interconnect for SOFCs. However, for RF magnetron sputtered $LaCrO_3$ films, the microstructural development of amorphous films during annealing has not been reported, despite the importance of the crystallization processes as one of the factors determining the final properties of the film. SEM and TEM were employed to study the microstructure of the La-Cr-O thin films deposited by RF magnetron sputtering on the stainless steel substrate. The "as-deposited" La-Cr-O thin film was found to be X-ray amorphous with no visible spots in electron diffraction pattern, but after annealing in air, the film transforms first to $LaCrO_4$ monazite type monoclinic phase at 495-530°C and further to the orthorhombic $LaCrO_3$ perovskite at 800°C. In certain cases the second transition to the $LaCrO_3$ perovskite occurred at 700°C. Formation of the porous self-organized $LaCrO_3$ structure was studied by SEM and *in-situ* high temperature TEM. The effect of environment (air, vacuum, hydrogen) on the microstructural evolution of La-Cr-O thin films during annealing is reported.

Key words: Self-Organized Structure/$LaCrO_3$ Perovskite/SOFC/SEM/*In-situ* High Temperature TEM/Electron Diffraction/Amorphous To Crystalline Phase Transition

N. Sammes et al. (eds.), Full Cell Technologies: State and Perspectives, 355-371.
© *2005 Springer. Printed in the Netherlands.*

1. INTRODUCTION

The $LaCrO_3$ based perovskites are considered promising materials in a variety of advanced energy system applications. These include being used as a ceramic interconnect for Solid Oxide Fuel Cells (SOFC) [1], as a material for heat exchangers [2], and as a catalysts for fuel oxidation [3]. Recently, the $LaCrO_3$ perovskites have also been considered for use as an anticorrosive coating for lower cost metallic interconnects used in Intermediate Temperature (IT) SOFCs. [4,5], Because the typical operating temperature for SOFC has been reduced to 650-750°C, the metallic interconnect has become a feasible substitute to the brittle and expensive $LaCrO_3$ based perovskite ceramics that were used as an interconnect material in High Temperature (HT) SOFCs. Still, even at 650-750°C, most prospective metallic interconnects such as Cr rich Ni or Fe based alloys are subject to severe oxidation in the aggressive SOFC environment, especially at the cathode side. Additionally, Cr evaporation is a problem on the cathode side. In the presence of water vapor, Cr evaporates primarily as the hydroxide which can then migrate and contaminate the cathode, significantly decreasing the overall SOFC performance [6-8]. One strategy to overcome these problems is to develop a corrosion resistant coating to improve the oxidation resistance and long term stability of the SOFC devices, while at the same time maintaining good electrical conductivity.

Magnetron sputtering is a very expedient method for depositing a variety of materials as thin films [9-11]. For $LaCrO_3$ based films deposited by magnetron sputtering, the films are typically amorphous when deposited on unheated substrates. However, the films can be transformed to the $LaCrO_3$ perovskite structure by additional annealing. During such annealing in air, two steps occur 1) La-Cr-O will transform first to the oxygen rich $LaCrO_4$ monazite type monoclinic phase at 495-530°C and 2) upon further heating the desired $LaCrO_3$ orthorhombic perovskite will be formed at 800-830°C. In certain cases, the formation of the $LaCoO_3$ orthorhombic phase was reported to occur at 700°C because of the heat pre-treatment of the thin film [4].

In this work results of SEM and TEM study of microstructural evolution of La-Cr-O $\rightarrow$ $LaCrO_4$ $\rightarrow$ $LaCrO_3$ thin films are reported. A microstructural analysis of the film development of pure $LaCrO_3$ as a function of temperature and environment is performed. The formation of the nanostructured self-organized microstructure of the $LaCrO_3$ thin film is demonstrated as a result of the two consecutive phase transitions which occurred during the X-ray amorphous film annealing.

2. EXPERIMENTAL TECHNIQUES

Thin films were deposited by RF magnetron sputtering onto Cr-containing stainless steel (SS) substrates. High-chromium ferritic Fe-23Cr steel coupons (SS 446) with the following chemical composition: Fe (73 wt %), Cr (23 wt %), Mn (1.5 wt %), Ni (0.3 wt %), Si (1.0 wt %), Al (1.0 wt%), C (0.2 wt %) were used as a substrate material. Stainless steel substrates (10x10x5 mm) were polished with a diamond spray to a mirror surface. The substrates were coated in the rf sputtering mode under $8 \times 10e^{-3}$ Torr Ar^+. The substrate temperature was 25°C at the beginning of deposition. The target to substrate distance was 5cm. After pre-sputtering the target for 30 min., the substrates were moved into position under the target and then remained stationary. Light green La-Cr-O films were deposited as a result of 8-hour magnetron sputtering at 500 Watts of power [4]. An "as deposited" film thickness, measured by phase shift technology using an interferometric surface profiler, was 800 nm. EDS analysis of the "as-deposited" film composition gave a La/Cr ratio of 56.54/43.46 at.%. After sputtering, the SS sample with deposited film was annealed at 500°C, 600°C, and 800°C for 1 hour using a controlled heating rate of 30°C per minute. After dwelling at annealing temperature for 1 hour the samples were cooled down at the same rate as during the heating. The samples were annealed both in air and in forming gas ($5wt\%H_2/95wt\%N_2$).

Plain view specimens of the films were prepared using a standard sample preparation technique. The as deposited and annealed at 800°C samples were cut and polished from a SS side down to ~200μm thickness. Further polishing was performed until the thickness of the sample became <30μm. The disks were dimple-ground in the center area from the SS side and Ar^+ ion-beam thinned to electron transparency using a Gatan ion-miller. The least damaging conditions of 4kV beam at a low incidence angle of 10° were used for samples preparation. The microstructure of La-Cr-O ceramics has been studied using A JEOL JEM-3010 TEM operating at 300kV and an environmental SEM. Both surface morphology and microstructure of the thin films have been investigated. High temperature *in-situ* TEM experiments were performed using a Gatan Model 652 double tilt heating stage, with a tantalum furnace, to heat perovskite foils up to 800°C at a controlled heating rate of 1 or 5°C per minute. The oxygen partial pressure in the microscope column is about 5×10^{-8} Pa, therefore the perovskite sample is in a highly reducing atmosphere during the high temperature experiments [12].

3. RESULTS AND DISCUSSION

3.1. SEM OF La-Cr-O $\rightarrow$ LaCrO$_4$ $\rightarrow$ LaCrO$_3$ STRUCTURES

The phase transformations of the amorphous "as deposited" La-Cr-O films to the LaCrO$_4$ monoclinic phase at 600°C and LaCrO$_3$ orthorhombic phase at 800°C was confirmed by XRD [13]. It was reported recently that in the case of La-Cr-O compounds, the amorphous to the monoclinic monazite type LaCrO$_4$ phase transition occurs at 495-530°C. The reaction is exothermic [14]. During this La-Cr-O to LaCrO$_4$ phase transition, strong oxygen absorption from the air occurs. The reaction of the decomposition of LaCrO$_4$ to LaCrO$_3$ perovskite further occurs at 780-840°C. The reaction LaCrO$_4$ (monoclinic) $\rightarrow$ LaCrO$_3$ (orthorhombic) + 0.5O$_2$ is endothermic and TG data showed a weight decrease that corresponds to the oxygen removal from the lattice [14]. The LaCrO$_4$ decomposition is considered to be composed of the four elementary reactions: a) nucleation and growth of LaCrO$_3$; b) a phase-boundary reaction between LaCrO$_4$ and LaCrO$_3$; c) diffusion of oxygen species through the LaCrO$_3$ layer; and d) evolution of O$_2$ gas. As one of the possible mechanisms of the oxygen evolution, the absorption of the oxygen species on the LaCrO$_3$ surface prior to deoxygenation was also proposed. The molecular volume and theoretical density of LaCrO$_4$ are 82.30 Å^3 molecule^{-1} and 5.15 g/cm^3 respectively, and for LaCrO$_3$ these values are 58.58 Å^3 molecule^{-1} and 6.77 g/cm^3 respectively [15]. Therefore, the loss of oxygen in LaCrO$_4$ leads to about 30% decrease of molecular volume, which corresponds to about 10% decrease in the grain diameter of LaCrO$_3$. Thus, two transitions, one from amorphous to monoclinic LaCrO$_4$ monazite type phase, and a second one from LaCrO$_4$ to orthorhombic LaCrO$_3$ perovskite phase, were detected during annealing of La-Cr-O thin films.

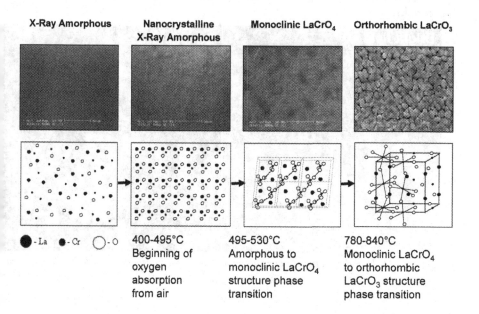

| X-Ray Amorphous | Nanocrystalline X-Ray Amorphous | Monoclinic LaCrO₄ | Orthorhombic LaCrO₃ |

● - La   ● - Cr   ○ - O

| 400-495°C Beginning of oxygen absorption from air | 495-530°C Amorphous to monoclinic $LaCrO_4$ structure phase transition | 780-840°C Monoclinic $LaCrO_4$ to orthorhombic $LaCrO_3$ structure phase transition |

Figure 1. SEM micrographs and a schematic presentation of the structure of the La-Cr-O thin film after magnetron sputtering and annealing at 500, 600, and 800°C for 1 hour in air.

The La-Cr-O, $LaCrO_4$ and $LaCrO_3$ thin films microstructures as a function of annealing temperatures are shown in Fig.1. The "as deposited" film was a very smooth uniform surface. The "as-deposited" film contained approximately 20at% oxygen as measured by EDS analysis. After annealing at 500, 600 and 800°C the oxygen amount was significantly increased (~ 60-80at%). After annealing at 500°C the first signs of structural transition can be seen.

Local structural features with sizes of 50-100 nm emerge from the smooth "as-deposited" surface as a result of the annealing of the La-Cr-O thin film at 500°C. An increase in oxygen content also accompanied this morphology change, but the films were still amorphous to x-ray diffraction.

After annealing at 600°C for 1 hour in air, the formation of the $LaCrO_4$ phase with a small amount (up to 5wt%) of La_2CrO_6 secondary phase was confirmed by XRD [13]. Fine grain structure of $LaCrO_4$ phase can be seen in Fig. 1. Annealing of the films, at 800°C for 1 hour in air, then lead to the subsequent formation of the orthorhombic $LaCrO_3$ perovskite. The porosity of the $LaCrO_3$ film was a result of the two phase transitions; from amorphous to $LaCrO_4$ followed by another transition to the $LaCrO_3$ perovskite phase.

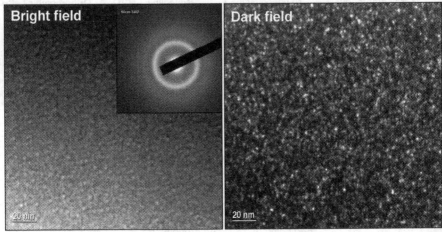

Figure 2. Bright and dark field images along with a selected area electron diffraction pattern of "as-deposited" La-Cr-O thin film.

3.2. TEM OF THE La-Cr-O THIN FILM

The bright and dark field images along with SAED of "as-deposited" La-Cr-O thin film are shown in Fig. 2. While no diffraction spots and only a broad amorphous ring can be seen as a SAED pattern, local structural inhomogeneities, such as short-range to medium range clusters (15-20Å) with no characteristics of a crystalline phase, can still be observed using dark field imaging. These local short- to medium-range clusters occur during the deposition of the La-Cr-O thin film by magnetron sputtering. Such structures are more stable than the homogeneous amorphous phase, but it has no characteristics of a crystalline phase, such as visible peaks in X-ray diffraction or diffraction spots in the transmission electron diffraction pattern. Such local structural inhomogeneities are one of the parameters that controls further heterogeneous nucleation and crystal growth processes during amorphous to crystalline transition [16]. Heating initiates long-range ordering (crystallization) of the structure that is dominated by nucleation and grain growth. All these processes reduce the free energy of the material, and, therefore, provide a thermodynamic driving force for further long range ordering.

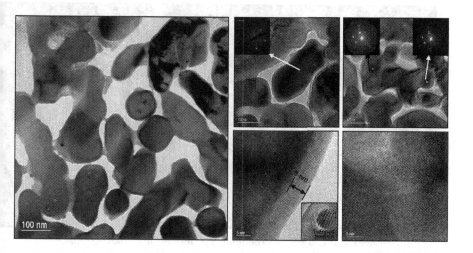

Figure 3. TEM images of the nanostructured self-organized LaCrO₃ thin film with an orthorhombic perovskite structure.

3.3. TEM OF THE LaCrO$_3$ THIN FILM

The bright field images of the LaCrO$_3$ orthorhombic perovskite thin film are shown in Fig. 3. A highly porous nanocrystalline structure with single crystalline grains of 100-300 nm grain size was formed as a result of the annealing of La-Cr-O thin film at 800°C in air for 1 hour (Fig.3A, B). As one can see the grains have very little direct contact between each other, and and are primarily contacted through an intergrain boundary phase. More detailed analysis revealed that this intergrain boundary phase is an amorphous layer on the surface of each grain with a thickness of about 5nm (Fig, 3C). The origin of this amorphous layer is not clear since it might be a residue of the parent amorphous structure that was retained after annealing, residual material not able to crystallize because it is kinetically frustrated by being of a dissimilar composition to the crystalline phases present, or the amorphous layer may be formed during the sample preparation during ion milling.

It can be seen from the bright field images that a number of grain types exist. While the majority of the grains are relatively round, irregular grains that appear to be the result of several smaller grains growing together exist, as well as an unusual grain that has a "tomato-slice" appearance (Fig.4). SAED patterns taken from a single irregular grain shows that the grain orientation is the same from two different parts of the grain, as one can see in Fig.3D. Each of the grain consists of one single domain with lattice fringes of 4Å. Rounded heavily twinned precipitates are also found within the single grains (Fig. 3D insert).

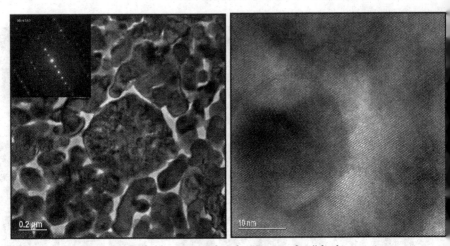

Figure 4. Bright field micrographs of a "tomato-slice" looking grain.

While the majority of the grains have a round dendritic morphology, another type of grains has been found to exist. A "tomato-slice" looking grains has much more complicated structure (Fig.4). The "tomato slice" is possibly the La_2CrO_6 phase, because X-ray analysis detected traces of this phase after annealing at 800°C for 1 hour in air. However, further analysis is required to determine the exact phase of the "tomato slice" grains.

EDS analysis of the $LaCrO_3$ grains and intergranular phases revealed that the La, Cr, and O elements are the major components of the dendritic grains (Fig.5). Some amount of Fe can be also detected because of the existence of the ferritic substrate. Cu is detected because of the copper grid that supports a thin film sample. However, in a number of selected intergranular locations both Si and Al elements were present in rather significant quantities. These two elements (Si and Al) are present as an impurity in the chromium rich ferritic alloy. The migration of Si and Al into the perovskite films and segregation to the grain boundaries can be expected to significantly increase the resistance of the films and therefore decrease cell/stack performance.

Si present in the substrate also creates another problem (Fig. 6). As can be seen in the SEM cross section, a continuous SiO_2 layer forms beneath Cr_2O_3 layers when the metallic 446 interconnect coated with 0.2 μm porous $LaCrO_3$ thin film is annealed at 900°C for 100 hours. Formation of such insulating layer is also detrimental for the electrical performance of the SOFC. Though the cross section analysis of structure of oxide layers did not reveal formation of alumina layer, there is still a significant amount of Al detected by EDS.

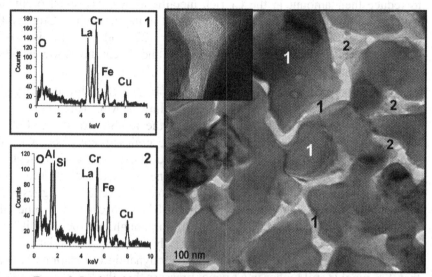

Figure 5. Bright field micrograph and EDS analysis of LaCrO₃ dendritic grains and interboundary phase.

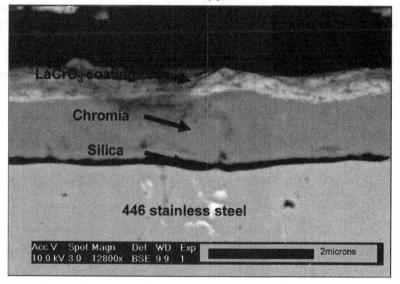

Figure 6. Backscattered cross section SEM image of 446 SS metallic interconnect material coated with 0.2 μm LaCrO₃ thin film after annealing at 900°C for 100 hours in air.

Further work is required to determine the location and composition of the Al containing phases and their effect on the resistivity of the material. One of the possibilities of diminishing the influence of the Si and Al impurities is to reduce their amount in the Cr rich metallic alloys. The most promising metallic alloy which is present on the market at the moment is Crofer APU alloy with a basic composition of Fe, Cr (22wt%), Mn (0.4wt%), Ti (0.05wt%), La (0.09wt%). The amount of Si and Al can be as low as 0.005 and 0.007, respectively.

Another type of impurity that is present in 446 SS alloy and can also affect the structure of the film after annealing is Mn. This leads to the formation of Mn rich grains on the surface of the 466 SS coated with $LaCrO_3$ thin film (Fig. 7). The composition of these grains is consistent with the $(Mn,Cr)_3O_4$ spinel phase. The formation of the spinel is not as detrimental as the continuous silica layer formation mentioned earlier. Indeed, alloys that form this spinel as the first stable native oxide have been suggested as possible alternatives to the chromia formers. This is because the spinel also has some conductivity when the layer is thin enough, and because it has a lower vapor pressure of Cr evaporation (mostly as $CrO_2(OH)_2$) relative to the normal Cr_2O_3 native oxide. However, thermodynamic calculations indicate that the spinel is still higher in Cr vapor pressure than the $LaCrO_3$ perovskite. Thought, the $LaCrO_3$ film is porous and potentially can not provide a complete corrosion protection for the metallic interconnect against oxidation in the aggressive SOFC environment, a certain degree of protection against a formation of oxide layers on the metallic interconnect surface still exists. As one can see from Fig. 7, the formation of large oxide grains is observed on the surface without $LaCrO_3$ coating, such as grooves or edges, however formation of such large grains is visibly decreased for the surface coated with porous $LaCrO_3$ perovskite coating.

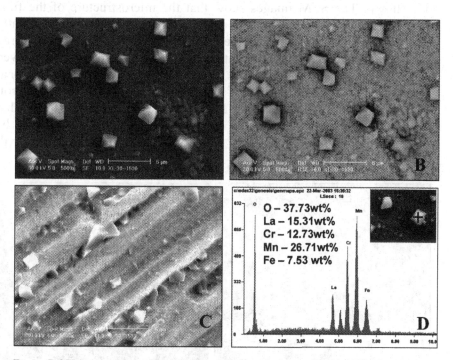

Figure 7. SEM micrographs of the LaCrO₃ thin film after annealing at 800°C for 100 hours. (A) and (C) are SEI images and (B) is a backscattered image. (D) is the EDS analysis of Mn rich grain that grown on the LaCrO₃ surface as a result of high temperature treatment.

3.4. EFFECT OF THE OXYGEN PARTIAL PRESSURE ON THE FORMATION OF LaCrO₄ AND LaCrO₃ PHASES

The effect of the environment on the formation of $LaCrO_4$ and $LaCrO_3$ phases has been studied in [15]. It was shown that at high oxygen partial pressure (pure O_2) $LaCrO_4$ will form at 650°C, while at low oxygen partial pressure in inert atmosphere (pure N_2) the formation of $LaCrO_3$ orthorhombic perovskite will occur during the high temperature treatment of the amorphous precursor. Therefore, we hypothesized that if we perform the annealing of X-ray amorphous La-Cr-O film in the reducing atmosphere, we will be able to avoid the formation of the intermediate oxygen rich $LaCrO_4$ phase, and the $LaCrO_3$ orthorhombic phase will be obtained without formation of the porosity. Forming gas (5%H_2/95%N_2) was used to anneal La-Cr-O thin film at 800°C for 1 hour. The microstructures of the $LaCrO_3$ thin film after annealing is shown in Fig. 8. For comparison the microstructure of the La-Cr-O film annealed in air at the same conditions is also shown. The SEM images show that the microstructure of the film annealed in forming gas is pronouncedly different, with a significantly smaller crystallite size than when the film is annealed in air. However, X-ray analysis revealed formation of an intermediate $LaCrO_4$ monazite phase even when annealing in forming gas. It is possible that the compositional inhomogeneity of the "as deposited" film, as seen in figure 2, may play a roll in allowing for the formation of the $LaCrO_4$ phase on a localized level. That is, despite annealing in a reducing environment, localized regions may be close enough to the composition of $LaCrO_4$ such that this phase will nucleate.

Figure 8. SEM micrographs of the $LaCrO_3$ thin film after annealing at $800^\circ C$ for 1 hour in air (A) and in forming gas ($5\%H_2/95\%N_2$) (B).

Microstructural Characterization Of La-Cr-O Thin Film
Deposited By RF Magnetron Sputtering On The Stainless Steel
Interconnect Materials For SOFC Application

367

3.5. HIGH TEMPERATURE *IN-SITU* TEM OF La-Cr-O THIN FILM

Figure 9. SAED of La-Cr-O thin film taken during the annealing at in-situ high temperature experiment in TEM microscope column.

In order to verify the mechanism of LaCrO$_3$ orthorhombic perovskite phase formation, the high temperature *in-situ* TEM of the amorphous La-Cr-O thin film was performed. The foil of La-Cr-O thin film with a thickness of about 100nm was prepared, which was surrounded by a much thicker sample that consisted of both 446SS and La-Cr-O thin film on the top of SS. At 500°C the film was amorphous, with only two weak electron diffraction spots visible (Fig. 9A). During the heating the first significant signs of the crystalline phase appeared at 700°C within the amorphous matrix (Fig. 9B). At 756°C the film fully transformed to the crystalline phase (Fig.9C). Such transformation did not occur everywhere, and an amorphous ring could be observed in a vast majority of the visible film areas. Therefore the amorphous phase is very stable up to very high temperatures in vacuum. No intermediate phase, such as LaCrO$_4$, has been observed during the heating. The one possible reason is that there is oxygen desorption from the amorphous film in the high vacuum (5x10^{-8} Pa) that exists inside the microscope column.

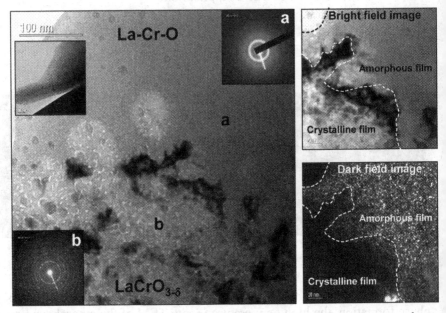

*Figure 10. A La-Cr-O sample after heating at 760°C for ~1 hour in the microscope column.
The pictures were taken at RT after cooling.*

The amorphous phase remained stable even after a full day of heating as can be seen in Fig 10a where significant areas of the film still can be shown remain amorphous as determined by SAED. The interface between amorphous and crystalline areas can be seen (Fig. 10). The crystalline phase has small 10-15 nm crystallites, which overlap with each other. Some nanoporosity is also present with pore sizes of 5-10 nm (Fig 11).

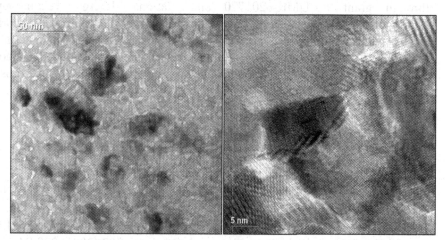

Figure 11. A LaCrO₃ perovskite after heating at 760°C for ~1 hour in reducing environment. The oxygen partial pressure in the microscope column is 5*10⁻⁸ Pa during the experiment.

4. SUMMARY

The microstructure of La-Cr-O thin film was characterized by SEM and TEM. The "as-deposited" La-Cr-O thin film was X-ray amorphous, and no electron diffraction spots can be seen in SAED pattern. However, it consists of short range clusters with a size of 1-2nm. As a result of annealing at different temperatures in air, several distinct film structures were formed. Annealing at 600°C in air lead to the formation of $LaCrO_4$ monoclinic structure; after annealing in air at 800°C the orthorhombic $LaCrO_3$ perovskite structure was formed. Since the monazite phase is less dense than the perovskite phase, the $LaCrO_4$ to $LaCrO_3$ phase transition leads to a porous morphology of $LaCrO_3$ perovskite phase. Annealing in forming gas (5%H2/95%N2) leads to a different microstructure with smaller crystallites, however the intermediate $LaCrO_4$ phase is not avoided completely.

A thin residual amorphous layer was found between grains in the annealed films. EDS analysis of these regions showed that the amorphous material contains Si and Al. SiO_2 was also formed as a continuous sublayer for films annealed in air for longer times. Since the source of these detrimental elements is the substrate (SS446), we conclude that interconnect metallic alloys with lower Al and Si impurity content are needed in order to prevent formation of insulating SiO_2 and Al_2O_3 layers.

ACKNOWLEDGEMENTS

The work at Drexel was supported by the National Science Foundation through grant # DMR-0201770 and National Energy Technology Laboratory, US Department of Energy under contract # 239811. TEM work at UIC was supported by Army Center for Nanoscience and Nanomaterials NC A&T State University. The authors would like to acknowledge Scot Laney, Univ. of Pittsburgh, Material Science and Engineering, for the cross section SEM image (Figure 6).

REFERENCES

[1]   W.Z. Zhu, S.C. Deevi, Mat Sci. Eng. A 348 1-2 (2003) 227-243.

[2]   D.B. Meadowcroft, P.G. Meier, A.C. Warren, Energy Convers., 12 (1972) 145-151.

[3]   J. Sfeir, P. Buffat, P. Moeckli, N. Xanthopoulos, R. Vasquez, H.J. Mathieu, J. van herle, K.R. Thampi, J. Catalysis, 202 (2001) 229-244.

[4]   C. Johnson, R. Gemmen, N. Orlovskaya, Composites, Part B: Engineering, 35 2 (2004) 167-172.

[5]   N. Orlovskaya, A. Coratolo, C. Johnson, R. Gemmen, J. Am. Ceram. Soc., (2004) in press.

[6]   D.-H. Peck, M. Miller, K. Hilpert, Solid State Ionics, 143 (2001) 401-412.

[7]   P. Kofstad, Proceedings of the Second European SOFC Forum, (1006)479-490.

[8]   J. Urbanek, M. Miller, H. Schmidt, K. Hilpert, Proceedings of the Second European SOFC Forum, (1006) 503-512.

[9]   D. O. Klenov, W. Donner, L. Chen, A. J. Jacobson, S. Stemmer, J. Mat. Res., 18 1 (2003) 188-194.

[10] K. Wakisaka, H. Kado, S. Yoshikado, Electroceramics in Japan VI, Key Enegineering Materials, 248 (2003) 121-124.

[11] E.L. Brosham, R. Mukundan, D. Brown, Q.X. Jia, R. Lujan, F.H. Garzon, Solid State Ionics, 166 (2004) 425-440.

[12] N. Orlovskaya, Y. Gogotsi, A. Nicholls, Proceedings of 7th Int. Symp. On Solid Oxide Fuel Cells, Eds. Yokokawa, H., Singhal, S.C., (2001) 624-630.

[13] N. Orlovskaya, C. Johnson, R. Gemmen, unpublished results.

[14] K. Azegami, M. Yoshinaka, K. Hirota, O. Yamaguchi, Mat. Res. Bull., 33 2 (1998) 341-348.

[15] A. Furusaki, H. Konno, R. Furuichi, Thermochiem Acta, 253 (1995) 253-264.

[16] H. Touir, J. Dixmier, K. Zellama, J. Morhange, P. Elkaim, J Non-Crystal Solids, 227-230 (1998) 906-910.

CaTiO$_3$-LaCrO$_3$-CaCrO$_3$ and CaTiO$_3$-LaCrO$_3$-La$_{2/3}$TiO$_3$ QUASI-TERNARY SYSTEMS

L. VASYLECHKO[1], V. VASHOOK[2], U. GUTH[2]

[1] "L'viv Polytechnic" National University, Lviv Ukraine
[2] Dresden University of Technology, Dresden, Germany

Abstract: Phase equilibria in the CaTiO$_3$–LaCrO$_3$–CaCrO$_3$ and CaTiO$_3$–LaCrO$_3$–La$_{2/3}$TiO$_3$ quasi-ternary systems and binary systems bounding them have been studied by means of high-resolution powder diffraction technique by using X-ray and synchrotron radiation. Based on the results of X-ray phase and structural analysis of the La$_{1-x}$Ca$_x$Cr$_{1-y}$Ti$_y$O$_3$ and La$_{1/3(3-2x-y)}$Ca$_x$Cr$_{1-y}$Ti$_y$O$_3$ samples, synthesized by a ceramic technique in air at 1620 K, the isothermal sections of corresponding phase diagrams are constructed. Extended solid solution with orthorhombic perovskite-like GdFeO$_3$-type of structure (space group *Pbnm*) and a wide homogeneity range is formed in the CaTiO$_3$–LaCr^{III}O$_3$–CaCr^{IV}O$_3$ quasi-ternary system. The region of existence of the solid solution at used experiment conditions reaches up to 0.6÷0.7 molar fractions of CaCr^{IV}O$_3$. Four kinds of solid solutions with different type of the *A*-cation deficient perovskite structures (space groups *Pbnm*, *Imma*, *I4/mcm* and *Cmmm* or *P2/m*) are formed in the CaTiO$_3$–LaCrO$_3$–La$_{2/3}$TiO$_3$ quasi-ternary system, depending on the level of deficiency of the structures. Concentration dependencies of the lattice parameters of the perovskite-like phases in the CaTiO$_3$–La$_{2/3}$TiO$_3$ and LaCrO$_3$–La$_{2/3}$TiO$_3$ pseudo-binary systems are presented. Stabilization of the La$_{2/3}$TiO$_3$ structure by the low-level substitution in both *A*- and *B*-cation sites and the structural peculiarities of the Ca- and Cr-stabilized La$_{2/3}$TiO$_3$ are discussed. Continuous phase transitions from orthorhombic (*Cmmm*) and monoclinic (*P2/m*), to tetragonal (*P4/mmm*) structure in the Ca- and Cr-stabilized La$_{2/3}$TiO$_3$ structures are described.

Key words: Mixed Titanate-Chromite/Perovskite/Phase Diagram/Phase Transition

N. Sammes et al. (eds.), Full Cell Technologies: State and Perspectives, 373-380.
© 2005 Springer. Printed in the Netherlands.

1. INTRODUCTION

Perovskite-type oxides formed in the $CaTiO_3$–$LaCrO_3$–$CaCrO_3$ and $CaTiO_3$–$LaCrO_3$–$La_{2/3}TiO_3$ quasi-ternary systems are of great interest, because of their high chemical stability and high electrical conductivity within broad ranges of temperature and oxygen partial pressure. Mixed occupation of the A- and B-sites by varying of the La:Ca and Cr:Ti ratios coupled with the formation of A-cation vacancies gives the possibility for the precise modification of the structural and physical-chemical properties, that are important for the optimization of electrode materials using in the different electrochemical devices (SOFC, oxygen sensors, oxygen pumps, oxygen permeable membranes, etc).

In the present work we report the results of the study of the interaction between components in the $CaTiO_3$–$LaCrO_3$–$CaCrO_3$ and $CaTiO_3$–$LaCrO_3$–$La_{2/3}TiO_3$ quasi-ternary systems. Phase relationships and structure peculiarities of the phases formed in these systems are described and discussed.

2. EXPERIMENTAL

$La_{1-x}Ca_xCr_{1-y}Ti_yO_3$ and $La_{(3-2x-y)/3}Ca_xCr_{1-y}Ti_yO_3$ samples were prepared by a standard ceramic solid state technique from the powders of La_2O_3, $CaCO_3$, Cr_2O_3 and TiO_2 with a 99.9 % purity. Details of the samples preparation are reported in [1-3]. X-ray phase and structural analysis were performed by means of the high-resolution powder diffraction (HRPD) technique using a Siemens D5000 X-ray diffractometer (Cu K_α- radiation, $\theta/2\theta$ - scanning mode, step width of $0.02°$, counting time per step – 7 s). High temperature behavior of the structures in the temperature range 293 to 1273 K in air was studied by means of HRPD technique and synchrotron radiation. *In situ* diffraction experiments were carried out at the Hamburg Synchrotron Radiation Laboratory HASYLAB by using powder diffractometer at beamline B2, equipped with STOE furnace. All crystallographic calculations have been performed by using WinCSD program package.

3. RESULTS AND DISCUSSION

Based on the results of X-ray phase and structural analysis of the $La_{1-x}Ca_xCr_{1-y}Ti_yO_3$ samples synthesized by a ceramic technique in air at 1620 K, the isothermal sections of the $CaTiO_3$–$LaCr^{III}O_3$–$CaCr^{IV}O_3$ quasi-ternary system has been constructed [3]. Extended solid solution with perovskite-like $GdFeO_3$-type of structure (space group *Pbnm*) is formed in this system in a wide homogeneity range. The existence area of the solid solution at used experiment conditions reaches up to $0.6\div0.7$ molar fractions of $CaCr^{IV}O_3$.

Lattice parameters and interatomic distances of the samples within the solid solution range change monotonically with the composition, in accordance with change of average radii both of A-and B-cations. The structural peculiarities of the solid solution formed in CaTiO$_3$–LaCr^{III}O$_3$–CaCr^{IV}O$_3$ quasi-ternary system are described in detail in [3].

The interaction of the components in the CaTiO$_3$–LaCrO$_3$–La$_{2/3}$TiO$_3$ quasi-ternary system is much more complicated. Four kinds of the $(1-x)$(CaTiO$_3$) $-$ x(La$_{2/3}$TiO$_3$) solid solutions with different type of A-cation deficient perovskite structures are formed in the CaTiO$_3$–La$_{2/3}$TiO$_3$ pseudo-binary system. Similar to CaTiO$_3$, the samples with x=0÷0.4 display the GdFeO$_3$-type structure. A-cation vacancies in the structures of these samples are statistically distributed on Ca(La) cation positions. At the intermediate concentration of La$_{2/3}$TiO$_3$ (x=0.4÷0.7) the solid solution with body-centered orthorhombic (pseudo-tetragonal) *Imma* structure is formed. Further increasing content of La$_{2/3}$TiO$_3$, and following A-cation deficiency led to tetragonal structure (space group *I4/mcm*). In the La$_{2/3}$TiO$_3$-rich range of the system (x=0.8÷0.96) the ordered perovskite-like structure is formed. The alternation of fully and partially (~40 %) occupied layers of A-cations is observed in these structures. The changing with La$_{2/3}$TiO$_3$ content lattice parameters of the perovskite phases formed in CaTiO$_3$–La$_{2/3}$TiO$_3$ system are presented in Fig. 1.

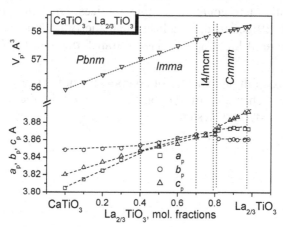

Figure 1. Concentration dependencies of the lattice parameters in the CaTiO$_3$–La$_{2/3}$TiO$_3$ system. Lattice parameters and cell volumes are reduced to perovskite-like cell.

In the $(1-x)$(LaCrO$_3$)$-x$(La$_{2/3}$TiO$_3$) pseudo-binary system, two kinds of the perovskite-like structures are found. LaCrO$_3$-rich compositions have the GdFeO$_3$-type structure, whereas in the La$_{2/3}$TiO$_3$-rich region of the system the solid solution with the ordered A-cation vacancies is formed. Both these perovskite phases co-exist in a wide concentration range of $x=0.2\div0.8$. An extended two-phase perovskite region is observed in the $LaCrO_3-La_{2/3}TiO_3$ system, in contrast to $CaTiO_3-La_{2/3}TiO_3$ one (Fig. 2).

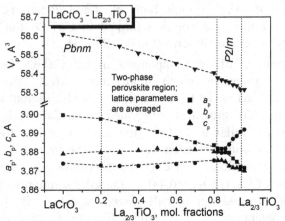

Figure 2. Concentration dependencies of the lattice parameters in the $LaCrO_3-La_{2/3}TiO_3$ system. Lattice parameters and cell volume are reduced to perovskite-like cell.

In the $CaTiO_3-LaCrO_3$ pseudo-binary system the solid solution with orthorhombic *Pbnm* structure is formed. Because $LaCrO_3$ and $CaTiO_3$ have different ratios of cell parameters within the same $GdFeO_3$-type structure ($a>b$ and $a<b$, respectively), some of the samples from this system display the metrically tetragonal or cubic lattices. *In situ* high-temperature synchrotron diffraction data indicate that the orthorhombic (pseudo-cubic) *Pbnm* structure of $La_{0.5}Ca_{0.5}Cr_{0.5}Ti_{0.5}O_3$ transforms into the *Imma* one above 1070 K.

The nature of the interaction of components in the pseudo-binary systems predetermines character of the phase equilibria in the $CaTiO_3-LaCrO_3-La_{2/3}TiO_3$ quasi-ternary system (Fig. 3).

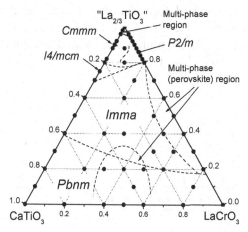

Figure 3. Isothermal section of the $CaTiO_3$–$LaCrO_3$–$La_{2/3}TiO_3$ *phase diagram.*

The formation of the extended solid solution with the $GdFeO_3$-type structure is observed in the region nearby to the $CaTiO_3$–$LaCrO_3$ pseudo-binary system. For this region the low level of statistically distributed *A*-cation vacancies (less as 7–13 %) is typical. The intermediate level of *A*-cation vacancies (23–7 %) favours the formation of structures with the body-centered orthorhombic *Imma* structure. Co-existence of two perovskite phases is observed in some samples between *Pbnm* and *Imma* structural fields (Fig. 3). At the concentration of *A*-cation vacancies ~ 23–28 % the formation of the solid solution with tetragonal structure (space group *I4/mcm*) is observed. Both, *Imma* and *I4/mcm* phases were found among the Ca-containing samples only. They do not exist in the $LaCrO_3$–$La_{2/3}TiO_3$ pseudo-binary system and in the adjoined to it part of the ternary system.

In the $La_{2/3}TiO_3$-rich corner of the phase diagram the perovskite-like structures with the alternation of fully and partially occupied layers of *A*-cations are formed. The "pure" $La_{2/3}TiO_3$ structure is thermodynamically unstable and can be stabilized with the various low-level of substitutions for La and Ti on the *A*- and *B*-sites, respectively [4-7]. There are controversies in the literature concerning the structure of $La_{2/3}TiO_3$. Different authors report *Pban* [4], *Cm2m* [5] or *Cmmm* [6] structural models for the "pure", Li- and Sr-stabilized structures of $La_{2/3}TiO_3$, respectively. In the $CaTiO_3$–$LaCrO_3$–$La_{2/3}TiO_3$ quasi-ternary system the $La_{2/3}TiO_3$ structure is stabilized by the low-level substitutions of Ca (4–20 at. %) and/or Cr (6–14 at. %) for La and/or Ti, respectively. In our previous publication [2] it was mentioned, that Ca-stabilized $La_{2/3}TiO_3$ structure could be described either in orthorhombic $2a_p \times 2a_p \times 2a_p$ lattice (space group *Cmmm*) or in monoclinic $\sqrt{2}a_p \times 2a_p \times \sqrt{2}a_p$ one (space group *P2/m*). Precise analysis of the structures from the $La_{2/3}TiO_3$-rich part of the ternary system proves that the Ca-stabilized $La_{2/3}TiO_3$ is orthorhombic (*Cmmm*), whereas for the Cr-stabilized samples the monoclinic deformation is evident. The graphical results of the Rietveld refinement of the $La_{0.64}Ca_{0.04}TiO_3$ and $La_{0.707}Ti_{0.88}Cr_{0.12}O_3$ structures are shown in Fig. 4, for examples. The refined structural parameters of these structures are given in Tab. 1.

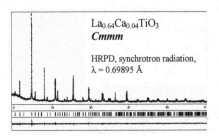
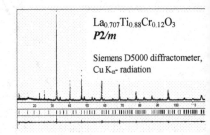

Figure 4. Graphical results of the Rietveld refinement of $La_{0.64}Ca_{0.04}TiO_3$ and $La_{0.707}Ti_{0.88}Cr_{0.12}O_3$ structures at 298 K.

From the extrapolation of the lattice parameters of the $(1-x)(CaTiO_3)$–$x(La_{2/3}TiO_3)$ and $(1-x)(LaCrO_3)$–$x(La_{2/3}TiO_3)$ solid solutions the cell parameters of "pure" $La_{2/3}TiO_3$ could be predicted: a_0=7.750 Å, b_0=7.721 Å, c_0=7.794 Å for *Cmmm* model and a_m=5.470 Å, b_m=7.794 Å, c_m=5.471 Å, β= 90.22° for *P2/m* one (Fig. 5). Taking into account the relationships between *Cmmm* and *P2/m* lattices ($a_m = c_m = \frac{1}{2}(a_0^2 + b_0^2)^{1/2}$, $b_m = c_0$, $\beta = 2arctg(a_0/b_0)$), an excellent coincidence between the values of the lattice parameters in both structural model was obtained. Since the predicted values of a_m and c_m are the same, and value of β is equal $2arctg(a_0/b_0)$, it may be concluded, that "pure" $La_{2/3}TiO_3$ has orthorhombic structure.

Table 1. Structural parameters of $La_{0.64}Ca_{0.04}TiO_3$ and $La_{0.707}Ti_{0.88}Cr_{0.12}O_3$.

Lattice parameters	Atoms, sites	x	y	z	B_{iso}	G
$La_{0.64}Ca_{0.04}TiO_3$, space group *Cmmm*, $R_I = 0.0747$. $R_P = 0.1504$						
	La1/Ca, 4g	0.2521(4)	0	0	0.67(2)	0.923(3)/ 0.077(3)
a=7.74664(6)	La2, 4h	0.2576(8)	0	1/2	0.98(5)	0.373(2)
b=7.72100(6)	Ti, 8n	0	0.2517(11)	0.2605(2)	0.77(4)	1.001(4)
c=7.78409(6)	O1, 4i	0	0.264(2)	0	1.2(4)	1.00(1)
V=465.58(1)	O2, 4j	0	0.224(2)	1/2	1.9(4)	0.99(1)
	O3, 4k	0	0	0.214(2)	1.1(4)	0.98(2)
	O4, 4l	0	1/2	0.260(2)	2.0(4)	0.99(2)
	O5, 8m	1/4	1/4	0.2320(9)	2.0(4)	1.01(1)
$La_{0.707}Ti_{0.88}Cr_{0.12}O_3$, space group *P2/m*, $R_I = 0.0611$, $R_P = 0.1595$						
	La1, 2m	0.252(2)	0	-0.2506(14)	0.75(8)	0.903(5)
	La2, 2n	0.253(4)	1/2	0.741(2)	0.94(14)	0.507(3)
a=5.4767(1)	Ti/Cr, 4o	0.256(3)	0.2627(5)	0.251(3)	0.76(14)	0.88(5)/ 0.12(5)
b=7.7680(2)	O1, 2n	0.297(7)	1/2	0.279(11)	1.4(14)	1.03(3)
c=5.4866(1)	O2, 2m	0.235(10)	0	0.293(7)	2.0(13)	0.97(3)
β=90.045(3)	O3, 2l	1/2	0.233(4)	1/2	1.4(16)	0.98(5)
V=233.42(1)	O4, 2k	0	0.256(5)	1/2	1.7(15)	0.97(6)
	O5, 2j	1/2	0.237(4)	0	1.5(15)	1.00(6)
	O6, 2j	0	0.279(4)	0	1.8(16)	0.99(5)

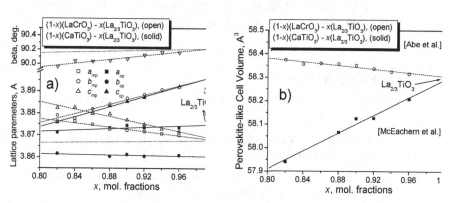

Figure 5. Perovskite lattice parameters and cell volumes of $(1-x)(La_{2/3}TiO_3)-x(CaTiO_3)$ and $(1-x)(La_{2/3}TiO_3)-x (LaCrO_3)$ solid solutions.

In situ HT investigation of the $La_{0.64}Ca_{0.04}TiO_3$ and $La_{0.6}Ca_{0.1}TiO_3$ structures revealed the continuous phase transformations from orthorhombic to tetragonal structure (space group P4/mmm) above the temperatures 704 and 784 K, respectively (Fig. 6, a). Approximately in the same temperature range the transition from monoclinic to tetragonal structure occurs in the Cr-stabilized $La_{2/3}TiO_3$ structure (Fig. 6, b). Possibly, this transition passes through the formation of intermediate orthorhombic phase in a narrow temperature range. The study of the thermal behaviour of the other structures in the $CaTiO_3$–$LaCrO_3$–$La_{2/3}TiO_3$ quasi-ternary system is now in progress.

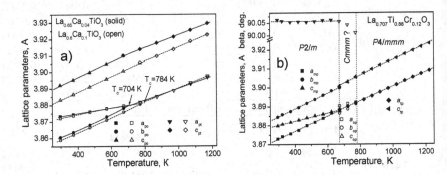

Figure 6. Temperature dependencies of the lattice parameters in $La_{0.65}Ca_{0.04}TiO_3$, $La_{0.6}Ca_{0.1}TiO_3$ (a) and $La_{0.707}Ti_{0.88}Cr_{0.12}O_3$ (b) structures

ACKNOWLEDGMENTS

The work was supported in parts by DFG (Project Gu-1-4), WTZ (UKR 01/12), Ukrainian Ministry of Education and Science (Projects "Cation" and M/85-2003) and ICDD Grant-in-aid program.

REFERENCES

[1] Vashook, L. Vasylechko, J. Zosel, U. Guth, Solid State Ionics. 159(3-4) (2003) 279.

[2] Vashook, L. Vasylechko, M. Knapp, H. Ullmann, U. Guth, J. Alloys and Compounds. Vol. 354(1-2) (2003) 13.

[3] Vashook, L. Vasylechko, J. Zosel, W. Gruner, H. Ullmann, U. Guth, J Solid State Chem., (2004) to be published.

[4] M.J. MacEachern, H. Dabkowska, J.D Garrett et al. Chem. Mat. 6 (1994) 2092.

[5] J. Sanz, J.A. Alonso, A. Varez, M.T. Fernandez-Diaz, J. Chem. Soc. Dalton Trans. (2002) 1406.

[6] C.J. Howard, Zh. Zhang J. Phys.: Condens. Matter. 15 (2003) 4543.

[7] M. Abe, K. Uchino, Mater. Res. Bull. 9 (1974) 147.

SOFC POWDER SYNTHESIS BY THE ORGANIC STERIC ENTRAPMENT METHOD

W.M. KRIVEN, B.R. ROSCZYK

Materials Science and Engineering Department, University of Illinois at Urbana-Champaign, Urbana, IL 61801, USA, (t) +1 217 333 5258 (f) 1 217 333 2736, kriven@uiuc.edu

Abstract: The organic steric entrapment was used to synthesize homogeneous powders of strontium doped lanthanum magnesium gallate ($La_{0.8}Sr_{0.2}Ga_{0.83}Mg_{0.17}O_{3-\delta}$), abbreviated "LSGM"; samaria doped cerium oxide ($Ce_{0.8}Sm_{0.2}O_{1.9}$), abbreviated "SDC"; strontium iron cobaltite ($SrFeCo_{0.5}O_x$), abbreviated "SFC". Powders were characterized for specific surface area and particle size distribution. Sintered pellets were examined by SEM and EDS and they showed homogeneous microstructures, which in the case of LSGM and SFC were multiphasic. The steric entrapment powder synthesis method is simple and relatively cheap. The resulting powders generally have a high surface area, good homogeneity and are highly sinterable.

Key words: Oxide Powder Synthesis/Organic Steric Entrapment/Doped Lanthanum Gallate/Samarium Doped Cerium Oxide/Strontium Iron Cobaltite

1. INTRODUCTION

As the demand for more pure and complex ceramic materials increases, there has been interest in chemical synthesis methods that tend to produce powders that are finer, more reactive and compositionally homogeneous. The traditional powder synthesis method, solid-state reaction, consists of mixing component powders at high temperature and relying on the increased diffusion to form desired compounds. The high temperature and large diffusion distances result in large particle sizes, more inhomogeneous powders and high energy costs.

N. Sammes et al. (eds.), Full Cell Technologies: State and Perspectives, 381-394.
© 2005 Springer. Printed in the Netherlands.

1.1. THE ORGANIC-INORGANIC STERIC ENTRAPMENT METHOD OF POWDER SYNTHESIS

The concept of organic steric entrapment was the focus of this work and can be divided into three categories: the aqueous polyvinyl alchohol (PVA) method, the non-aqueous ethylene glycol (EG) or polyethylene (PEG) method and the nano-sol modified, PVA method. The method of organic steric entrapment was developed in our laboratory over the past 10 years and has resulted in numerous publications [1-16].

1.1.1. THE AQUEOUS POLYVINYL ALCOHOL METHOD

Polyvinyl alcohol (PVA) is a cheap, commercially available polymer that may be produced by hydrolysis of polymerized vinyl acetate. PVA may have varying degrees of hydrolysis resulting from only partial replacement (up to about 87% for partially hydrolyzed PVA) of the acetate groups (-$COOCH_3$) by the hydroxyl groups (-OH) along the polymer molecule. The PVA ceramic powder synthesis method uses PVA dissolved in an aqueous solution to not only chelate with the ions in solution, as in the case of glycine and other organic molecules, but to physically prevent precipitation by sterically inhibiting interaction among dissolved ions and retarding their precipitation from aqueous solution. The positively charged cations are thought to become sterically entrapped by the polymer coils and their mobility is lessened by weak hydrogen bonding from the hydroxyl (-OH) groups. The steric effect is supported by the fact that the molar ratio of the total dissolved cation valence charge to the negatively charged hydroxyl groups can be varied from 2:1 to 12:1. Therefore, the valence charge of the cations is much greater than the available chelation sites. The amount of polymer can be adjusted to provide the minimum amount of polymer needed to stabilize and separate the metal ions.

When compounds are dissolved in solution, they may mix at the molecular level such that powders made by soft solution methods become relatively homogeneous powders and can be synthesized at low temperatures. If heated to temperatures only high enough that the organic components are decomposed, but not providing enough energy to allow crystallization, amorphous powders may result. Upon further heating, the powders can be crystallized, and generally form the most thermodynamically stable phase due to the intimate mixing of ions. The intimate mixing is especially useful in the synthesis of complex material systems consisting of several components.

Various powders have been produced by the PVA method including single phase calcium-aluminate and yttrium aluminate. Gulgun et al. [3] produced amorphous, fine powders at temperatures as low as 650 $^{\circ}C$ with crystalline powders after heat treatment at 900 $^{\circ}C$ for 1 h. The nano-sized cordierite powder prepared by Lee and Kriven [2] had a specific surface area after attitor milling of 181 m^2/g. During the heating of the polymer-nitrate solution, NO_x gases were evolved from the decomposition of the nitrates in solution and caused the precursor powder to be very porous and aerogel-like in nature with a high surface area that can be easily milled to a sub-micron sized powder. Other significant materials such as yttrium aluminum garnet (YAG) [4] and components of portland cement [6] have been produced by the PVA method.

1.1.2. THE POLYETHYLENE GLYCOL AND ETHYLENE GLYCOL METHODS

One limitation of the PVA method is that the source chemicals must be water-soluble. To broaden the applications but maintain the concept of steric entrapment, the polyethylene glycol (PEG) and the ethylene glycol (EG) methods were developed to take advantage of source chemicals that were soluble in alcohols. PEG, $H(OCH_2CH_2)_nOH$, is a linear or branched neutral polyether diol with a hydroxyl group at the end of the molecule. PEG, like PVA, also prevents segregation or precipitation of cations in solution by steric entrapment and chelation. To take advantage of the alcohol soluble, or non-aqueous method, a duplex zirconia/alumina composite has been synthesized using the PEG method and an alcohol-soluble zirconate source (zirconium 2,4-pentanedionate [8]. As in the PVA method, the properties of the synthesized powder will be dependent on the type and weight ratio of added PEG.

To increase the mixing effects and interaction with the hydroxyl groups, the PEG polymer may be replaced by its monomer, ethylene glycol (EG, $OH(CH_2)OH$). The EG method also has the advantage of the use of alcohol soluble source chemical such as titanium isopropoxide. Nano-sized titania, calcium titanate [17,18] and barium titanate [9] have been synthesized by the EG method at relatively low organic weight percentages.

1.1.3. NANO SOLS FOR MIXED AQUEOUS/NON-AQUEOUS ROUTES

When the synthesis of complex compounds requires precursor sources, some of which are soluble in aqueous solutions while others are only soluble in non-aqueous solutions, it will be difficult or impossible to find a common solvent for both types of solutions, without incurring significant expense. For this case, a nano-sol "shortcut" has been developed, based on the premise that diffusion distances through a nanoparticle are insignificantly short. A choice of medium is made such that most of the cation sources dissolve in one medium (e.g., nitrates dissolve in an aqueous medium), and the remaining cations are added as solid nano particles, for the solution to coat during drying and calcinations. A successful example has been the synthesis of lead zirconate titanate (PZT)), where the lead zirconate precursors were dissolved in aqueous medium, and a commercial titania nanosol (containing 45% anatase (TiO_2 particles) and supplied by Altair Technologies, Reno, Nevada, USA)) were dispersed in the organic-inorganic mixture prior to calcination and conversion to crystalline powders of PZT [12].

1.2. OBJECTIVE OF THIS RESEARCH

Since complicated multicomponent ceramic materials with precise levels of doping are required for some of the new materials that show promise in solid oxide fuel cells, we aim to use the relatively low-cost, simple steric entrapment method to synthesize several solid oxide fuel cell materials, and evaluate some of their characteristics. We have selected lanthanum gallates doped with strontium and magnesium, samarium doped ceria and a promising non-stoichoimetric composition composed of strontium iron and cobalt.

2. AQUEOUS SYNTHESIS AND CHARACTERIZATION OF SOLID OXIDE FUEL CELL MATERIALS

2.1. EXPERIMENTAL METHODS

The flow chart summarizing the powder synthesis method is seen in Fig.

2.1.1. POWDER SYNTHESIS

2.1.1.1 Doped lanthanum gallate (LSGM)

The composition $La_{0.8}Sr_{0.2}Ga_{0.83}Mg_{0.17}O_{3-\delta}$ (LSGM) was produced by adding stoichoimetric amounts of lanthanum nitrate hexahydrate (99.99%), gallium (III) nitrate hydrate (99.9%), magnesium nitrate hexahydrate (99%), strontium nitrate (99+%) (Aldrich, Milwaukee, WI) to deionized water and allowing full dissolution. A 5 wt% polyvinyl alcohol (PVA) solution (partially hydrolyzed, Celvol 205S, Celanese Chemicals) was added in a 4 to 1 ratio of dissolved positive cation charge to the negative PVA (-OH) functional end groups. The solution was allowed to stir for about 45 minutes for steric entrapment of the metal cations, and then heated over a hot plate while continuously being stirred at about $130°C$ until all of the water had evaporated and only a yellow resin remained with little violent release of NO_x gases. The resin was dried overnight in an oven at ~90°C and ground with mortar and pestle.

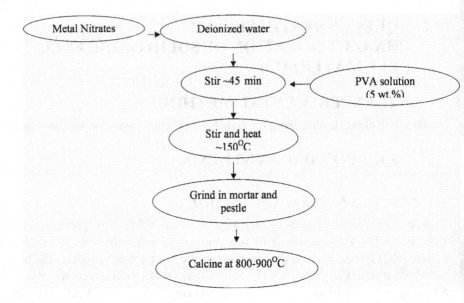

Figure 1. Flow chart for mixed oxide powder synthesis by the organic, steric entrapment
method.

2.1.1.2 Samaria Doped Ceria (SDC)

$Ce_{0.8}Sm_{0.2}O_{1.9}$ (SDC) was made in a similar manner as LSGM using cerium (III) nitrate hexahydrate (99%) and samarium (III) nitrate hexahydrate (99.9%)(Aldrich, Milwaukee, WI). A cation to PVA ratio of 4:1 was also used. The powder reacted violently with combustion of the dried resin while on the hot plate. A fine yellow ash resulted.

2.1.1.3 Strontium iron cobalt oxide (SFC)

$SrFeCo_{0.5}O_x$ (SFC) was also produced by the PVA method with iron (III) nitrate nonahydrate (99.99%), cobalt (II) nitrate hexahydrate (98%) and strontium nitrate (99+%) (Aldrich, Milwaukee, WI) with a 4:1 catio:PVA ratio. The resulting dried resin was black/gray in color.

2.1.2. POWDER CHARACTERIZATION AND PROCESSING

Synthesized powders were calcined and characterized by standard techniques of thermogravimetric analysis, BET specific surface area measurements, and particle size determination by sedigraphy, as well as X-ray diffraction, and scanning electron microscopy (SEM) equipped with energy dispersive spectroscopy (EDS) for microchemical analyses. To obtain powder compacts, the milled powder was passed through a 100 and 325 mesh sieve, and then pressed into disks in a 1/4" cylindrical metal die by uniaxial dry pressing. The green disks were then cold isostatically pressed (CIP) at 60,000 psi for about 10 min, placed in a platinum crucible and fired in air at various times and temperatures in a $MoSi_2$ heating element furnace.

3. RESULTS

3.1. POWDER SYNTHESIS AND CHARACTERIZATION

The calcinations conditions, specific surface area measurement and particle size of the powders are listed in Table 1. The sintered densities of materials are summarized in Table 2.

Table 1. Specific surface area and particle size distributions

SOFC Material	Calcination Temperature (°C)/1h	Specific Surface Area (m²/g)	Median Particle Size ($\Box$m)
$La_{0.8}Sr_{0.2}Ga_{0.83}Mg_{0.17}O_{3-\delta}$ (LSGM)	800	34.69	0.64 ± 0.51
$Ce_{0.8}Sm_{0.2}O_{1.9}$ (SDC)	800	18.51	0.37 ± 0.71
$SrFeCo_{0.5}O_x$ (SFC)	900	36.00	0.91 ± 0.69

$^{\phi}$*measured by BET nitrogen gas adsorption*

$^{\gamma}$*after attrition milling for 1 h, measured by sedimentation*

Table 2. Sintered densities of selected SOFC materials

Material	Sintering Temperature/Time (°C/h)	Bulk Density (g/cm³)	% Theoretical Density
LSGM	1300/3	6.18	93.13
	1450/4	6.36	95.80
SDC	1300/4	5.41	75.27
	1400/4	6.83	95.06
	1500/4	7.05	98.16
SFC	1000/5	3.62	--
	1100/5	4.97	--
	1200/5	5.09	--

3.2. X-RAY DIFFRACTOMETRY

The LSGM calcined powder was amorphous, but after sintering at 1400°C crystalline peaks matching a primative cubic structure with a~3.9304 Å were found. The SDC was also found to be single phase with a face-centered cubic structure similar to CeO_2 but with an increased lattice parameter, [a], from 5.41134 Å of CeO_2 to 5.4407 Å due to the doping of the larger Sm ions.

Kim et al. [19] asserted that the maximum temperature at which $SrFeCo_{0.5}O_x$ was stable as a single phase was 900 °C. Our results seem to somewhat agree with this assertion. The powder calcined at 800 °C was composed mainly of the perovskite $Sr(Co_{0.5}Fe_{0.5})O_3$ with trace peaks of spinel $CoFe_2O_4$ that persisted to 900 °C. At a temperature of 1100 °C the $Sr(Co_{0.5}Fe_{0.5})O_3$ phase had decomposed into at least two phases which may be strontium iron oxide and strontium cobalt oxide but they could not be definitely identified due to the poor resolution given by the crushed SFC pellets.

3.3. SCANNING ELECTRON MICROSCOPY

Samples for electron microscopy were chosen from the sintering temperatures and times that provided high density (Table 2). The theoretical density of the SFC materials could not be calculated because of the unknown phase content, although it was evident that a sintering temperature above 1000 °C was necessary for high density. The SDC pellet sintered at 1500°C for 5h was thermally etched at 1300°C for 30 min to accentuate the grain boundaries. Grain sizes of 1-2 $\square$m were observed, and the material was single phase. The sintered LSGM had grain sizes ranging from 1-5 µm (Fig. 2.). Unfortunately, while the pellets sintered at 1450 °C did reach a high density, there was a problem with cracking that would reduce the density and conductivity. The microstructure of the sintered SFC pellets could be characterized as three phases: long rods or plates, a dark gray and light gray phase (Fig. 3).

4. DISCUSSION

4.1. LSGM

LSGM produced by the PVA method was found to have high surface area and a sub-micron particle size that sintered to >95 % theoretical at 1450 °C, which is 50 °C lower than that required by powders produced by solid-state reaction. LSGM powders calcined at 800 °C were fairly amorphous with no remaining organic material, but may have some reduction in surface area after crystallization due to pre-sintering.

The microstructure of the sintered LSGM (seen in Fig. 2) consisted of irregular small grains (> 5 μm) with a possible second phase present at the grain boundaries. Solid state reaction methods have yielded grain sizes up to ~20 μm for a similar composition. If a second phase was present at the grain boundaries, there was no evidence from x-ray diffraction. Huang et al. [20-22] also found impurity phases at the grain boundaries at low Mg doping levels of y = 0.05 and higher doping y > 0.30. The composition studied for this work used a Mg doping of y =0.17 which was less than the maximum solubility limit of y=0.2 [23-24].

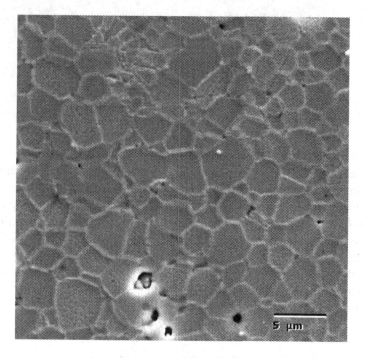

Figure 2. SEM micrograph of LSGM pellet sintered at 1450 °C for 4 h and thermally etched at 1300 °C for 30 min

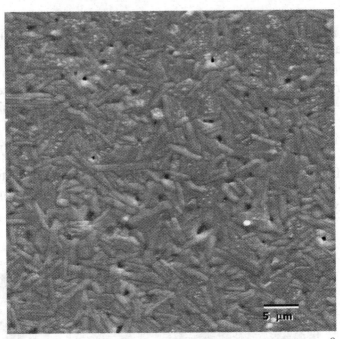

Figure 3. SFC pellet sintered at 1200 °C for 5 h and thermally etched at 1300 °C for 30 min

If there was a grain boundary phase, most likely it may be due to the level of Sr doping. The Sr was doped at a level of x=0.2 which was higher than the solubility limit found by previous work. Furthermore, LSGM powders made from sol-gel methods [25] with lower Sr doping (x=0.1) did not have a glassy impurity phase ($SrLaGaO_4$). The powder had finer particle size and greater surface area, but was only able to reach 93 % density at a sintering temperature of 1370 °C. This was explained in part by the absence of the glassy phase present in powders produced by solid state reaction. Therefore, the glassy grain boundary phase may aid in sintering and result in improved electrical properties.

4.2. SAMARIUM DOPED CERIUM

The samarium doped cerium powder produced by the PVA method was a sub-micron powder with a high surface area. As explained in previous sections, the PVA resin was heated to a higher temperature on the hot plate to allow the explosive decomposition of the nitrates in a more contained manner. This resulted in precursor powders that were crystalline and nearly free of organic material. One should note that the surface area of the SDC is lower than that of the other SOFC materials possibly due to presintering of the particles at the combustion temperature. This process is in fact very similar to the glycine-nitrate method except that the PVA does not necessarily support its own combustion and must be heated to complete the reaction. Although the specific surface area produced by the glycine-nitrate method (about 71 m^2/g) was greater than that produced by the PVA method in this study, it depended heavily on the fuel (organics) to nitrate ratio which was more optimized in the case of glycine-nitrate. The surface area of SDC powder may be optimized by a systematic study with varying ratios of PVA to nitrates.

4.3. SDC

The SDC powder produced by the PVA method was able to reach 95 % density at a sintering temperature of 1400 ^{O}C, much lower than that achieved by solid state reaction. This is also close to the sintering temperatures achievable with the glycine-nitrate method, but required less organics and gave a higher yield of powder.

4.4. STRONTIUM IRON COBALT OXIDE

Kim et al. [19] suggested that the maximum temperature at which $SrFeCo_{0.5}O_x$ may be stable as a single phase is around 900 ^{O}C, although attainment of equilibrium is slow at that temperature. Due to the intimate mixing of ions in solution, a single phase is expected at low calcination temperatures, but the SFC powder produced by the PVA method was a mixture of the perovskite $Sr(Fe_{0.5}Co_{0.5})O_3$ with a cobalt iron oxide minor phase at calcination temperatures up to 900 ^{O}C. This may have occurred due to the higher decomposition temperatures required for strontium nitrate (~500 ^{O}C) and the evidence of strontium nitrate crystals in the precursor powder diffraction pattern. This non-homogeneous mixing of Sr may have caused a strontium-depleted phase at low temperature calcination resulting in the perovskite and cobalt iron oxide phase mixture. More organic material for steric entrapment of the strontium ions may be needed to synthesize $SrFeCo_{0.5}O_x$ at low temperatures.

At temperatures of 1100 ^{O}C, it is evident that other phases have developed, but due to poor resolution of the x-ray data of the crushed sintered pellets, the exact phase composition could not be determined. From microstructural examination using backscattered electron imaging, there does seem to be a three phase composition similar to that found by Kim et al. that includes the perovskite $Sr(Co_{1-x}Fe_x)O_{2.5-\delta}$ in the form of rods or needles with $SrFe_{1.5-x}Co_xO_{3.25-\delta}$ and $(Co,Fe)_3O_4$ visible as gray and black regions, respectively. From Fig. X, the fraction of the rod shaped perovskite seems larger than the fraction present in solid state synthesis techniques this may be due to the more intimate mixing of ions in solution leading to a more thermodynamically favorable microstructure. SEM/EDS evidence provides support for the existence of a cobalt-rich phase that is less rich in both Sr and Fe.

There was also a problem of cracking of the SFC pellets during sintering in air. This problem has been observed in other $Sr(FeCo)O_x$ materials [26] and it was proposed that is was caused by oxidation of the perovskite phase (Fe^{3+} to Fe^{4+}) accompanied by an increase in unit cell dimensions. Our samples were composed of a high fraction of perovskite phase with a relatively small particle size causing any change in the unit cell of the perovskite to create stresses leading to the development of cracks. Sintering in a non-oxidizing atmosphere such as nitrogen may prevent oxidation and cracking.

5. CONCLUSIONS

Using steric entrapment, solid oxide fuel cell materials were successfully produced by an aqueous synthesis method. The powders were more homogeneous and required lower processing temperatures than did powders made by more conventional methods. Properties of the powder and sintered microstructures were comparable with those properties attained using more complex methods such as sol-gel and with less organics than methods such as the glycine-nitrate method. Further work is needed to fully optimize the PVA to cation ratio to produce the most favorable powder characteristics.

ACKNOWLEDGEMENTS

This work was partially funded by the United States Air Force Office of Scientific Research under grant number F49620-01-1-500. It was carried out in the Center for Microanalysis of Materials, of the Frederick Seitz Materials Research Laboratory, University of Illinois at Urbana-Champaign, which is partially supported by the U.S. Department of Energy under grant DEFG02-91-ER45439.

REFERENCES

[1] Gulgun M.A., Kriven W.M. A simple solution polymerizaion route for oxide powder synthesis. Ceramic transactions, 1995, v. 62, p.57-66

[2] Lee, S.J., Kriven W.M. Crystallization and densification of nano-size amorphous cordierite powder prepared by a solution-polymerization

route. Journal of the American ceramic society, 1998, v.81, N10, p.2605-2612.

[3] Gulgun, M.A., Nguyen, M.H., Kriven W. M. Polymerized organic-inorganic synthesis of mixed oxides. Journal of the American ceramic society, 1999, v.82, N.3, p.555-560.

[4] Nguyen, M. H., Lee S.J., Kriven W.M. Journal of materials research, 1999, v.14, N.8, p.3417-3426.

[5] Lee S. J., Kriven W. M. Preparation of ceramic powders by a solution-polymerizaiton route employing PVA solution. Ceramic engineering and science proceedings, 1998, v.19, N.4, p.469-476.

[6] Lee S.J., Benson E.A., Kriven W.M. Preparation of Portland cement components by PVA solution polymerization. Journal of the American ceramic society, 1999, v.82, N.8, p. 2049-2055.

[7] Lee S.J., Biegalski M.D., Kriven W.M. Ceramic engineering and science proceedings, 1999, v.20, N.3, p.11-18.

[8] Lee S.J., Kriven,W.M. A submicron-scale duplex zirconia and alumina composites by polymer complexation processing. Ceramic engineering and science proceedings, 1999, v.20, N.3, p. 69-76.

[9] Lee S.J., Biegalski M.D., Kriven W.M. Journal of materials engineering, 1999, v.14, N.7, p.3001-3006.

[10] Kriven W.M., Lee S.J., Gulgun M.A., Nguyen M.H., Kim D.K. (invited review paper). Synthesis of oxide powders via polymeric steric entrapment in Innovative processing/synthesis: ceramics, glasses, composites III. Published as Ceramic transactions, 2000, v.108, p.99-110.

[11] Kim D.K., Kriven W.M. Synthesis and characterization of YAG materials and alumina-YAG composites. Journal of the American Ceramics Society 2005, submitted.

[12] Rosczyk B.R., Lee S.J., Kriven W.M. Preparation of titanate powders by an ethylene glycol method. Journal of the American ceramics society, 2005, submitted.

[13] Lee S.J., Lee Ch.H., Kriven W.M. Synthesis of low-firing anorthite powder by the steric entrapment route. Ceramic engineering and science proceedings, 2002, v.23, N.3, P.33-40.

[14] Kriven W.M., Rosczyk B.R., Kremeyer, Song B., Chen W. Transformation toughening of a calcium zirconate matrix by dicalcium silicate under ballistic impact. Ceramic engineering and science proceedings, 2003, v.24, N.3, p.383-388.

[15] Rosczyk B.R., Kriven W.M., Mason T.O., Solid oxide fuel cell materials synthesized by an organic steric entrapment method, 2003, v.24, N.3, p.287-292.

[16] M. Gordon, J. Bell and W. M. Kriven, Comparison of Naturally and Synthetically-Derived, Potassium-Based Geopolymers, Ceramic Transactions, vol. xxx edited by J. P. Singh, N. P. Bansal and W. M. Kriven (2004), in press.

[17] Lee S.J., Lee Ch.H. Fabrication of nano-sized titanate powder via a polymeric steric entrapment route and planetary milling process. Journal of the Korean ceramic society, 2002, v.39, N.4, p.336-340.

[18] Lee S.J., Kim Y.C., Hwang J.H., An organic-inorganic solution technique for fabrication of nano-sized $CaTiO_3$ powder. Journal of Ceramic Processing Research, 2004, v.5, N.3, p.223-226.

[19] Kim S. Y. et al., Determination of oxygen permeation kinetics in a ceramic membrane based on the composition $SrFeCo_{0.5}O_{3.25-\square}$. Solid state ionics, 1998, v.109, p.3-4.

[20] Huang K., Tichy R.S., Goodenough J.B. Superior perovskite oxide-ion conductor; strontium- and magnesium-doped $LaGaO_3$: phase relationships and electrical properties. Journal of the American ceramic society, 1988, v.81, N.10, p.2565-2575.

[21] Huang K., Tichy R.S., and Goodenough J.B. Superior perovskite oxide-ion conductor; strontium- and magnesium-doped $LaGaO_3$ II. Ac impedance spectroscopy. Journal of the American ceramic society, 1998, v.81, N.10, p.2576-2580.

[22] Huang K., Tichy R.S., and Goodenough J.B. Superior perovskite oxide-ion conductor; strontium- and magnesium-doped $LaGaO_3$ III. Performance tests of single ceramic fuel cells. Journal of the American ceramic society, 1998, v.81, N.10, p.2576-2580.

[23] Ishihara T., Matsuda H., Takita Y. Doped $LaGaO_3$ perovskite type oxide as a new oxide ionic conductor. Journal of the American chemical society, 1994, v.116, N.5, P.3801-3803.

[24] Ishihara T., Honda M, Shibayama T. Intermediate temperature solid oxide fuel cells using a new $LaGaO_3$ based oxide ion conductor. I. Domped $SmCoO_3$ as a new cathode material. Jounral of the electrochemical society, 1998, v. 145, N.9, p.3177-3183.

[25] Huang K., Feng M., Goodenough J.B. Sol-gel synthesis of a new oxide-ion conductor Sr- and Mg-doped $LaGaO_3$ perovskite. Journal of the American ceramic society, 1996, v.79, N.4, p.1100-1104.

[26] Kleveland K.E., Sintering behavior, microstructure and phase composition of $Se(Fe,Co)O_3$ ceramics. Journal of the American ceramic society, 2000, v.83, N.12, p.3158-3164.

NANOSCALED OXIDE THIN FILMS FOR ENERGY CONVERSION

IGOR KOSACKI

Metals and Ceramics Division, Oak Ridge National Laboratory, Oak Ridge, TN 37830

Abstract: This paper discusses the electrical transport properties of oxygen conductors and how they may be optimized by controlling their microstructure or thickness in the nanometer range. The parameters determining the ionic conductivity in cubic stabilized zirconia are reviewed and the case for enhanced ionic conductivity in yttria, (YSZ) and scandia (ScSZ) stabilized zirconia is considered. The obtained results show that properties of nanoscaled oxides can be greatly enhanced. This is attributed to a significant contribution from grain boundary/interface conductivity when the grain size or the film thickness is decreased below 100nm. These observations can have important implications for the development of nanostructured electrochemical devices with enhanced performance.

1. INTRODUCTION

The most important aspects of the study of oxygen conductors are the abilities to enhance their ionic conductivity and reaction kinetics. Both features are essential for the development of electrochemical devices including fuel cells, gas sensors and ionic membranes. These devices have the potential to deliver high economic and ecological benefits; however to achieve satisfactory performance, it is necessary to optimize the ionic conductivity of the solid electrolytes.

N. Sammes et al. (eds.), Full Cell Technologies: State and Perspectives, 395-416.
© 2005 Springer. Printed in the Netherlands.

Stabilized zirconia is an excellent material for such applications due to its high level of ionic conductivity and a desirable chemical stability in both oxidizing and reducing atmospheres [1-4]. The ionic conductivity in this system is attributed to mobile oxygen vacancies, which are created by acceptor doping. The conductivity has an extrinsic character and is controlled by oxygen vacancies generated by acceptors - $[A'_{Zr}] = 2 [V_o^{..}]$. However, there is an optimum of doping level (typically 15-20 atomic %) beyond which the conductivity decreases [1, 3, 4].

A number of studies have also shown that the ionic conductivity of stabilized zirconia depends on the size of acceptors: the conductivity tends to be highest with a minimization of the conductivity activation energy for those acceptors whose ionic radius closely matches that of Zr^{4+} host cation [1, 4]. This observation suggests that the ionic transport can be related to the defect energy, which is defined by the interaction between the acceptors and oxygen sublattice. This interaction results in the barrier for ion hopping between equivalent positions in the lattice. Therefore, there has been considerable effort to optimize the ionic conductivity of stabilized zirconia by controlling the acceptors concentration and their size [1, 3].

An alternative solution to enhance the ionic conductivity involves increasing the mobility of the ion species. A number of recent studies have shown that this can be promoted when the material microstructure is in the nanometer range. This is attributed to grain boundary and interfacial effects, which can exhibit orders of magnitude greater diffusivity than that of the lattice [5-7]. To accentuate these phenomena, the material microstructural features should be reduced to the value where the volume fraction of grain boundary region represents a few percent of the total, which corresponds to grain sizes below 100nm [6-8]. The influence of microstructure on ionic transport has been recently reported for nanocystalline CaF_2 [9, 10], CeO_2 [7, 11, 12] and stabilized cubic zirconia, ZrO_2: (Y or Sc-doped) thin films [7, 8, 13-15]. For example, a fifty (50)-fold increase in the ionic conductivity has been observed in 20nm grain sized YSZ thin films as compared to that of conventional microcrystalline ceramics or bulk single crystals [7]. Such enhancement can be attributed to either the greatly reduced impurity concentration at grain boundaries [16] or to the dominance of size-dependent defect equilibria [6, 7, 11, 12] as the grain size is reduced.

The ionic conductivity of YSZ can be also enhanced by the introduction of high density of dislocations [17] or interfaces that act as rapid diffusion paths for oxygen vacancies. Such an idea has been discussed for $BaF_2/CaF_2/BaF_2$ superlattices where a substantial increase of ionic conductivity was observed [18]. In this system a progressive increase in the conductivity was correlated with the increase of interfacial density.

Observed effects were interpreted by internal fluoride ion transfer using a space charge model [9, 18].

The point that still eludes researchers is whether the observed enhancement of the ionic conductivity in solid electrolytes is attributed to enhanced mobility, to changes in the defect equilibria or to an increased contribution of the surface/interface conductivity. It is thus important to gain a fundamental understanding of the factors leading to fast oxygen transport such as atomistic migration parameters (hopping energy, relaxation time), grain boundary properties and surface exchange effects. Each effect is dependent on the microstructure which complicates the resolution of the source of the observed enhancement of ionic conductivity.

Another important issue in the research of nanomaterials is the ability to determine the electrical conductivity related to surface/interface effects what will help in better understanding ionic transport in grain boundary and bulk areas of solid electrolytes. This can be achieved when the thin films with controlled grain size or the thickness in the nanometer range will be evaluated [19]. The knowledge of these relationships is very important for optimization of the ionic transport and the development of new nanoscaled materials which conductivity will not be limited by lattice diffusion. The understanding of the surface and grain boundary diffusivity is needed to better engineer these fascinating materials and their applications in a range of important technological devices.

2. LATTICE DIFFUSION - LIMITED IONIC CONDUCTIVITY IN MICROCRYSTALLINE YSZ

The ionic transport in solids is attributed to the hopping of ionic carriers between the equivalent positions in the crystal lattice. This mechanism is known as lattice diffusion and depends on the jumping distance and frequency of moved ions. The understanding of the influence of these factors on the ionic conductivity is very important for the development of material with enhanced ionic transport. The question of what is the limit of ionic conductivity in solids will be addressed by analyzing the ionic transport in cubic stabilized zirconia systems with different acceptor dopants.

The ionic conductivity of oxygen conductor can be written as [5, 20]:

$$\sigma = 2 \cdot e \cdot [V_o^{\cdot\cdot}] \cdot \mu_{V_O} \tag{1}$$

where $e, [V_o^{\cdot\cdot}]$ and μ_{V_O} are the elemental charge, concentration and mobility of oxygen vacancies respectively. The temperature dependence of the conductivity arises from the behavior of the concentration and mobility of oxygen vacancies. In the extrinsic regime, the concentration of oxygen vacancies is related to the acceptor concentration - $[A']$ and is not temperature dependent:

$$[V_o^{..}] = 1/2[A']$$ (2)

In general, this dependence can be more complex since it can be influenced by the relative amounts of extrinsic defects and impurities, as well by association between defects. In this case, we should use the effective concentration of acceptors which is - $[A'_{eff}] \leq [A']$.

The temperature dependence of the ionic mobility is described in the terms of ionic species moving across the energy barrier formed between two equivalent positions in the lattice [5, 20]:

$$\mu_{V_o} = \frac{a \cdot e \cdot v \cdot r^2}{kT} \cdot \exp\left(\frac{\Delta S_m}{k}\right) \cdot \exp\left(\frac{-\Delta H_m}{kT}\right)$$ (3)

where, a is a numerical factor, which depends on the lattice type and conduction mechanism, v is the characteristic vibrational frequency and r is a characteristic jump distance. ΔS_m and ΔH_m are the entropy and enthalpy related to the ion hopping proces respectively. The combination of equations (1-3) yields the expression for the extrinsic ionic conductivity:

$$\sigma \cdot T = \frac{a \cdot e^2 \cdot v \cdot r^2}{k} \cdot [A'_{eff}] \cdot \exp\left(\frac{\Delta S_m}{k}\right) \cdot \exp\left(\frac{-\Delta H_m}{kT}\right)$$ (4)

For the comparison with experimental data, a more useful form of Eq.4 can be written as:

$$\sigma \cdot T = \sigma_o \cdot \exp\left(\frac{E_{act}}{kT_o}\right) \cdot \exp\left(\frac{-E_{act}}{kT}\right)$$ (5)

As can be seen, the pre-exponential factor has two contributions. One,

$\sigma_o = \dfrac{a \cdot e^2 \cdot v \cdot r^2}{k} \cdot [A'_{eff}]$ is related to lattice ion diffusion and its value should be determined mainly by the hopping distance and the concentration of oxygen vacancies (Eq.4), while the second, $\exp(E_{act}/kT_o)$ describes the configurational entropy with some characteristic temperature, T_o related to a transition to state of zero entropy. In general, this factor can be dependent upon the microstructure and defect concentration. Note that the conductivity activation energy, E_{act} in extrinsic regime, is related to ion hopping enthalpy - $E_{act} = \Delta H_m = \Delta S_m \cdot T_o$.

Equation (5) describes the relationship between the log of the pre-exponential factor and the activation energy - $\log \sigma_o (E_{act})$, which should be linear and is known as the Mayer-Neldel rule [21, 22]. This relationship was first observed for semiconductors and recently for ionic conductors [21, 23], as well as for organic thin film transistors [24]. The conductivity plots determined for a family of similar specimens obeying this rule converge to a point at temperature T_0, which is constant within a class of related materials. The value of T_0 should be related to the material melting point where the hopping energy at $T = T_0$ approaches zero (Eq.5).

Equation (5), can be used to examine the temperature dependent conductivity of ZrO_2 stabilized by different acceptors - Y, Yb, Sc, Ca, Sm, Gd, Nd with a the same dopants level of 15-20 atomic%. These data have been reported by many authors [1-4]; however the separation from the pre-exponential factor of the part related to configurational entropy can explain the conductivities obtained for different acceptors. In this case σ_O (Eq.4) should depend only on the atomistic factors determined by ZrO_2 lattice and is a constant for all stabilized zirconia specimens.

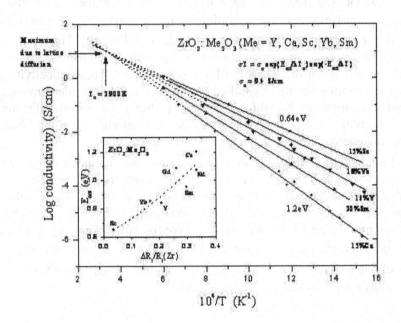

Figure 1. Temperature dependence of ionic conductivity of microcrystalline acceptor stabilized zirconia. Data taken from [1-4]. Solid lines represent the result of fitting the model (Eq.5) to experimental results [8]. The insert is the dependence of the conductivity activation energy as a function of acceptor radius.

Fig.1 shows the temperature dependence of the ionic conductivity determined for different acceptors introduced into ZrO_2. As can be seen ionic conductivity is dependent on the dopants and tends to be highest for S with the lowest conductivity activation energy of 0.64 eV. On the other hand, the lowest conductivity value and highest activation energy of 1.2 eV characterize those specimens doped with Ca. The observed changes in the electrical conductivity activation energy can be correlated with the ionic radii of the acceptors substituted of Zr^{4+} ions. Among the acceptors examined, Sc^{3+} (0.87Å) has the radii closest to that of Zr^{4+} (0.84Å), which corresponds to the lowest activation energy, see insert in Fig.1. Note that Eq.5 describes the temperature dependence of the ionic conductivity very well and yields a value for T_0 of 2900K, which is close to the reported melting point of 2953K for microcrystalline YSZ [2]. This suggests that maximum ionic conductivity value of ~10 S/cm (Fig.1) should be observed for acceptor-doped ZrO_2 at T_0 temperature and would be independent of the dopant type. Experimental data obtained at temperatures below 1000°C are in good agreement with presented predictions. This illustrates a limit of the ionic conductivity, which can be reached for microcrystalline stabilized zirconia and explains why the conductivity of this system was practically not increased during the study of this material for last 40 years.

The above model of the ionic conductivity is based on the assumption that the transport of oxygen ions is described by lattice diffusion mechanism. This diffusion is limited by the ion jump distance, determined by ZrO_2 lattice constant. The enhancement of the ionic conductivity in acceptor-doped ZrO_2 ceramics can be explained by a decrease in the activation energy dependent on the size of acceptor. This behavior is consistent with experimental results reported for specimens, which microstructure is in the micrometer range and concentration of grain boundaries is negligible. To reach higher level of ionic conductivity, it is necessary to develop materials which diffusivity will be not limited by the lattice mechanism.

3. IONIC CONDUCYIVITY OF NANOCRYSTALLINE YSZ CERAMIC THIN FILMS

The introduction of interfaces/grain boundaries has been proved to be a powerful way to enhance the ionic transport, due to higher diffusivity of grain boundary region, which is often orders of magnitude greater than lattice diffusivity [6, 7, 25]. Since the volume fraction of the grain boundary depends upon the microstructure, it is possible to obtain a material whose lectrical conductivity is controlled by interfaces and grain boundaries. It has been shown that such effect can be achieved when the grain size is less than 100 nm, which should occur when the fraction of grain boundaries is $\geq 1\%$ [25].

The influence of the microstructure on electrical transport and non-stoichiometry was reported for CeO_2, ZrO_2:Ca bulk [12, 15, 16] and YSZ and ScSZ nanocrystalline thin films [7, 13, 14]. Bulk specimens are usually prepared by pressure-densified processing (1.1GPa, 600°C) from powders of 5nm crystallite size, while nanocrystalline thin films were obtained by the lymeric precursor spin coating technique, which has been proved to be particularly useful for the formation of dense nanocrystalline films at temperatures about 1000°C lower than those required to prepare micrometer-sized specimens [26]. The uniform distribution of grain size as well as the ability to control the thickness and microstructure of the films prepared by the polymeric precursor method allow comprehensive structure/property studies to be made on nanocrystalline materials [7].

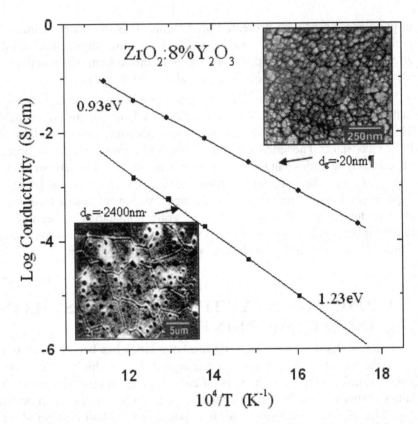

Figure 2. Temperature dependence of the electrical conductivity determined for nanocrystalline thin film and microcrystalline bulk YSZ. The insert shows the SEM images of both specimens.

Fig. 2 presents the comparison between the electrical conductivities of micro- and nanocrystalline YSZ specimens and the enhancement in the ionic conductivity has been observed for 20nm grain sized films [7, 14]. The nanocrystalline YSZ exhibits about two orders of magnitude greater conductivity (at 400°C) as compared to microcrystalline specimens [7]. This is associated with a decrease in the activation energy from 1.23 to 0.93eV and also by an increase in the preexponential factor, σ_0 from 9.4 to 240S/cm for micro- and nanocrystalline specimens respectively [7, 8]. Such a large increase in σ_0 can not be explained by changes in the lattice hopping distance r, which should be similar for micro- and nanocrystalline specimens. In the case of nanocrystalline materials, the increased conductivity is probably attributed to higher mobility of ionic species in the grain boundary area. Presumably, the conductivity enhancement results from the fact that nanocrystalline YSZ is characterized by a higher percentage of the ions residing in the grain boundary area. The grain boundary phase appears to have two key characteristics necessary for enhance ionic diffusion: higher defect densities (increase of σ_o factor) and higher mobility of ionic species (decrease E_{act}). This correlates with the tracer diffusion studies performed on YSZ single crystal. Recently reported data showed that surface exchange coefficient is about 2 orders of magnitude larger that lattice diffusion [27]. A complete explanation of the observed enhancement in the electrical conductivity of nanocrystalline stabilizes zirconia requires more studies using different acceptor dopants.

4. NON-STOICHIOMETRY OF NANOCRYSTALLINE STABILIZED ZIRCONIA

Interesting information regarding the non-stoichiometry and defect equilibria in nanocrystalline materials can be obtained from the study of electrical conductivity as a function of the oxygen partial pressure. These measurements can provide information about the contribution of the electronic and ionic conductivities to the electrical transport. An example of the influence of the microstructure and the type of acceptors on the electrical transport was discussed in a recently reported study performed for Sc- and Y-stabilized zirconia [7, 13, 14].

Figure 3 presents the relationship between the electrical conductivity and oxygen partial pressure determined for 20nm grain sized YSZ and ScSZ thin films at 900°C. In this figure the electronic conductivity estimated from data reported for microcrystalline YSZ [3, 28] and ScSZ [29] are also presented. As can be seen, the electrical conductivity of nanocrystalline specimens shows a different P_{O2} behavior than the microcrystalline materials. The conductivity of nanocrystalline films exhibits a P_{O2}-independent region and anomalous decrease at high and intermediate oxygen activity region respectively. These regions can probably be attributed to extrinsic ionic conductivity while at low P_{O2}, the electrical conductivity of ScSZ follows the $P_{O2}^{-1/4}$ relationship typical for extrinsic electronic conductivity. The ionic conductivity of nanocrystalline ScSZ certainly has ahigher value than nano- and microcrystalline YSZ (Fig.3).

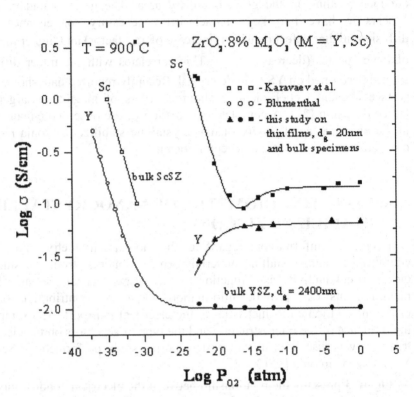

Figure 3. The comparison between conductivities determined at 900°C for nanocrystalline thin films and microcrystalline bulk specimens of YSZ and ScSZ as a function of P$_{O2}$ [13]. Open symbols represent electronic conductivity data for bulk microcrystalline specimens [28, 29].

The electronic conductivity of nanocrystalline ScSZ dominates the electrical transport at P$_{O2}$ values below 10^{-18} atm, while for YSZ, becomes significant only at oxygen partial pressure less than 10^{-25} atm. This observation strongly suggests the presence on nonequivalent positions of oxygen vacancies distributed in ZrO$_2$ matrix when different acceptors are introduced. Also there may be an interaction between defects resulting in defect association.

The anomalies in electrical conductivity observed for nanocrystalline specimens can be explained by a model in which a defect pair formation was proposed [13, 14, 30]. The association between oxygen vacancies and acceptor dopants has been considered, which results in the formation of a donor - $(A' - V_O^{\cdot\cdot} - e)^x$ and an acceptor - $(A' - h)^x$ center, (A' denotes Y or Sc). In this case, acceptor doped ZrO_2 can be considered being a compensated semiconductor with donor and acceptor levels whose concentration in ionized and unionized states can be described by Fermi statistics. Based on this model and assuming that the total concentration of dopants is related to the sum of acceptors and donors, which can be in ionized and unionized states, an expression for the ionic and electronic conductivities dependent upon oxygen activity has been obtained [13, 30]. This model predicts a decrease of ionic conductivity at both low- and high-oxygen activity due to an increase number of unionized donors, which oxygen vacancies can not contribute to the ionic transport [13].

Since the electronic conductivity of nanocrystalline ScSZ becomes significant in reducing atmosphere (Fig.3), its contribution to the electrical transport should be considered. This can be discussed based on the dependence of the ionic transference number, $t_i = \sigma_i / (\sigma_i + \sigma_e)$ as a function of temperature and oxygen activity. Such information is important for the development of ScSZ solid electrolyte for Solid Oxide Fuel Cells. Figure 4 presents the relationship between the ionic transference number and oxygen activity, which has been determined based on the presented conductivity measurements and the defect model [13].

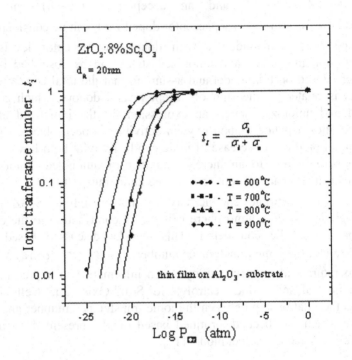

Figure 4. The ionic transference number as a function of oxygen activity determined for nanocrystalline ScSZ thin films at different temperatures [13].

As can be seen (Fig.4), nanocrystalline ScSz is an excellent ionic conductor at $P_{O2} > 10^{-12}$atm. At the P_{O2} region of 10^{-12} to 10^{-20}atm, the material exhibits a mixed ionic-electronic conductivity with t_i varying from 1 to 0.1, while at $P_{O2} < 10^{-20}$atm, ScSZ is the electronic conductor. These results strongly suggest some limitation for using ScSZ as a solid electrolyte in reducing atmospheres. The observed enhancement of the electronic conductivity in nanocrystalline ScSZ is very unique and strongly related to the microstructure controlled non-stoichiometry. The electrical conductivity measurements recently performed for bulk microcrystalline (1.6μm) specimens showed only very small changes with oxygen activity that suggests dominant ionic transport [13]. This behavior is similar to observed for YSZ, which is good ion conductor at $P_{O2} > 10^{-25}$atm.

The observed enhancement of the electronic conductivity in nanocrystalline ScSZ is probably related to the fact that Sc forms a shallower donor level, $(Sc'_{Zr} - V_o^{\cdot\cdot} - e)^x$ compared with deep donor, $(Y'_{Zr} - V_o^{\cdot\cdot} - e)^x$ formed in YSZ. The ionization energy of such donor in ScSZ has been determined from the temperature dependence of the minimum conductivity, where $\sigma_i = \sigma_e$ observed for the P_{O2} dependent behavior (Fig.3). This yields a donor ionization energy of 0.9eV [8, 13], which is much lower than 2.2eV reported for YSZ [28]. The positions of the energy levels determined for YSZ and ScSZ are presented in Fig.5 [8]. The ionic conductivity due to the high level of acceptor concentration, $10^{21}cm^{-3}$, is extrinsic in character and therefore the enhancement, which has been observed for nanocrystalline specimens, has to be related to the mobility factor.

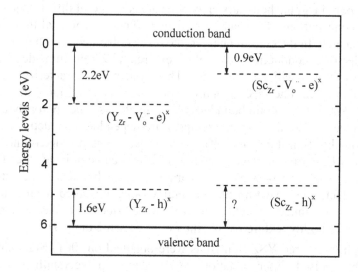

Figure 5. The diagram of defect energy levels determined for YSZ and ScSZ [8].

The electrical conductivity of ScSZ is strongly related to the microstructure and for nanocrystalline specimens can be dominated by electronic transport in the temperature range of 600-900°C and $P_{O2} < 10^{-18}$atm, while microcrystalline specimens show only ionic conductivity [7, 13]. The results of the study of electrical transport in both nanocrystalline YSZ and ScSZ thin films are showing that their non-stoichiometry can be controlled by the microstructure and in addition that the electrical conductivity is also influenced by the interaction between lattice defects and type of acceptor dopants. The large increase in electronic conductivity observed for nanocrystalline ScSZ suggests that it may find use as a buffer layer on the anode site of a solid electrolyte in SOFC, which can be YSZ or microcrystalline ScSZ. Such compatibility between the electrode and electrolyte should reduce the internal resistance and finally improve the SOFC performance.

5. ELECTRICAL CONDUCTIVITY OF HIGHLY TEXTURED YSZ THIN FILMS

The discussed above results regardnig the electrical conductivity of nanocrystalline YSZ and ScSZ were obtained on ceramic polycrystalline materials where the negative effects of blocking grain boundaries can influence the electrical transport. Very important for better understanding ionic transport in grain boundary and bulk areas is the ability to separate those two contributions. This can be achieved when the epitaxial thin films will be evaluated. In the current chapter, the results obtained in a study of the in-plane electrical conductivity of highly textured YSZ thin films deposited on MgO substrates will be discussed. This system is characterized by a single film/substrate interface, which can provide very rapid ionic transport with the absence of any grain boundary blocking effects. The approach taken here is to characterize the in-plane temperature-dependent conductivity as a function of the film thickness. This study was able to determine the conditions where the surface/interface effects can dominate the bulk properties. It was also possible to separate of grain boundary and lattice related conductivities and find the maximum value attributed to interfacial diffusivity. The initial results obtained in a study of the electrical conductivity in highly textured YSZ films were presented in [19].

The highly textured YSZ thin films were obtained on (001) MgO single crystals by pulsed laser ablation of a dense polycrystalline cubic ZrO_2:9.5%Y_2O_3 target using a pulsed KrF (248-nm wavelength) extimer laser. This technique was able to obtain high quality of YSZ films with varied thickness from 2μm to 15nm. The structure of epitaxial YSZ films was characterized by transmission electron microscopy (TEM), x-ray diffraction (XRD) and x-ray texture maping. The high quality of the YSZ films is evident in lattice imaging of YSZ/MgO cross sections as seen for 15nm thick film in Fig.6. Examinations of the films with different thickness reveled YSZ structures with nearly perfect crystalline structures that were well aligned with the (001) single crystal MgO substrate. This alignment of the two lattices is also evident in the cube-on-cube diffraction pattern and the small (< 2°) mosaic spread of the (001) YSZ diffraction spot. The TEM and x-ray studies do confirm the highly textured growth of YSZ thin films which appear to have negligible grain boundaries parallel to the substrate. This ensures that in-plane geometry current flow is not limited by grain boundaries [19].

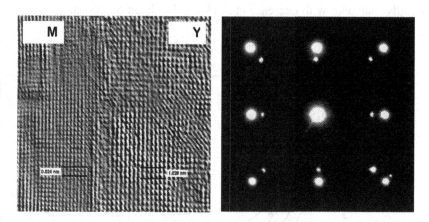

Figure 6. Lattice structure and orientation of YSZ thin films. TEM image of YSZ/MgO cross section (left) and the electron diffraction pattern along (001) direction (right) [19].

The electrical conductivity measurements were performed on highly textured YSZ thin films using impedance spectroscopy and the in-plane electrode configuration [19, 31]. The YSZ films with thickness ranging from 60nm to 2μm exhibit identical temperature-dependent conductivities, which mimicked the response of the cubic YSZ single crystals. The strong correlation between the conductivity of single crystals and these YSZ films is not surprising in view of the strong cube-on-cube texture of the films. The associated activation energies below 650°C averaged ~ 1.1eV for both the single crystals and these films. The activation energies decreased somewhat in the higher temperature region (> 650°C) and reached values of 0.9eV for single crystal and 1.0eV for the thicker films. The decrease in the activation energy was considered to be related to oxygen-vacancy trapping by the acceptor dopants, which results in the defect association $(Y'_{Zr} - V_o^{..})^.$ at lower temperatures [13, 30]. The enhancement in the electrical conductivity was observed for the 29nm thick film and is nearly tenfold greater than that of the thicker films below 650°C and threefold above 650°C. An even greater effect is seen for the 15nm thick film, which exhibits tenfold increases in the conductivity and 150-fold at 800 and 400°C respectively [31]. In this case, the activation energy remains 0.62eV and is lower than the value of 0.9eV reported for oxygen diffusivity in YSZ single crystals [17, 19, 31]. The observed enhancement in the electrical conductivity of nanometer thick films can be associated with YSZ/MgO interface, which contribution increases with decreasing film thickness. The effect of film thickness on the electrical conductivity at selected temperatures is highlighted in Fig.7. The transition from a conductivity, which is independent of thickness to one that rapidly increases with decrease in thickness has been observed.

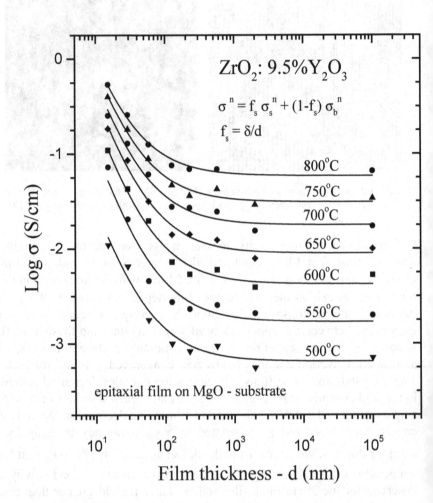

Figure 7. Thickness-dependent electrical conductivity of YSZ thin films determined at different temperatures. Solid lines represent the results of fitting the experimental data to the model [31]

This transitionin in the thickness-dependent conductivity coincides with the shift from a constant activation energy for films thickness > 60nm to one that changes with temperature as the thickness is reduced [31]. These observations clearly show that ionic conductivity of YSZ can exhibit a substantial nanoscale effect when the film thickness is reduced below 60nm. The observed enhancement in the conductivity can be attributed to

YSZ/MgO interface region or to free surface effects. An earlier study of YSZ single crystals by isotope exchange depth profiling and low energy ion scattering techniques, has been shown that the surface is enriched in yttrium as well as impurities (SiO_2, CaO, Na_2O) that segregate from the bulk [32] and cause a reduction in the surface conductivity. These results suggest that the surface conductivity can be less than of the interface, and the interface conductivity dominates the electrical transport in nanoscale films.

The observed relationship between the conductivity and film thickness can be explained as a superposition of YSZ/MgO interface conductivity, (σ_s) and YSZ lattice, (σ_b) contributions. An analysis of these contributions to the total conductivity, (σ) has been performed using a rule of mixtures model, which is usually applied to describe the conductivity of two phase materials [33, 34]:

$$\sigma^n = \sigma_s^n f_s + \sigma_b^n \cdot (1 - f_s) \qquad (6)$$

where, f_s is the volume fraction occupied by the interfacial conduction path and is represent by δ / d where δ is the thickness of the interface, and d is the film thickness. Equation (6) describes a two conduction phase system where the conductivity is dependent on the volume content of each phases. The parameter n depends on the manner in which these conductors are connected and for real material can be a function of morphology, roughness, strain and impurity segregation effects [33]. Its value can change within - 1 < n < 1 [35].

Equation (6) was used to describe the thickness dependence of the conductivity observed for highly textured YSZ thin films. It was found that keeping n = 0.31 an excellent fit to all experimental data has been obtained (solid lines in Fig.7). This value is consistent with n = 1/3 obtained by theoretical calculations for a 3-D ion diffusion model [35]. This suggests that ionic transport in the YSZ/MgO interface can occur in more that one lattice plane, which would be consistent with the hopping of oxygen ions between equivalent oxygen vacancy positions. To achieve this, a few atomic layers would be involved creating a conduction region at the interface and an additional energy barrier for ion movement across the interface. Next, the thickness, (δ) of the conductive interface zone was estimated to be ~1.6nm. This value correlates well with that obtained for a HfO_2/Si heterostructure using transmission electron microscopy combined with high-resolution x-ray photoelectron spectroscopy, where a value of 1-1.6nm was found for the interface thickness [36]. A similar value (1nm) was also reported for grain boundary thickness determined for ZrO_2:Ca by impedance spectroscopy [16]. In addition, values for the thickness of space charge regions associated with grain boundaries are of the order of 1 – 2nm, were reported for polycrystalline YSZ [37].

Finally, using these estimated n and δ values, one can separate the bulk and interface components of the temperature dependent conductivities via Eq.6. Obtained results are presented in Fig.8. As can be seen, the interface related conductivity is about 3-4 orders of magnitude larger than that of the lattice, and its activation energy of 0.45eV is about half that of the lattice.

This behavior illustrates the potential to enhance the ionic conductivity when electrical transport involves interfacial effects. The present estimates of the interfacial conductivity contribution in epitaxial YSZ thin films correlate well with the enhancement in grain boundary oxygen diffusivity obtained by tracer diffusion and secondary ion mass spectroscopy (SIMS) [38]. An enhancement of the ^{18}O diffusivity in the grain boundaries of YSZ was reported to be more that three orders of magnitude greater than the diffusivity in single crystals.

The ability to reach ionic conductivities of the level of 100S/cm in nanoscaled YSZ thin films is quite unique and is not possible to obtain in conventional materials as their diffusivity is limited by the lattice [7, 31]. The increase of the conductivity of epitaxial YSZ thin films offers new opportunity for oxygen conductors whose properties can be effectively controlled by the thickness and epitaxy level of these materials.

6. CONCLUSIONS

In the present paper the question of how the microstructure can influence the electrical transport of nanocrystalline stabilized zirconia was discussed. The reviewed results obtained from the study of nanocrystalline YSZ and ScSZ thin films have shown the possibility of the enhancement of ionic or electronic conductivity.

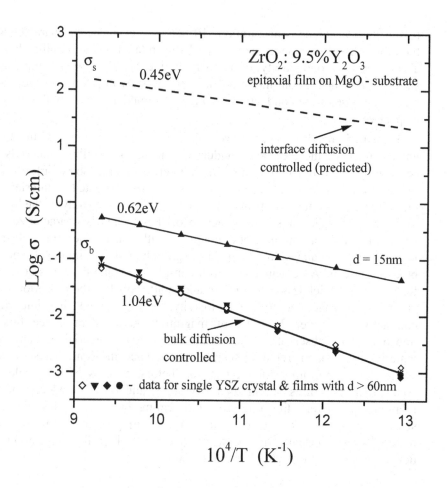

Figure 8. Temperature dependence of bulk- and interface-limited electrical conductivities determined for YSZ thin films [31].

This depends on the type of acceptor and the energy level related to the defect associations. The electrical transport of YSZ and ScSZ is determined by grain-size–dependent non-stoichiometry and is attributed to grain boundary effects, controlled by material microstructure.

The study of electrical conductivity as a function oxygen partial pressure is very helpful to obtain the information related to the defects thermodynamics and their relation with the material band structure. Analysis of these data in terms of the defect model enabled the determination of the electronic and ionic conductivities and correlate they with material microstructure. The ionic conductivity of nanocrystalline YSZ and ScSZ due to the high acceptor concentration, $10^{21}cm^{-3}$, is extrinsic in character and therefore observed enhancement, is probably related to the mobility factor. The electrical conductivity of nanocrystalline ScSZ is very unique and strongly depends upon the oxygen activity showing ionic, mixed and electronic type of the conductivity at high, intermediate and low P_{O2} regions respectively.

Presented results have shown that it is possible to achieve an enhancement of the electrical conductivity in nanocrystalline materials by controlling their defect equilibria and microstructure. Some information related to the maximum value of the conductivity attributed to interface or grain boundary controlled diffusivity was obtained in the study of the highly textured YSZ thin films. The structure of these films is characterized by negligible dislocation network coupled with only one film/substrate interface, which can be a model of single grain boundary with the absence of blocking effects. An enhancement in the in-plane conductivity was observed for films with thickness below 60nm and is attributed to the transition from lattice to interface controlled diffusivity. It was found that interfacial conductivity is over 3 orders of magnitude larger than lattice limited transport. This observation shows the tremendous potential of nanoscale materials with nanostructured features to enhance the ionic transport and offers new opportunity for oxygen conductors. Studies of nanoscale ion conducting thin films are in their beginning and the knowledge of the relationship between the properties and scaling factor (thickness or grain size) seems to be a key factor to the understanding of the nature of these materials and consequently to the realization of their full impact on the development of new electrochemical devices.

ACKNOWLEDGEMENTS:

The author wishes to thank Prof. H. U. Anderson, Dr's T. Suzuki and P. Jasinski from University of Missouri-Rolla for their collaboration in the study of nanosrystalline oxide thin films. The collaboration with Dr's C. M. Rouleau, P. F. Becher and D. H. Lowndes in the study of highly textured YSZ thin films is greatly acknowledged.

Research partly sponsored by the Division of Materials Sciences and Engineering, U. S. Department of Energy, under Contract DE-AC05-00OR22725 with UT-Battelle, LLC.

REFERENCES

[1] Advances in Ceramics Vol. **3**, Science and Technology of Zirconia. A.H.Heuer and L.W. Hobs (Eds.), The American Ceramics Society, Columbus, Ohio 1981.

[2] Subbarao E.C., in Ref.[1], pp. 1 - 24.

[3] Baumard J.F. and P. Abelard, Advances in Ceramics Vol. **12** , Science and Technology of Zirconia II, N. Clausen, M. Ruhle and A.H. Heuer (Eds.), The American Ceramics Society, Columbus, Ohio 1983, pp. 555 - 571.

[4] Minh N. Q., J. Am. Ceram. Soc. 1993, **76**, 563 – 575.

[5] Philibert J., Atom Movements, Diffusion and Mass Transport in Solids, Monographies De Physique, Paris 1991.

[6] Sigiel R. W., in Materials Interfaces, Wolf D. and Yip S., eds. Chapman & Hall, New York 1992, pp. 431 – 460.

[7] Kosacki I. and Anderson H.U., in Encyclopedia of Materials: Science and Technology, Elsevier Science Ltd., New York 2001, vol.**4**, pp. 3609 - 3617.

[8] Kosacki I. and Anderson H.U., Electrochemical Society Proceedings, Vol. **2001-28**, The Electrochemical Society Inc. Pennington, NJ 2001, pp. 238 - 251.

[9] Maier J., Solid State Ionics 2000, **131**, 13 - 22.

[10] Puin W., S. Rodewald, R. Ramlau, P. Heitjans and J. Maier, Solis State Ionics 2000, **131**, 159 - 164.

[11] Suzuki T., I. Kosacki, H.U. Anderson and Ph. Colomban, J. Am. Ceram. Soc. 2001, **84**, 2007 - 2014.

[12] Chiang Y. M., I. Kosacki, H. L. Tuller, and J. Y. Ying, Appl. Phys. Lett. 1996, 69, 185 – 187.

[13] Kosacki I., H. U. Anderson, Y. Mizutani and K. Ukai, Solid State Ionics 2002, **152-153**, 431 - 438.

[14] Kosacki I., T. Suzuki, V. Petrovsky and H. U. Anderson, Solid State Ionics 2000, **136-137**, 1225 - 1233.

[15] Mondal P., A. Klein, W. Jaegermann and H. Hahn, Solid State Ionics 1999, **118**, 331 - 339.

[16] Aoki M., Y-M. Chiang, I. Kosacki, J-R. Lee, H. L. Tuller and Y. J. Liu, J. Am. Ceram. Soc. 1996, **79**, 1169 - 1180.

[17] Otsuka K., A. Kuwabara, A. Nakamura, T. Yamamoto, K. Matsunga and Y. Ikuhara, Appl. Phys. Lett. 2003, **83**, 877 – 879.

[18] Sata N., K. Eberman, K. Eberl and J. Maier, Nature 2000, **408**, 946 - 949.

[19] Kosacki I., C. M. Rouleau, P. F. Becher and D. H. Lowndes, Electrochem. Solid State Lett. 2004, **7**, 11.

[20] Sorensen O. T., ed. Nonstoichiometric Oxides, Academic Press, New York, 1981.

[21] Nowick A. S., W-K Lee and H. Jain, Solid State Ionics 1988, **28-30**, 89 – 95.

[22] Dyre J. C., J. Phys. C: Solid State Physics 1986, **19**, 5655.

[23] Honke D. K., J. Phys. Chem. Solids 1980, **41**, 777.

[24] Meijer E. J., M. Matters, P. T. Herwig, D. M. de Leeuw and T. M. Klapwijk, Appl. Phys. Lett. 2000, **76**, 3433 – 3435.

[25] Siegel R. W.and G. C. Hadjipanayis, eds. Nanophase Materials Synthesis-Properties-applications, Kluwer, Dodrecht, 1994.

[26] Anderson H. U., C. C. Chen and M. M. Nasrallah, U. S. Patent No.5, 494, 700, Feb. 1996.

[27] Manning P.S., J. D. Sirman, R. A. De Souza and J. A. Kilner, Solid State Ionics 1997, **100**, 1 - 10.

[28] Park J. H. and R. N. Blumenthal, J. Electrochem. Soc. 1989, **136**, 2868.

[29] Karavaev Yu. N., A. D. Neuiman and S. F. Palguev, Electrochimiya 1987, **23**, 121.

[30] Kosacki I., V. Petrovsky and H. U. Anderson, J. Electroceramics 2000, **4**, 243 - 249.

[31] Kosacki I., C. M. Rouleau, P. F. Becher and D. H. Lowndes Solid State Ionics 2004, in press.

[32] Ridder van M., R. G. van Welzenis, H. H. Brongerman and U. Kreissig, Solid State Ionics 2003, **158**, 67 – 77.

[33] Uvarov A. F., Solid State Ionics 2000, **136-137**, 1267 - 1272.

[34] Yuan S.L., Z.Y. Li, G. Peng, C. S. Xiong, Y. H. Xiong and C. Q. Tang, Appl. Phys. Lett. 2001, **79**, 90 – 92.

[35] B. I. Shklovskii, A. L. Efros, "Electronic Properties of Doped Semiconductors", Springer-Verlag, Berlin Heidelberg, New York 1984.

[36] Lee J. C., S. J. Oh, M. Cho and C. S. Hwang, Appl. Phys. Lett. 2004, **84**, 1305 - 1307.

[37] Guo X., and J. Maier, J. Electrochem. Soc. 2001, **148** (3) E121 - E126.

[38] U. Brossman, G. Knoner, H. E. Schaefer and R. Wurschum, Rev. Adv. Mater. Sci. 2004, **6** 7 – 11.

Authors Index

Keyword Index

9 781402 03497